MATH/STAT
Y0-BSD-008
Berkeley
LIBRARY OF THE UNIVERSITY OF CALIFORNIA

Pitman Research Notes in Mathematics Series

Submission of proposals for consideration
Suggestions for publication, in the form of outlines and representative samples, are invited by the Editorial Board for assessment. Intending authors should approach one of the main editors or another member of the Editorial Board, citing the relevant AMS subject classifications. Alternatively, outlines may be sent directly to the publisher's offices. Refereeing is by members of the board and other mathematical authorities in the topic concerned, throughout the world.

Preparation of accepted manuscripts
On acceptance of a proposal, the publisher will supply full instructions for the preparation of manuscripts in a form suitable for direct photo-lithographic reproduction. Specially printed grid sheets are provided and a contribution is offered by the publisher towards the cost of typing. Word processor output, subject to the publisher's approval, is also acceptable.

Illustrations should be prepared by the authors, ready for direct reproduction without further improvement. The use of hand-drawn symbols should be avoided wherever possible, in order to maintain maximum clarity of the text.

The publisher will be pleased to give any guidance necessary during the preparation of a typescript, and will be happy to answer any queries.

Important note
In order to avoid later retyping, intending authors are strongly urged not to begin final preparation of a typescript before receiving the publisher's guidelines and special paper. In this way it is hoped to preserve the uniform appearance of the series.

Longman Scientific & Technical
Longman House
Burnt Mill
Harlow, Essex, UK
(tel (0279) 26721)

Titles in this series

1. Improperly posed boundary value problems
A Carasso and A P Stone
2. Lie algebras generated by finite dimensional ideals
I N Stewart
3. Bifurcation problems in nonlinear elasticity
R W Dickey
4. Partial differential equations in the complex domain
D L Colton
5. Quasilinear hyperbolic systems and waves
A Jeffrey
6. Solution of boundary value problems by the method of integral operators
D L Colton
7. Taylor expansions and catastrophes
T Poston and I N Stewart
8. Function theoretic methods in differential equations
R P Gilbert and R J Weinacht
9. Differential topology with a view to applications
D R J Chillingworth
10. Characteristic classes of foliations
H V Pittie
11. Stochastic integration and generalized martingales
A U Kussmaul
12. Zeta-functions: An introduction to algebraic geometry
A D Thomas
13. Explicit *a priori* inequalities with applications to boundary value problems
V G Sigillito
14. Nonlinear diffusion
W E Fitzgibbon III and H F Walker
15. Unsolved problems concerning lattice points
J Hammer
16. Edge-colourings of graphs
S Fiorini and R J Wilson
17. Nonlinear analysis and mechanics: Heriot-Watt Symposium Volume I
R J Knops
18. Actions of fine abelian groups
C Kosniowski
19. Closed graph theorems and webbed spaces
M De Wilde
20. Singular perturbation techniques applied to integro-differential equations
H Grabmüller
21. Retarded functional differential equations: A global point of view
S E A Mohammed
22. Multiparameter spectral theory in Hilbert space
B D Sleeman
24. Mathematical modelling techniques
R Aris
25. Singular points of smooth mappings
C G Gibson
26. Nonlinear evolution equations solvable by the spectral transform
F Calogero
27. Nonlinear analysis and mechanics: Heriot-Watt Symposium Volume II
R J Knops
28. Constructive functional analysis
D S Bridges
29. Elongational flows: Aspects of the behaviour of model elasticoviscous fluids
C J S Petrie
30. Nonlinear analysis and mechanics: Heriot-Watt Symposium Volume III
R J Knops
31. Fractional calculus and integral transforms of generalized functions
A C McBride
32. Complex manifold techniques in theoretical physics
D E Lerner and P D Sommers
33. Hilbert's third problem: scissors congruence
C-H Sah
34. Graph theory and combinatorics
R J Wilson
35. The Tricomi equation with applications to the theory of plane transonic flow
A R Manwell
36. Abstract differential equations
S D Zaidman
37. Advances in twistor theory
L P Hughston and R S Ward
38. Operator theory and functional analysis
I Erdelyi
39. Nonlinear analysis and mechanics: Heriot-Watt Symposium Volume IV
R J Knops
40. Singular systems of differential equations
S L Campbell
41. N-dimensional crystallography
R L E Schwarzenberger
42. Nonlinear partial differential equations in physical problems
D Graffi
43. Shifts and periodicity for right invertible operators
D Przeworska-Rolewicz
44. Rings with chain conditions
A W Chatters and C R Hajarnavis
45. Moduli, deformations and classifications of compact complex manifolds
D Sundararaman
46. Nonlinear problems of analysis in geometry and mechanics
M Atteia, D Bancel and I Gumowski
47. Algorithmic methods in optimal control
W A Gruver and E Sachs
48. Abstract Cauchy problems and functional differential equations
F Kappel and W Schappacher
49. Sequence spaces
W H Ruckle
50. Recent contributions to nonlinear partial differential equations
H Berestycki and H Brezis
51. Subnormal operators
J B Conway

52 Wave propagation in viscoelastic media
F Mainardi
53 Nonlinear partial differential equations and their applications: Collège de France Seminar. Volume I
H Brezis and J L Lions
54 Geometry of Coxeter groups
H Hiller
55 Cusps of Gauss mappings
T Banchoff, T Gaffney and C McCrory
56 An approach to algebraic K-theory
A J Berrick
57 Convex analysis and optimization
J-P Aubin and R B Vintner
58 Convex analysis with applications in the differentiation of convex functions
J R Giles
59 Weak and variational methods for moving boundary problems
C M Elliott and J R Ockendon
60 Nonlinear partial differential equations and their applications: Collège de France Seminar. Volume II
H Brezis and J L Lions
61 Singular systems of differential equations II
S L Campbell
62 Rates of convergence in the central limit theorem
Peter Hall
63 Solution of differential equations by means of one-parameter groups
J M Hill
64 Hankel operators on Hilbert space
S C Power
65 Schrödinger-type operators with continuous spectra
M S P Eastham and H Kalf
66 Recent applications of generalized inverses
S L Campbell
67 Riesz and Fredholm theory in Banach algebra
B A Barnes, G J Murphy, M R F Smyth and T T West
68 Evolution equations and their applications
F Kappel and W Schappacher
69 Generalized solutions of Hamilton-Jacobi equations
P L Lions
70 Nonlinear partial differential equations and their applications: Collège de France Seminar. Volume III
H Brezis and J L Lions
71 Spectral theory and wave operators for the Schrödinger equation
A M Berthier
72 Approximation of Hilbert space operators I
D A Herrero
73 Vector valued Nevanlinna Theory
H J W Ziegler
74 Instability, nonexistence and weighted energy methods in fluid dynamics and related theories
B Straughan
75 Local bifurcation and symmetry
A Vanderbauwhede
76 Clifford analysis
F Brackx, R Delanghe and F Sommen
77 Nonlinear equivalence, reduction of PDEs to ODEs and fast convergent numerical methods
E E Rosinger
78 Free boundary problems, theory and applications. Volume I
A Fasano and M Primicerio
79 Free boundary problems, theory and applications. Volume II
A Fasano and M Primicerio
80 Symplectic geometry
A Crumeyrolle and J Grifone
81 An algorithmic analysis of a communication model with retransmission of flawed messages
D M Lucantoni
82 Geometric games and their applications
W H Ruckle
83 Additive groups of rings
S Feigelstock
84 Nonlinear partial differential equations and their applications: Collège de France Seminar. Volume IV
H Brezis and J L Lions
85 Multiplicative functionals on topological algebras
T Husain
86 Hamilton-Jacobi equations in Hilbert spaces
V Barbu and G Da Prato
87 Harmonic maps with symmetry, harmonic morphisms and deformations of metrics
P Baird
88 Similarity solutions of nonlinear partial differential equations
L Dresner
89 Contributions to nonlinear partial differential equations
C Bardos, A Damlamian, J I Díaz and J Hernández
90 Banach and Hilbert spaces of vector-valued functions
J Burbea and P Masani
91 Control and observation of neutral systems
D Salamon
92 Banach bundles, Banach modules and automorphisms of C*-algebras
M J Dupré and R M Gillette
93 Nonlinear partial differential equations and their applications: Collège de France Seminar. Volume V
H Brezis and J L Lions
94 Computer algebra in applied mathematics: an introduction to MACSYMA
R H Rand
95 Advances in nonlinear waves. Volume I
L Debnath
96 FC-groups
M J Tomkinson
97 Topics in relaxation and ellipsoidal methods
M Akgül
98 Analogue of the group algebra for topological semigroups
H Dzinotyiweyi
99 Stochastic functional differential equations
S E A Mohammed

100 Optimal control of variational inequalities
V Barbu
101 Partial differential equations and dynamical systems
W E Fitzgibbon III
102 Approximation of Hilbert space operators. Volume II
C Apostol, L A Fialkow, D A Herrero and D Voiculescu
103 Nondiscrete induction and iterative processes
V Ptak and F-A Potra
104 Analytic functions – growth aspects
O P Juneja and G P Kapoor
105 Theory of Tikhonov regularization for Fredholm equations of the first kind
C W Groetsch
106 Nonlinear partial differential equations and free boundaries. Volume I
J I Díaz
107 Tight and taut immersions of manifolds
T E Cecil and P J Ryan
108 A layering method for viscous, incompressible L_p flows occupying R^n
A Douglis and E B Fabes
109 Nonlinear partial differential equations and their applications: Collège de France Seminar. Volume VI
H Brezis and J L Lions
110 Finite generalized quadrangles
S E Payne and J A Thas
111 Advances in nonlinear waves. Volume II
L Debnath
112 Topics in several complex variables
E Ramírez de Arellano and D Sundararaman
113 Differential equations, flow invariance and applications
N H Pavel
114 Geometrical combinatorics
F C Holroyd and R J Wilson
115 Generators of strongly continuous semigroups
J A van Casteren
116 Growth of algebras and Gelfand–Kirillov dimension
G R Krause and T H Lenagan
117 Theory of bases and cones
P K Kamthan and M Gupta
118 Linear groups and permutations
A R Camina and E A Whelan
119 General Wiener–Hopf factorization methods
F-O Speck
120 Free boundary problems: applications and theory, Volume III
A Bossavit, A Damlamian and M Fremond
121 Free boundary problems: applications and theory, Volume IV
A Bossavit, A Damlamian and M Fremond
122 Nonlinear partial differential equations and their applications: Collège de France Seminar. Volume VII
H Brezis and J L Lions
123 Geometric methods in operator algebras
H Araki and E G Effros
124 Infinite dimensional analysis–stochastic processes
S Albeverio
125 Ennio de Giorgi Colloquium
P Krée
126 Almost-periodic functions in abstract spaces
S Zaidman
127 Nonlinear variational problems
A Marino, L Modica, S Spagnolo and M Degiovanni
128 Second-order systems of partial differential equations in the plane
L K Hua, W Lin and C-Q Wu
129 Asymptotics of high-order ordinary differenti equations
R B Paris and A D Wood
130 Stochastic differential equations
R Wu
131 Differential geometry
L A Cordero
132 Nonlinear differential equations
J K Hale and P Martinez-Amores
133 Approximation theory and applications
S P Singh
134 Near-rings and their links with groups
J D P Meldrum
135 Estimating eigenvalues with *a posteriori/a pri* inequalities
J R Kuttler and V G Sigillito
136 Regular semigroups as extensions
F J Pastijn and M Petrich
137 Representations of rank one Lie groups
D H Collingwood
138 Fractional calculus
G F Roach and A C McBride
139 Hamilton's principle in continuum mechanics
A Bedford
140 Numerical analysis
D F Griffiths and G A Watson
141 Semigroups, theory and applications. Volum
H Brezis, M G Crandall and F Kappel
142 Distribution theorems of L-functions
D Joyner
143 Recent developments in structured continua
D De Kee and P Kaloni
144 Functional analysis and two-point differentia operators
J Locker
145 Numerical methods for partial differential equations
S I Hariharan and T H Moulden
146 Completely bounded maps and dilations
V I Paulsen
147 Harmonic analysis on the Heisenberg nilpote Lie group
W Schempp
148 Contributions to modern calculus of variation
L Cesari
149 Nonlinear parabolic equations: qualitative properties of solutions
L Boccardo and A Tesei
150 From local times to global geometry, control physics
K D Elworthy

151 A stochastic maximum principle for optimal control of diffusions
U G Haussmann
152 Semigroups, theory and applications. Volume II
H Brezis, M G Crandall and F Kappel
153 A general theory of integration in function spaces
P Muldowney
154 Oakland Conference on partial differential equations and applied mathematics
L R Bragg and J W Dettman
155 Contributions to nonlinear partial differential equations. Volume II
J I Díaz and P L Lions
156 Semigroups of linear operators: an introduction
A C McBride
157 Ordinary and partial differential equations
B D Sleeman and R J Jarvis
158 Hyperbolic equations
F Colombini and M K V Murthy
159 Linear topologies on a ring: an overview
J S Golan
160 Dynamical systems and bifurcation theory
M I Camacho, M J Pacifico and F Takens
161 Branched coverings and algebraic functions
M Namba
162 Perturbation bounds for matrix eigenvalues
R Bhatia
163 Defect minimization in operator equations: theory and applications
R Reemtsen
164 Multidimensional Brownian excursions and potential theory
K Burdzy
165 Viscosity solutions and optimal control
R J Elliott
166 Nonlinear partial differential equations and their applications. Collège de France Seminar. Volume VIII
H Brezis and J L Lions
167 Theory and applications of inverse problems
H Haario
168 Energy stability and convection
G P Galdi and B Straughan
169 Additive groups of rings. Volume II
S Feigelstock
170 Numerical analysis 1987
D F Griffiths and G A Watson
171 Surveys of some recent results in operator theory. Volume I
J B Conway and B B Morrel
172 Amenable Banach algebras
J-P Pier
173 Pseudo-orbits of contact forms
A Bahri
174 Poisson algebras and Poisson manifolds
K H Bhaskara and K Viswanath
175 Maximum principles and eigenvalue problems in partial differential equations
P W Schaefer
176 Mathematical analysis of nonlinear, dynamic processes
K U Grusa
177 Cordes' two-parameter spectral representation theory
D F McGhee and R H Picard
178 Equivariant K-theory for proper actions
N C Phillips
179 Elliptic operators, topology and asymptotic methods
J Roe
180 Nonlinear evolution equations
J K Engelbrecht, V E Fridman and E N Pelinovski
181 Nonlinear partial differential equations and their applications. Collège de France Seminar. Volume IX
H Brezis and J L Lions
182 Critical points at infinity in some variational problems
A Bahri
183 Recent developments in hyperbolic equations
L Cattabriga, F Colombini, M K V Murthy and S Spagnolo
184 Optimization and identification of systems governed by evolution equations on Banach space
N U Ahmed
185 Free boundary problems: theory and applications. Volume I
K H Hoffmann and J Sprekels
186 Free boundary problems: theory and applications. Volume II
K H Hoffmann and J Sprekels
187 An introduction to intersection homology theory
F Kirwan
188 Derivatives, nuclei and dimensions on the frame of torsion theories
J S Golan and H Simmons
189 Theory of reproducing kernels and its applications
S Saitoh
190 Volterra integrodifferential equations in Banach spaces and applications
G Da Prato and M Iannelli
191 Nest algebras
K R Davidson
192 Surveys of some recent results in operator theory. Volume II
J B Conway and B B Morrel
193 Nonlinear variational problems. Volume II
A Marino and M K Murthy
194 Stochastic processes with multidimensional parameter
M E Dozzi
195 Prestressed bodies
D Iesan
196 Hilbert space approach to some classical transforms
R H Picard
197 Stochastic calculus in application
J R Norris
198 Radical theory
B J Gardner
199 The C* – algebras of a class of solvable Lie groups
X Wang

Critical points at infinity in some variational problems

Longman Scientific & Technical,
Longman Group UK Limited,
Longman House, Burnt Mill, Harlow
Essex CM20 2JE, England
and Associated Companies throughout the world.

Copublished in the United States with
John Wiley & Sons, Inc., 605 Third Avenue, New York, NY 10158

First published 1989

AMS Subject Classifications: 21, 37, 39, 45

ISSN 0269-3674

British Library Cataloguing in Publication Data
Bahri, A.
Critical points at infinity in some variational problems.
1. Calculus of variations
I. Title.
515'.64
ISBN 0-582-02164-2

Library of Congress Cataloging-in-Publication Data
Bahri, A.
Critical points at infinity in some variational problems
p. c. -- (Pitman research notes in mathematics series, ISSN 0269-3674: 182)
Bibliography: p.
ISBN 0-470-21147-4
1. Calculus of variations. 2. Critical point theory (Mathematical analysis)
3. Differential equations.
Partial, I. title.
II. Series.
QA316.B242 1988
515'.64--dc19 88-12264 CIP

Printed and bound in Great Britain
by Biddles Ltd, Guildford and King's Lynn

Contents

Introduction

0.1 We consider in this work a scalar type non compact variational problem and we raise the question of proving the existence of "true" critical points for the considered functional, in a framework where the Palais-Smale condition does not hold.

The problem may be stated in general terms: Suppose that we are given, on a Hilbert space (or manifold) E, a function $f: E \to R$, which we assume to be C^2 for sake of simplicity.

We want to find solutions in E of the equation:

$$\begin{cases} \partial f(x) = 0 \\ x \in E \end{cases} \quad \text{where } \partial f \text{ is the gradient of } f. \tag{1}$$

The classical variational theory provides us with a simple principle which allows us to detect and also count the number of solutions of (1), assuming a compactness condition on the gradient flow:

Let $a \leq b$; $f^a = \{x \in E \text{ s,t } f(x) \leq a\}$; $f^b = \{x \in E \text{ s. t. } f(x) \leq b\}$; $f_a = \{x \in E \text{ s. t } f(x) \geq a\}$ $f_b = \{x \in E \text{ s. t } f(x) \geq b\}$.

The following statements describe the variational principle under its simplest form (also its most crucial form).

Proposition 0.1: (see [01] for example). If f satisfies the Palais-Smale condition in $f^{-1}([a,b])$ and if (1) has no solution in $f^{-1}([a,b])$, then f^a is a retract by deformation of f^b (rsp f_b is a retract by deformation of f_a).

Corollary 0.2: Under the hypotheses of Proposition 0.1, the homotopy and homology of f^a (rsp f_a) and f^b (rsp f_b) are the same.

Corollary 0.3: Under the hypotheses of Proposition 0.1, (1) has a solution x_0 such that $f(x_0) \in [a,b]$ if the homotopy (or homology) groups of f^a (rsp f_a) and f^b (rsp f_b) differ.

These results may be considerably improved and the difference of topology in the level sets of f induced by the solutions of (1) may be computed, under additional hypotheses on f.

However, there's a key-hypothesis which is the Palais-Smale condition. We restate it here:

Definition 0.4: A C^2 functional $f:E \to \mathbb{R}$ satisfies the Palais-Smale condition on $f^{-1}([a,b])$ if any sequence (x_n) such that $a \leq f(x_n) \leq b$ and $\partial f(x_n) \to 0$ is precompact.

In practice there are many examples of functionals violating the Palais-Smale condition. The two following examples, on real funtions of real variables, are already interesting.

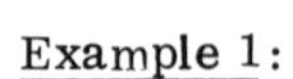

Example 1:

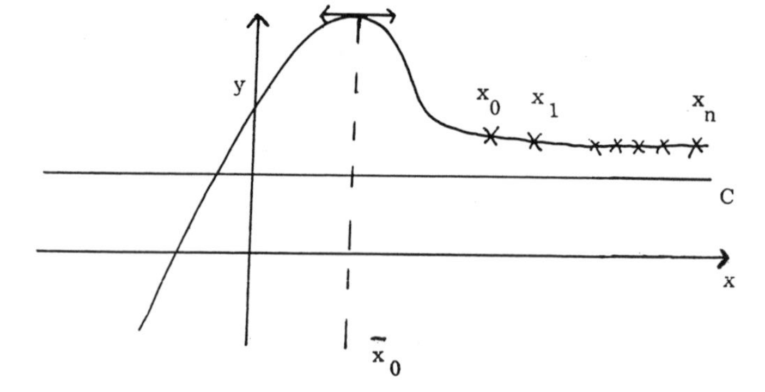

This drawing displays a sequence (x_n) ("a critical point at infinity") such that $\partial f(x_n) \to 0$ and $f(x_n) \to C$ which does not have any convergent subsequence as x_n tends to $+\infty$. As may be easily verified, $f^{C+\varepsilon}$ is not connected while $f^{C-\varepsilon}$ is connected, for $\varepsilon > 0$ small enough. Thus, this critical point at infinity induces a difference of topology in the level sets of f.

If one wants to prove the existence, by a global type argument, of $\bar{x}_0$ (the "true" critical point), then one should take into account this critical point at infinity, which plays the role of a minimum.

The following example is simply the graph of $\tilde{f} = -f + K$ where f is the function of example 1 and K is an irrelevant constant.

Example 2:

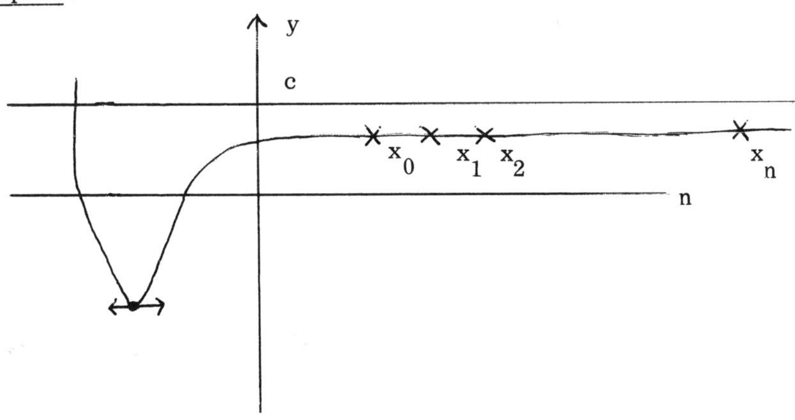

Obviously, the Palais-Smale condition does not hold in a neighbourhood of c. However, for $\varepsilon > 0$ small enough, $\tilde{f}^{c+\varepsilon}$ is contractible as well as $\tilde{f}^{c-\varepsilon}$. Hence, no difference of topology has been introduced by this critical point at infinity in the "decreasing" level sets. On the contrary, if one considers $\tilde{f}_{c+\varepsilon}$ and $\tilde{f}_{c-\varepsilon}$, the difference of topology is the same as that in example 1.

This fact, in the classical variational theory (in finite dimension also), is surprising: a non degenerate critical point (and even a degenerate one under some hypotheses) induces, in finite dimension, a difference of topology for the "increasing" as well as for the "decreasing" level sets. As examples 1 and 2 display it, this is no longer true for critical points at infinity.

We thus see that, if we no longer assume that the Palais-Smale condition holds, the classical variational theory does not apply, even in some extended way. We need to come back to the basic elements of this theory, to the deformation along flow-lines and to the computation of the differences in topology which are induced even if these flow-lines are non compact.

0.2 Some examples of non compact variational problems in nonlinear analysis, in geometry and physics

The category of non compact variational problems is wide. There are geometric problems, as the Yamabe problem or the Kazdan-Warner equation, the problem of finding harmonic maps of a given degree, the Yang-Mills equations, the nonlinear wave equation under non rationally dependent boundary conditions, etc.

The non-compactness bears different features in each of these cases. Nevertheless, there are some very strong common points between these problems, as it is shown throughout the existing literature (e.g. [02], [03], [04]), at least if we except the non-linear wave equation, where the non-compactness is related to the small divisors-problem. Similarly, there are the non linear elliptic differential equations with supercritical exponent (as, for example, the equation $\Delta u + u^7 = 0$ on an open Ω of $\mathbb{R}^3$).

We further develop here the non-compactness in Yamabe-type equations, including the Kazdan-Warner problem. We refer the reader to A. Bahri, "Pseudo-Orbits of contact forms" [05] for another framework where the non-compactness arises.

<u>The non-compactness in Yamabe-type equations and in the Kazdan-Warner problem</u>: We will describe the non-compactness of these equations for open sets Ω of S^3. Generalizations to the n-dimensional situation are completed in this work. Generalizations to manifolds are completed in Bahri-Brezis [08].

Let $S^3 = \{x \in \mathbb{R}^4 \text{ s.t } |x| = 1\}$, c the standard metric on S^3 and Ω a regular open set of S^3. Let K be a positive function of class C^2 on $\overline{\Omega}$.

We look for a function $u: \overline{\Omega} \to \mathbb{R}$ satisfying

$$\begin{cases} -8\Delta u + 6u = K(x)\, u^5 \\ u > o \text{ in } \Omega \\ \text{if } \Omega \neq S^3, \quad u = o|_{\partial\Omega}. \end{cases} \tag{2}$$

When Ω is not S^3, we will assume that K is constant. When Ω is S^3, (2) has the following geometric interpretation: Does there exist a metric g on S^3, conformal to c, such that the scalar curvature of (S^3, g) is $K(g = u^4 c)$. The same equation exists on any compact Riemannian manifold. If K is constant, this is the Yamabe equation; otherwise, this is the Kazdan-Warner problem.

The Yamabe equation has been solved by T. Aubin ([06]) for the case of the non locally conformally flat manifolds (for the precise statements, see [06]); the other cases have been solved by R. Schoēn [07], who uses the positive mass theorem.

The proof of R. Schoēn relies on a sharp analysis which is present also in our study of the non compactness for such equations. However, we point out here that our method allows us also to solve cases of actual non compactness (i.e. the Palais-Smale condition does not hold, even in a neighbourhood of the minimum) as for example, in the Kazdan-Warner problem.

As for the existence results we are able to derive with such techniques on manifolds, we refer the reader to Bahri-Brezis [08].

As for the Kazdan-Warner equation, obstructions are known, due to Kazdan-Warner [09] and also Bourguignon-Ezin ([010]). These obstructions lead to counter examples (see e.g. [010]) which precisely violate the sufficient condition that we provide for the existence of a solution to (2). We assume here that K is strictly positive.

We describe the variational framework of problem (2): Let

$$I(u) = \frac{1}{\int_{\Omega} K(x)\, u^{6}\, dv} ,$$

where u belongs to $H_0^1(\Omega)$ if Ω is not S^3, u belongs to $H^1(S^3)$ otherwise

$$\left(\|u\|_{\Omega}^2 = \int_{\Omega} (8|\nabla u|^2 + 6u^2)\, dv \right).$$

We consider the functional $I(u)$ on the set

$$\Sigma = \{ u \in H_0^1(\Omega) \text{ s.t } \|u\|_{\Omega} = 1 \}.$$

It is easy to see that $\operatorname{Inf}_{\Sigma} I$ is not reached if $\Omega = S^3$ and K is not constant or if Ω is not S^3. A critical point of I on Σ yields a solution of (2), if it is a positive function.

Let d be the geodesic distance on (S^3, c). Let, for a belonging to S^3 and $\lambda > 0$, $\delta(a, \lambda)$ be the function on S^3:

$$\delta(a, \lambda)(x) = C \frac{\sqrt{\lambda}}{\sqrt{\lambda^2+1+(\lambda^2-1)\operatorname{Cos} d(a,x)}} \tag{3}$$

C is chosen so that $\|\delta(a, \lambda)\|_\Omega = 1$ (C is independent of λ and a). Let

$$\Sigma^+ = \{u \in \Sigma \text{ s.t } u \geq 0 \text{ in } \Omega\}.$$

For $\varepsilon > 0$ and $p \in \mathbb{N}$ given, we introduce:

$$\begin{cases} V(p,\varepsilon) = \{u \in \Sigma^+ \text{ s.t } \exists\, a_1, \_\_, a_p \in \Omega,\ \exists\, \lambda_1, \_\_, \lambda_p \in \\ ]0,+\infty[\text{ s.t } \left\| u - \frac{1}{\sqrt{S}} \sum_{i=1}^{p} \frac{1}{K(a_i)^{1/4}} \delta(a_i, \lambda_i) \right\|_\Omega \leq \varepsilon \\ \text{with } S = \sum_{i=1}^{p} \frac{1}{\sqrt{K(a_i)}},\ \lambda_i \geq \frac{1}{\varepsilon};\ \frac{\lambda_i}{\lambda_j} + \frac{\lambda_j}{\lambda_i} + d(a_i, a_j^2)\lambda_i\lambda_j \cdot \geq \frac{1}{\varepsilon} \\ \forall i \neq j;\ \text{if } \partial\Omega \neq \phi,\ \lambda_i d(a_i, \partial\Omega) \geq 1/\varepsilon \end{cases} \tag{4}$$

The set $V(p, \varepsilon)$ described by (4) has a simple interpretation: it is a neighbourhood of the critical points at infinity of the functional I on Σ^+. This means that any sequence (u_n) of Σ^+, such that $I(u_n)$ remains bounded and $\partial I(u_n) \to 0$ remains in a $V(p, \varepsilon)$ (with ε arbitrarily small and p upper-bounded) for n large enough.

The condition on the $\lambda_i (\lambda_i \geq \frac{1}{\varepsilon})$ means that the function u, when ε tends to zero, becomes closer and closer (in the H^1 norm) to functions of the type $\frac{1}{\sqrt{S}} \sum_{i-1}^{p} \frac{1}{K(a_i)^{1/4}} \delta(a_i, \lambda_i)$ which precisely have then "the tendency to leave the space H^1". This is displayed by the fact that $|\nabla w|^2$ and w^6 tend to a Dirac mass concentrated at the points a_i (which might also change with n) when λ_i tends to $+\infty$ (or ε tends to zero) $\left(w = \frac{1}{\sqrt{S}} \sum_{i=1}^{p} \frac{1}{K(a_i)^{1/4}} \delta(a_i, \lambda)\right)$.

The other condition, on the λ_i's mean that two concentration points a_i and

a_j for $j \neq i$, cannot come very close if the concentrations λ_i and λ_j do not become very different (in order to avoid interaction phenomena which are excluded by the fact that $\partial I(u_n) \to 0$).

The condition $d(a_i, \partial\Omega)\ \lambda_i \geq \frac{1}{\varepsilon}$ means that a concentration point has to be in Ω (relatively to λ_i) so that the total energy carried by $\delta(a_i, \lambda_i)$ remains on Ω. Thus, the introduction of the set $V(p,\varepsilon)$ is natural and we have the following result:

Proposition 0.5: Let (u_n) be a sequence of Σ^+ such that $\partial I(u_n) \to 0$, $(I(u_n))_n$ being bounded and such that u_n tends weakly to zero in $H_0^1(\Omega)$. There exists then a subsequence again denoted u_n, an integer p and a sequence $(\varepsilon_n)_n$ with $\varepsilon_n \to 0$ such that $u_n \in V(p, \varepsilon_n)\ \forall\, n$.

Remark 0.6: If $\partial I(u_n) \to 0$ and $(I(u_n))$ is bounded, then (u_n) is a bounded sequence in H_0^1. Hence (u_n) has a weak limit $\bar{u}$.

If $\bar{u}$ is non zero, then $\bar{u}$ is a critical point of I. We assume, in

$\|\bar{u}\|\Omega$

Proposition 0.5, that $\bar{u}$ is zero as we are looking for non zero solutions of (2). In the general case, Proposition 0.5 holds with

$$\frac{u_n - \bar{u}}{\|u_n - \bar{u}\|} \in V(p, \varepsilon_n)\ .$$

Proposition 0.5 is a crucial step in the understanding of the non-compactness. It relies on the work of Sacks and Uhlenbeck (02), Siu and Yau [03], P. L. Lions [011], M. Struwe [012], C. H. Taubes [013], Brezis-Coron [04].

However, Proposition 0.5 is not satisfactory from the variational point of view. Indeed, the following questions remain unanswered: Do these sequences (u_n) induce a difference of topology in the level sets of I? Are there some which would do so while others would not? What are the induced differences of topology? Can one compute this defect of compactness and use it, in a contradiction argument, in order to prove the existence of "true" solutions?

0.3 The results

0.3.1 Yamabe-type equations: The following Proposition provides us with a useful parametrization of the set $V(p, \varepsilon)$ which has been defined in (5). (This definition extends easily to the case where the dimension is n.)

Proposition 0.7: Let p be a non zero integer. For ε small enough and for u in $V(p, \varepsilon)$, the problem:

$$\operatorname{Min} \left\| u - \sum_{i=1}^{p} \alpha_i \delta(a_i, \lambda_i) \right\|_{S^n}, \quad \alpha_1, \_\_, \alpha_p > 0,$$

$a_1, \_\_, a_p \in \Omega^p;\ \lambda_1, \_\_, \lambda_p \in \,]0, +\infty[^p$ has a unique solution.

We then denote $a_i(u)$, $\lambda_i(u)$ and $\alpha_i(u)$ the solution of this minimization problem.

Let us first consider the case $\Omega \neq S^n$. Then, after stereographic projection, problem (2) (with $K(x) = C^{st}$) becomes:

$$\begin{cases} -\Delta u = u^{\frac{n+2}{n-2}} & \text{in } \Omega \\ u > 0 \text{ in } \Omega \text{ and } u = 0 / \partial\Omega. \end{cases} \tag{5}$$

Ω is now an open bounded regular subset of $\mathbb{R}^n$.

The existence results on equation of type (5) are joint work with J. M. Coron (A. Bahri-J. M. Coron [014]) and H. Brezis (A. Bahri-H. Brezis [08]). We state in this introduction the existence result for the case $\Omega \subset \mathbb{R}^n$. For the general case, the reader is referred to [08].

We then have:

Theorem 0.8: If Ω is a bounded, connected regular open set in $\mathbb{R}^n$ such that $\tilde{H}_\star(\Omega; \mathbb{Z}_z) \neq 0$, (5) has a solution.

The non-compactness phenomenon in equations of type (5) goes beyond the existence proofs as stated in Theorem 0.8.

To describe this phenomenon, we introduce some notations:

$$
(F)\quad\begin{cases}
\|u\| = \left(\int |\nabla u|^2 dx\right)^{1/2} \quad \Sigma = \{u \in H_0^1 / \|u\| = 1\} \\
\Sigma^+ = \{u \in \Sigma / u \geq 0\} \quad I(u) = \dfrac{1}{\left(\int |u|^{2n/n-2}\right)} \; ; \\
\delta(a,\lambda)(x) = C \left| \dfrac{\lambda}{1+\lambda^2 |x-a|^2} \right|^{\frac{n-2}{2}}, \\
C \text{ such that } \|\delta(a,\lambda)\| = 1.
\end{cases}
$$

For x in Ω, we define the function:

$$y \to H(x,y)$$

by

$$\Delta_y H(x,y) = 0 \text{ in } \Omega$$

$$H(x,y) = \frac{1}{|x-y|^{n-2}} \text{ on } \partial\Omega$$

for $a = (a_1, \_\_\_, a_p) \in \Omega^p$, we define the matrix $M(a)$ in $\mathbb{R}^{p^2}$ by:

$$
(g)\quad\begin{cases}
M_{ij}(a) = H(a_i, a_j) - \dfrac{1}{|a_i - a_j|^{n-2}} \; ; \; M_{ij}(a) = -\infty \text{ if } i \neq j \text{ and } a_i = a_j \\
M_{ii}(a) = H(a_i, a_i)
\end{cases}
$$

Let

$$
\begin{cases}
\rho(a) : \text{ the smallest eigenvalue of } M(a) \\
\rho(a) = -\infty \text{ if } a_i = a_j \text{ with } i \neq j
\end{cases}
\tag{6}
$$

The matrix $M(a)$ is clearly related to Green's function of the Laplacian on Ω ; this matrix has an equivalent on an arbitrary Riemannian manifold with or without boundary. In order to estimate the defects of compactness induced by the $V(p,\varepsilon)$, $\varepsilon \to 0$, we follow a gradient line in Σ^+

$$\frac{du}{ds} = -I'(u) \; ; \; u(0) \in \Sigma^+ \tag{7}$$

Let

$$\omega(u) = \frac{1}{\left(\int |u|^{2n/n-2}\right)^{\frac{n-2}{4}}}$$

We assume that there exists $p \in \mathbb{N}$ and $\varepsilon(s)$ small so that $u(s)$ belongs to $V(p, \varepsilon(s))$ for s large enough. We note $\lambda_i(s) = \lambda_i(u(s))$; $a_i(s) = a_i(u(s))$; $\rho(s) = \rho(a(s))$

Definition 0.9: We define the critical points at infinity of the variational problem (5) to be the orbits of the flow which remain in a $V(p, \varepsilon(s))$, where $\varepsilon(s)$ is a given function such that $\varepsilon(s) \to 0$ when $s \to +\infty$.

We then have:

Theorem 0.10: Assume that $\forall\, i \in [1, p]$, $\overline{\lim}\, d(a_i(s), \partial\Omega) > 0$. We then have $\overline{\lim}\, \rho(s) \geq 0$. If $\overline{\lim}\, \rho(s) > 0$, then $\rho(s)$ and $a(s)$ converge when $s \to +\infty$ and $\lambda_i(s) \sim C_i s^{\frac{1}{n-2}}$; $s > 0$.

Theorem 0.10 is derived from Theorem 0.11 (Chapter 4 of this book).

Theorem 0.11: For any $\delta > 0$, there exists $\varepsilon > 0$ and $s_0 > 0$ such that if $u(s) \in V(p, \varepsilon_0)$ for $0 \leq s \leq s_0$, $d(a_i(s), \partial\Omega) \geq \delta$ for $0 \leq s \leq s_0$, $d(a_i(s), a_j(s)) \geq \delta\ \forall\, i \neq j$ for $0 \leq s \leq s_0$, then for any $\bar{s} \geq s_0$ such that $u(s)$ remains in $V(p, \varepsilon_0)$ for $s \in (0, \bar{s})$, we have:

$$\frac{\dot{\lambda}_i}{\lambda_i \lambda_i^{\frac{n-2}{-2}}}(\bar{s}) = \frac{\bar{\alpha}\,\omega(u)}{2\lambda_i \alpha_i} \Big\{ \omega(u)^{4/n-2} \alpha_i^{\frac{n+2}{n-2}} \frac{H(a_i, a_i)}{\lambda_i^{\frac{n-2}{2}}}$$

$$-\omega(u)^{4/n-2} \left\{ \sum_{j \neq i} \frac{1}{\lambda_i^{\frac{n-2}{2}}} \times \frac{\alpha_i^{4/n-2} \alpha_j + \alpha_j^{\frac{n+2}{n-2}}}{|a_i - a_j|^{n-2}} - \alpha_j^{\frac{n+2}{n-2}} H(a_i, a_j) \right. +$$

$$+ \sum_{j\neq i} \frac{\alpha_j}{\lambda_j^{\frac{n-2}{2}} |a_i - a_j|^{n-2}} + \frac{1}{\lambda_i^{\frac{n-2}{2}}} o\left(\Sigma \frac{1}{\lambda_k}\right)\Bigg\}$$

$$|\dot{a}_i(\bar{s})| \leq \frac{C}{\lambda_i} \left(\Sigma \frac{1}{\lambda_k^{n-2}}\right) \quad ; \ \bar{\alpha}, \ C \ \text{and} \ \bar{C} \ \text{are constant}$$

$$\dot{\alpha}_i(\bar{s}) = \_\bar{C}\, w(u)\, \alpha_i \left(1 \_ \alpha_i^{4/n-2} \omega(u)^{4/n-2} \int_{\mathbb{R}^n} \delta^{2n/n-2}\right) + 0 \left(\Sigma \frac{1}{\lambda_k^{n-2}}\right)$$

$$\int_{\mathbb{R}^n} |\nabla v|^2 dx(\bar{s}) \leq K \Sigma \frac{1}{\lambda_i^{n-2}} \quad \text{where } v = u - \sum_{i=1}^{p} \alpha_i \delta(a_i, \lambda_i)$$

The preceding formula allows us to describe the difference in topology induced by these critical points at infinity. Namely, considering:

$$C_p = \{(a_1, \_\_, a_p) \in \Omega^p \ \text{s, t} \ M(a) \geq 0\} \tag{8}$$

$$\partial C_p = \{(a_1, \_\_, a_p) \in \Omega^p \ \text{s, t} \ M(a) \geq 0 \ \text{and} \ \rho(a) = 0\}$$

σ_p acts freely on C_p.

Let $b_p = (p+1) S^{2/n-2}$, where $S = \int_{\mathbb{R}^n} \delta^{2n/n-2}$

Let, for $c \in \mathbb{R}$, $I^c = \{u \in \Sigma^+ \ \text{s. t} \ I(u) \leq c\}$

For $p \in \mathbb{N}^\star$, $\Delta_{p-1} = \{(\alpha_1, \_\_, \alpha_p); \ \alpha_i \geq 0 \ \sum_{i=1}^{p} \alpha_i = 1\}$

We then have:

<u>Theorem 0.12:</u> For $\varepsilon > 0$ small enough, the pair

$$(I^{b_p+\varepsilon} \cap V(p,\varepsilon), \ I^{b_p-\varepsilon} \cap V(p,\varepsilon))$$

is homotopically equivalent to

$$(C_p \times_{\sigma_p} \Delta_{p-1}, \ C_p \times \partial\Delta_{p-1} \cup_{\sigma_p} \partial C_p \times \Delta_{p-1}).$$

Theorem 04 again follows from Chapters 3, 4 and 5 of this book. It provides us with the topological difference of topology in the level sets. This formula (A. Bahri-J. M. Coron [015]) should be useful in order to prove Theorem 0.8 under more general conditions on Ω. There should also be a similar formula for Yamabe-type functionals on manifolds.

0.3.2 The Kazdan-Warner Problem: We are looking for sufficient conditions on K such that

$$\begin{cases} -8\Delta u + 6u = K(x)\, u^5 \\ u > 0 \end{cases} \tag{10}$$

has a solution.

We make the following assumptions on K:

K is a positive, C^2, function with non degenerate critical points $y_1, \_\_\_, y_m$ such that $(-8\Delta K + 6K)$, $(y_i) \neq 0 \ \forall_i = 1, \_\_\_, m$.

We then have:

Theorem 0.13: (A. Bahri-J. M. Coron [016][017]). Let k_i be the Morse index of K at y_i. Let $L = \{i/(-8\Delta K + 6K)(y_i) > 0\}$.

If $\sum_{i\in L} (-1)^{k_i} \neq 1$, (10) has a solution.

The proof of Theorem 0.13 relies on the precise study of the critical points at infinity of the variational problem:

In Chapters 3, 4, and 5 of this book, the phenomena taking place along flow-lines are more precisely studied in the case of $\Omega = S^n$.

We prove that, if $K \equiv 1$, then the Palais-Smale condition is satisfied on the flow-lines in the $V(p, \varepsilon)$'s. Similarly, if K is no longer constant, but if $n = 3$, the Palais-Smale condition is satisfied on the flow-lines in the $V(p, \varepsilon)$'s for $p \geq 2$.

Furthermore, for $p = 1$, we see that these flow-lines have to concentrate at the critical points of K. An expansion of K at y_i shows then that we must have $(-8\Delta K + 6K)(y_i) > 0$ and that the Morse index of I at the critical point at infinity defined by y_i is $3-k_i$ where k_i is the Morse index of K at y_i. Hence Theorem 0.13.

The problem is open for $n \geq 4$. However, there is a qualitative difference between the situation for $n = 4$ and the situation for $n \geq 5$.

Indeed, if we try to follow the same method for $n \geq 4$, we are led to consider s-uples $(y_{i_1}, \_\_, y_{i_s})$ of critical points of K and to define a matrix:

$$M(y_{i_1}, \_\_, y_{i_s}) = \begin{bmatrix} \ddots & & a_{rj} & \\ & \ddots & & \\ & a_{jj} & \ddots & \\ & & & \ddots \end{bmatrix} \tag{11}$$

where

$$a_{jj} = -\Delta K(y_{i_j})\ ; \quad a_{r_j} = \frac{-G(y_{i_j}, y_{i_r})}{[K(y_{i_j}) K(y_{i_r})]^{4/n-2}}$$

L is the conformal Laplacian: G is its Green function. We conjecture then that to any s-uple of critical points of K such that $M(y_{i_1}, \_\_, y_{i_s})$ is strictly positive (we assume that these matrices are not degenerate), there is an associated critical point at infinity for I of index $\sum_{j=1}^{s} (n-k_{i_j}) + s-1$. Following the argument for Theorem 0.13, we are led to compare

$$Q = \sum_{\substack{(y_{i_1}, -, y_{i_s}) \\ s\ M > 0}} (-1)^{s(n+1)-1-\sum_{j=1}^{s} k_{i_j}} \quad \text{and } 1.$$

As one can easily see, Q is always equal to 1 for $n \geq 5$. Thus, our method should lead to the following result on S^4: If Q is different from 1, K is the scalar-curvature function of a metric conformal to the standard one on S^4. On S^n, for $n \geq 5$, our method would provide a compactification of the variational problem but would not lead by such easy arguments to an existence theorem.

Conclusion: Following "Pseudo-orbits of contact forms" ([05]), this work shows that critical points at infinity are useful in existence proofs. In a certain way, the critical points at infinity are simpler to understand than the "true" critical points as they are more constrained than these. Thus, often, their study, although very technical, leads to existence proofs.

Conversely, the non compactness in variational problems, understood in the variational sense (i.e. difference of topology in the level sets) leads to new geometrical structures, which are of independent interest.

However, certain forms of non compactness are more complicated, such as those related to an equation such as $\Delta u + u^7 = 0$ on an open set Ω of $\mathbb{R}^3$.

References for introduction

[01] J. Milnor. Morse Theory. Annals of Math. Studies 51, Princeton Univ. Press, (1963).

[02] J. Sacks-K. Uhlenbeck. The existence of minimal immersions of 2-spheres. Annals of Math. 113 (1981), 1-24.

[03] Y. T. Siu-S. T. Yan. Compact Kahler manifolds of positive bisectional curvature. Inv. Mathematical 59 (1980), 189-204.

[04] H. Brezis-J. M. Coron. Convergence of solutions of H. systems or how to blow bubbles. Archive Rat. Mech. Anal. 89, 1 (1985), 21-56.

[05] A. Bahri. Pseudo-orbits of contact forms. Longman Scientific and Technical, 1988.

[06] T. Aubin. Metrique riemannienne et courture. J. Diff. Geo. 11 (1976), 573-598.

[07] R. Schoën. Conformal deformation of a Riemannian metric to constant scalar curvature. J. Diff. Geo. 20 (1984), 479-495.

[08] A. Bahri-H. Brezis. To appear.

[09] J. Kazdan-F. Warner. Remarks on some quasilinear elliptic equations. Comm. Pure Appl. Math. 38 (1975), 557-569.

[010] J. P. Bourguignon-J. P. Ezin. Scalar curvature functions on a conformal class of metrics and conformal transformations. Preprint.

[011] P. L. Lions. The concentration compactness principle in the calculus of variations, the limit case. Rev. Mat. Iberoamericana, 1, 1 (1985), 145-201 and 1, 21 (1985), 45-121.

[012] M. Struwe. A global compactness result for elliptic boundary problems involving nonlinearities. Math. Z. 187 (1984), 511-517.

[013] C. H. Taubes. Path connected Yang-Mills moduli spaces. J. Diff. Geo. 19 (1982), 337-392.

[014] A. Bahri-J. M. Coron. On a non linear elliptic equation involving the limit Sobolev exponent. To appear, Comm. Pure and Applied Math.

[015] A. Bahri-J. M. Coron. Sur une equation elliptique non lineaire avec l'exposant critique de Sobolev. C. R. Acad. Sc. Paris, 301, serie I (1985), 345-348.

[016] A. Bahri-J. M. Coron. Un theorie des points critiques à l'infini pour l'equation de Yamabe et le problème de Kazdan-Warner. C. R. Acad. Sc. Paris, 300, serie I (1985), n'15, 513-516.

[017] A. Bahri-J. M. Coron. To appear.

Part 1 The flow

1 Technical lemmas

Let α be strictly positive real.

In the following lemmas, X^α is the function $|X|^\alpha$ or $|X|^{\alpha-1}X$, indifferently; however, the same function is used in each formula, together with its corresponding derivatives.

Lemma 1.1: There exists a constant M, such that for any $(a,b) \in \mathbb{R}^2$:

$$|(a+b)^\alpha - a^\alpha - \alpha a^{\alpha-1} b| \leq M(|b|^\alpha + |a|^{\alpha-2} \inf(a^2, b^2)) \tag{1.1}$$

Lemma 1.2: There exists a constant M, such that for any $(a_1, -, a_p) \in \mathbb{R}^p$:

$$|(\Sigma a_i)^\alpha - \Sigma a_i^\alpha| \leq M \sum_{i \neq j} |a_i|^{\alpha-1} \inf(|a_i|, |a_j|) \tag{1.2}$$

Lemma 1.3: There exists a constant M, such that for any $(a_1, -, a_p) \in \mathbb{R}^p$:

$$|(\Sigma a_i)^\alpha - \Sigma a_i^\alpha - \alpha \sum_{i \neq j} a_i^{\alpha-1} a_j| \leq M(\sum_{i \neq j} \operatorname{Sup}(|a_i|^{\alpha-2}, |a_j|^{\alpha-2}) \inf(a_i^2, a_j^2) +$$

$$+ \sum_{i \neq j} \inf(|a_i|^{\alpha-1}, |a_j|^{\alpha-1}) \operatorname{Sup}(|a_i|, |a_j|)) \tag{1.3}$$

Lemma 1.4: There exists a constant M, such that for any $(a_1, -, a_p) \in \mathbb{R}^p$:

$$|(\Sigma a_i)^\alpha - \Sigma a_i^\alpha - \alpha a_{i_0}^{\alpha-1} (\sum_{i \neq i_0} a_i)| \leq M(\sum_{\substack{i \neq i_0 \\ i \neq k}} |a_i|^{\alpha-1} \inf(|a_i|, |a_k|) +$$

$$+ \sum_{i \neq i_0} |a_{i_0}|^{\alpha-2} \inf(a_{i_0}^2, a_i^2) + \sum_{i \neq i_0} \inf(|a_i|^{\alpha-1}, |a_{i_0}|^{\alpha-1}) (\sum_{i \neq i_0} |a_i|) \tag{1.4}$$

The proof of these lemmas just requires some thought and is thus left to the reader.

We now introduce, for $x_i \in \mathbb{R}^n$ and $\lambda_i > 0$:

$$\delta_i(x) = \delta(x_i, \lambda_i)(x) = C_0 \frac{\lambda_i^{\frac{n-2}{2}}}{(1+\lambda_i^2 |x-x_i|^2)^{\frac{n-2}{2}}} \tag{1.5}$$

With a suitable choice of C_0, δ_i satisfies:

$$\Delta \delta_i + \delta_i^{\frac{n+2}{n-2}} = 0 \text{ in } \mathbb{R}^n. \tag{1.6}$$

Let

$$\varepsilon_{ij} = \left(\frac{1}{\lambda_i/\lambda_j + \lambda_j/\lambda_i + \lambda_i\lambda_j |x_i - x_j|^2} \right)^{\frac{n-2}{2}} ; \; x_i \text{ and } x_j \in \mathbb{R}^n; \; \lambda_i > 0; \; \lambda_j > 0 \tag{1.7}$$

We then have the following estimates:

<u>Estimate 1:</u> $\int_{\mathbb{R}^n} \delta_i^{\frac{n+2}{n-2}} \delta_j = C_0^{2n/n-2} C_1 \varepsilon_{ij} + 0(\varepsilon_{ij}^{n/n-2})$; $C_1 > 0$ (E1)

$$\left(C_1 = \int_{\mathbb{R}^n} \frac{dx}{(1+|x|^2)^{\frac{n+2}{2}}} \right)$$

<u>Estimate 2:</u> $\int_{\mathbb{R}^n} \delta_i^{n/n-2} \delta_j^{n/n-2} = 0(\varepsilon_{ij}^{n/n-2} \log \varepsilon_{ij}^{-1})$ (E2)

<u>Estimate 3:</u> Let α, β, $\alpha > 1$, $\beta > 1$ such that $\alpha + \beta = \frac{2n}{n-2}$; $\theta = \inf(\alpha, \beta)$

$$\int_{\mathbb{R}^n} \delta_i^{\alpha} \delta_j^{\beta} = 0 \left(\varepsilon_{ij} (\log \frac{1}{\varepsilon_{ij}})^{\frac{n-2}{n}\theta} \right) \tag{E3}$$

<u>Estimate 4:</u> Let $\alpha > \beta > 1$; $\alpha + \beta = \frac{2n}{n-2}$

$$\int_{\mathbb{R}^n} \delta_i^{\beta} \inf(\delta_i^{\alpha}, \delta_j^{\alpha}) = 0(\varepsilon_{ij}^{n/n-2} \log \varepsilon_{ij}^{-1}) \tag{E4}$$

<u>Estimate 5:</u> There exists a constant C such that:

$$\left| \frac{\partial \delta_i}{\partial x_i} \right| \leq C \lambda_i \tag{E5}$$

$$\left|\frac{\partial \delta_i}{\partial \lambda_i}\right| \leq \frac{C}{\lambda_i} \tag{E6}$$

Proof of Estimates 1-5

The proof of Estimate 5 is a straightforward computation. (E4) is easily derived from (E2) as:

$$\delta_i^{\beta} \inf(\delta_i^{\alpha}, \delta_j^{\alpha}) \leq \delta_i^{n/n-2}\ \delta_j^{n/n-2} \tag{1.8}$$

We derive (E3) from (E2) as follows:

Assume $\theta = \alpha$ for instance.

Let

$$p_1 = \frac{1}{\alpha}\frac{n}{n-2}; \quad q_1 = \frac{p_1}{p_1 - 1} \tag{1.9}$$

Then

$$\int_{\mathbb{R}^n} \delta_i^{\alpha} \delta_j^{\beta} \leq \left(\int_{\mathbb{R}^n} \delta_i^{n/n-2}\ \delta_j^{n/n-2}\right)^{\frac{n-2}{n}\alpha} \left(\int_{\mathbb{R}^n} \delta_j^{(\beta-\alpha)q_1}\right)^{1/q_1} \tag{1.10}$$

by Hölder's inequality.

Now,

$$(\beta-\alpha)\, q_1 = 2\left(\frac{n}{n-2} - \alpha\right) \frac{n}{n-\alpha(n-2)} = \frac{2n}{n-2} \tag{1.11}$$

(E3) follows from (1.10) as $\int_{\mathbb{R}^n} \delta_j^{2n/n-2}$ is a constant independent of j.

Proof of (E1): Let

$$I = \int_{\mathbb{R}^n} \delta_j^{\frac{n+2}{n-2}}\ \delta_i \tag{1.12}$$

and

$$y = \lambda_i x \tag{1.13}$$

Let

$$\mu = \mathrm{Max}(\lambda_i/\lambda_j,\ \lambda_j/\lambda_i,\ \lambda_i\lambda_j|x_i - x_j|^2) \tag{1.14}$$

$$d_{ij} = x_i - x_j \, . \tag{1.15}$$

I is equal to:

$$I = C_0^{\frac{2n}{n-2}} \left(\frac{\lambda_j}{\lambda_i}\right)^{\frac{n-2}{2}} \int_{\mathbb{R}^n} \left(\frac{1}{1+|x|^2}\right)^{\frac{n+2}{2}} \left(\frac{1}{1+\left|\frac{\lambda_j}{\lambda_i} y+\lambda_j d_{ij}\right|^2}\right)^{\frac{n-2}{2}} dy \tag{1.16}$$

First assume:

$$\mu = \lambda_i / \lambda_j \, . \tag{1.17}$$

We then have:

$$1 + \left|\frac{\lambda_j}{\lambda_i} x + \lambda_j d_{ij}\right|^2 = (1+\lambda_j^2 |d_{ij}|^2) \left(1+2 \frac{\lambda_j}{1+\lambda_j^2 |d_{ij}|^2} d_{ij} \cdot \frac{\lambda_j}{\lambda_i} x + \right. \tag{1.18}$$

$$\left. + \left(\frac{\lambda_j}{\lambda_i}\right)^2 \frac{|x|^2}{1+\lambda_j^2 |d_{ij}|^2}\right)$$

Thus, we have, if $|x| \leq 1/4 \, \lambda_i / \lambda_j$

$$\left(\frac{1}{1+\left(\frac{\lambda_j}{\lambda_i} x+\lambda_j d_{ij}\right)^2}\right)^{\frac{n-2}{2}} = \left(\frac{1}{1+\lambda_j^2 |d_{ij}|^2}\right)^{\frac{n-2}{2}} \left(1-(n-2) \frac{\lambda_j}{1+\lambda_j^2 |d_{ij}|^2} d_{ij} \cdot \frac{\lambda_j}{\lambda_i} x \right.$$

$$\left. + \, 0 \left(\left(\frac{\lambda_j}{\lambda_i}\right)^2 |x|^2\right)\right) \tag{1.19}$$

Now:

$$\int_{|x| \leq 1/4 \, \lambda_i / \lambda_j} \frac{|x|^2}{(1+|x|^2)^{\frac{n+2}{2}}} dx = 0 \left(\operatorname{Log} \frac{\lambda_i}{\lambda_j} \right) \tag{1.20}$$

$$\int_{|x| \leq 1/4 \, \lambda_i / \lambda_j} \frac{dx}{(1+|x|^2)^{\frac{n+2}{2}}} = C_1 + 0 \left(\left(\frac{\lambda_j}{\lambda_i}\right)^2\right) \tag{1.21}$$

$$\int_{|x|\geq 1/4\lambda_i/\lambda_j} \frac{dx}{(1+|x|^2)^{\frac{n+2}{2}}} = 0\left(\left(\frac{\lambda_j}{\lambda_i}\right)^2\right) \tag{1.22}$$

$$\int \frac{x\,dx}{(1+|x|^2)^{\frac{n+2}{2}}} = 0 \tag{1.23}$$

From these estimates, we derive:

$$I = C_0^{2n/n-2}\, C_1 \left(\frac{\lambda_j}{\lambda_i(1+\lambda_j^2|d_{ij}|^2)}\right)^{\frac{n-2}{2}} + 0\left(\left(\frac{\lambda_j}{\lambda_i}\right)^{\frac{n+2}{2}} \log\frac{\lambda_j}{\lambda_i}\right) \tag{1.24}$$

Thus, under (1.17), when ε_{ij} goes to zero:

$$I = C_0^{2n/n-2}\, C_1\, \varepsilon_{ij} + 0\left(\varepsilon_{ij}^{\frac{n+2}{n-2}} \log\frac{1}{\varepsilon_{ij}}\right) \tag{1.25}$$

(1.25) is (E1) if $\mu = \lambda_i/\lambda_j$.
The case $\mu = \lambda_j/\lambda_i$ is similar as:

$$\int_{\mathbb{R}^n} \delta_i^{\frac{n+2}{n-2}}\, \delta_j = -\int_{\mathbb{R}^n} \Delta\delta_i\,\delta_j = -\int_{\mathbb{R}^n} \delta_i\, \Delta\delta_j = \int_{\mathbb{R}^n} \delta_i\,\delta_j^{\frac{n+2}{n-2}} \tag{1.26}$$

We are thus left with the case:

$$\mu = \lambda_i\lambda_j|d_{ij}|^2 \tag{1.27}$$

In this case, we write I as follows:

$$I = C_0^{2n/n-2} \int_{\mathbb{R}^n} \frac{1}{(1+|x|^2)^{\frac{n+2}{2}}}\; \frac{dx}{(\lambda_i/\lambda_j + |\sqrt{\lambda_j/\lambda_i}\,x + \sqrt{\lambda_i\lambda_j}\,d_{ij}|^2)^{\frac{n-2}{2}}} \tag{1.28}$$

Using (1.26), we may assume:

$$\lambda_j \leq \lambda_i \tag{1.29}$$

We have

$$\lambda_i/\lambda_j + \left|\sqrt{\lambda_j/\lambda_i}\,x + \sqrt{\lambda_i\lambda_j}\,d_{ij}\right|^2 = (\lambda_i/\lambda_j + \lambda_i\lambda_j|d_{ij}|^2) \times \tag{1.30}$$

$$\times\left\{1 + \frac{\lambda_j/\lambda_i|x|^2 + 2\lambda_j x.\,a_{ij}}{\lambda_i/\lambda_j + \lambda_i\lambda_j|d_{ij}|^2}\right\}$$

By the same arguments used in the case where (1.17) holds, we have:

$$\int_{|x|\le\frac{\sqrt{\mu}}{10}} \frac{1}{(1+|x|^2)^{\frac{n+2}{2}}} \frac{dx}{(\lambda_i/\lambda_j+|\sqrt{\lambda_j/\lambda_i}x+\sqrt{\lambda_i\lambda_j}d_{ij}|^2)^{\frac{n-2}{2}}} \tag{1.31}$$

$$= C_1\varepsilon_{ij} + 0(\varepsilon_{ij}^{n/n-2})$$

Let

$$B_1 = \{x \in \mathbb{R}^n / |x + \lambda_i d_{ij}| \le 1/10\,\lambda_i |d_{ij}|\} \tag{1.32}$$

$$B_2 = \{x \in \mathbb{R}^n / |x| \le \frac{\sqrt{\mu}}{10}\}. \tag{1.33}$$

and

$$L(x) = \frac{1}{(1+|x|^2)^{\frac{n+2}{2}} (\lambda_i/\lambda_j+|\sqrt{\lambda_i/\lambda_j}x+\sqrt{\lambda_i\lambda_j}d_{ij}|^2)^{\frac{n-2}{2}}} \tag{1.34}$$

By definition of B_1 and B_2, we have:

$$\int_{(B_1\cup B_2)^c} L(x)\,dx \le \frac{C}{\mu^{\frac{n-2}{2}}} \int_{\sqrt{\mu}}^{+\infty} \frac{r^{n-1}dr}{(1+r^2)^{\frac{n+2}{2}}} \tag{1.35}$$

Thus

$$\int_{(B_1\cup B_2)^c} L(x)\,dx = 0\left(\frac{1}{\mu^{\frac{n-2}{2}}}\right) = 0(\varepsilon_{ij}^{n/n-2}) \tag{1.36}$$

While, on B_1, we have:

$$|x| \geq 9/10\ \lambda_i |d_{ij}| \tag{1.37}$$

Thus

$$\int_{B_1} L(x)\,dx \leq \frac{C}{\lambda_i^{n+2} |d_{ij}|^{n+2}} \left(\frac{\lambda_j}{\lambda_i}\right)^{\frac{n-2}{2}} \int_0^{\lambda_i |d_{ij}|} \frac{r^{n-1}}{\left(1+\frac{\lambda_j^2}{\lambda_i^2} r^2\right)^{\frac{n-2}{2}}} dr \tag{1.38}$$

$$\leq \frac{C}{\lambda_i^{n+2} |d_{ij}|^{n+2}} \left(\frac{\lambda_i}{\lambda_j}\right)^{\frac{n+2}{2}} \lambda_j \lambda_i |d_{ij}|^2 = 0(\varepsilon_{ij}^{n/n-2})$$

(1.31), (1.36) and (1.38) yield (E1) in the case when $\mu = \lambda_i \lambda_j |d_{ij}|^2$ (the case $\lambda_i \leq \lambda_j$ is symmetric as already noted).
This completes the proof of (E1).

Proof of (E2) :

We consider

$$S = \int_{\mathbb{R}^n} \varepsilon_i^{n/n-2} \delta_j^{n/n-2} = C_0^{2n/n-2} \int_{\mathbb{R}^n} \frac{(\lambda_i \lambda_j)^{n/2}}{(1+\lambda_i^2 |x-x_i|^2)^{n/2} (1+\lambda_j^2 |x-x_j|^2)^{n/2}} \tag{1.39}$$

Let

$$a_{ij} = \frac{x_j - x_i}{2} \quad ; \quad Z = x - \frac{x_i + x_j}{2} \tag{1.40}$$

We have

$$S = C_0^{2n/n-2} \frac{1}{(\lambda_i \lambda_j)^{n/2}} \int_{\mathbb{R}^n} \frac{dz}{\left(\frac{1}{\lambda_i^2} + |z+a_{ij}|^2\right)^{n/2} \left(\frac{1}{\lambda_j^2} + |z-a_{ij}|^2\right)^{n/2}} \tag{1.41}$$

By symmetry arguments, we may assume:

$$\lambda_j \leq \lambda_i \tag{1.42}$$

Consider first the case:

$$\lambda_i\lambda_j|a_{ij}|^2 \geq C\,\lambda_i/\lambda_j; \quad \text{with } C \text{ a large constant} \tag{1.43}$$

Thus

$$\lambda_j^2|a_{ij}|^2 \geq C\,; \quad \lambda_i^2|a_{ij}|^2 \geq C \tag{1.44}$$

Then

$$S \leq \frac{1}{(\lambda_i\lambda_j)^{n/2}}\left\{\int_{|z+a_{ij}|\leq 1/\lambda_i}\frac{C_1\lambda_i^n}{(\frac{1}{\lambda_j^2}+|a_{ij}|^2)^{n/2}}\,dz + \right. \tag{1.45}$$

$$\left. + \int_{|z-a_{ij}|\leq 1/\lambda_j}\frac{C_1\lambda_j^n}{(\frac{1}{\lambda_i^2}+|a_{ij}|^2)^{n/2}}\,dz + \int_{\substack{|z+a_{ij}|\geq\frac{1}{\lambda_i}\\ |z-a_{ij}|\geq\frac{1}{\lambda_j}}}\frac{C_1\,dz}{|z+a_{ij}|^n|z-a_{ij}|^n}\right\}$$

with C_1 a suitable constant.
Indeed, if $|z+a_{ij}| \leq 1/\lambda_i$, hence by (1.44), if $|z+a_{ij}| \leq |a_{ij}|$, then $|z-a_{ij}| \geq |a_{ij}|$; and if $|z-a_{ij}| \leq \frac{1}{\lambda_j}$, hence by (1.44), if $|z-a_{ij}| \leq |a_{ij}|$, then $|z+a_{ij}| \geq |a_{ij}|$.
Consider:

$$\int_{\substack{|z+a_{ij}|\geq\frac{1}{\lambda_i}\\ |z-a_{ij}|\geq\frac{1}{\lambda_j}}}\frac{dz}{|z+a_{ij}|^n|z-a_{ij}|^n} \tag{1.46}$$

We proceed as follows:

$$\int_{\substack{|z-a_{ij}|\geq\frac{1}{\lambda_i}\\ |z-a_{ij}|\geq\frac{1}{\lambda_j}}}\frac{dz}{|z+a_{ij}|^n|z-a_{ij}|^n} \leq \int_{\frac{1}{\lambda_i}\leq|z+a_{ij}|\leq|a_{ij}|}\frac{dz}{|z+a_{ij}|^n|a_{ij}|^n} +$$

$$+ \int_{\frac{1}{\lambda_j} \le |z-a_{ij}| \le |a_{ij}|} \frac{dz}{|a_{ij}|^n |z-a_{ij}|^n} + \int_{\substack{|z+a_{ij}| \ge |a_{ij}| \\ |z-a_{ij}| \ge |a_{ij}|}} \frac{dz}{|z+a_{ij}|^n |z-a_{ij}|^n}$$

(1.47)

Again, we used here the fact that if $|z \pm a_{ij}| \le |a_{ij}|$, then $|z \mp a_{ij}| \ge |a_{ij}|$.

Now,

$$\int_{\substack{|z+a_{ij}| \ge |a_{ij}| \\ |z-a_{ij}| \ge |a_{ij}|}} \frac{dz}{|z+a_{ij}|^n |z-a_{ij}|^n} \le C_2 \int_{|z+a_{ij}| \ge |a_{ij}|} \frac{dz}{|z+a_{ij}|^{2n}} =$$

$$= \frac{C_2}{n+1} \frac{1}{|a_{ij}|^n} \times \mathrm{Vol}(S^{n-1}) \tag{1.48}$$

with C_2, a suitable constant.

Indeed, if $|z+a_{ij}| \ge |a_{ij}|$ and $|z-a_{ij}| \ge |a_{ij}|$, then $|z-a_{ij}| \ge \frac{1}{C_2^{1/n}} |z+a_{ij}|$, with C_2 large enough.

On the other hand,

$$\int_{\frac{1}{\lambda_i} \le |z+a_{ij}| \le |a_{ij}|} \frac{dz}{|z+a_{ij}|^n} = \mathrm{Log}\, \lambda_i |a_{ij}| \; ; \tag{1.49}$$

$$\int_{\frac{1}{\lambda_j} \le |z-a_{ij}| \le |a_{ij}|} \frac{dz}{|z-a_{ij}|^n} = \mathrm{Log}\, \lambda_j |a_{ij}|$$

(1.45), (1.47), (1.48) and (1.49) yield:

$$S \le C_3 \left[\frac{1}{\left(\frac{\lambda_i}{\lambda_j} + \lambda_i \lambda_j |a_{ij}|^2\right)^{n/2}} + \frac{1}{\left(\frac{\lambda_j}{\lambda_i} + \lambda_i \lambda_j |a_{ij}|^2\right)^{n/2}} + \frac{1}{(\lambda_i \lambda_j |a_{ij}|^2)^{n/2}} \times \right.$$
$$\left. \times \left[\mathrm{Log}(\lambda_i \lambda_j |a_{ij}|^2) + C\right] \right] \tag{1.50}$$

which yield (E2) under (1.43).

We now assume:

$$\lambda_j \leq \lambda_i \qquad \text{(symmetry arguments)} \tag{1.42}$$

$$\lambda_i \lambda_j |a_{ij}|^2 \leq C\, \lambda_i/\lambda_j \tag{1.51}$$

i. e.

$$\lambda_j^2 |a_{ij}|^2 \leq C \tag{1.52}$$

We then set in (1. 39) :

$$z = \lambda_i (x - x_i) \tag{1.53}$$

Hence:

$$S = \left(\frac{\lambda_j}{\lambda_i}\right)^{n/2} \int_{\mathbb{R}^n} \frac{dz}{(1+|z|^2)^{n/2} (1+|\frac{\lambda_j}{\lambda_i} z - 2a_{ij}\lambda_j|^2)^{n/2}} \tag{1.54}$$

which we split in two parts:

$$\left(\frac{\lambda_j}{\lambda_i}\right)^{n/2} \int_{|z| \leq 4C \frac{\lambda_i}{\lambda_j}} \frac{dz}{(1+|z|^2)^{n/2} (1+|\frac{\lambda_j}{\lambda_i} z - 2a_{ij}\lambda_j|^2)^{n/2}} = T \tag{1.55}$$

and

$$\left(\frac{\lambda_j}{\lambda_i}\right)^{n/2} \int_{|z| \geq 4C \frac{\lambda_i}{\lambda_j}} \frac{dz}{(1+|z|^2)^{n/2} (1+|\frac{\lambda_j}{\lambda_i} z - 2a_{ij}\lambda_j|^2)^{n/2}} = F \tag{1.56}$$

C defined in (1. 52)

We have:

$$T \leq C_1 \left(\frac{\lambda_j}{\lambda_i}\right)^{n/2} \operatorname{Log}(\lambda_i/\lambda_j) \;; \quad C_1 \text{ appropriate constant} \tag{1.57}$$

hence

$$T \leq 0(\varepsilon_{ij}^{n/n-2} \operatorname{Log} \varepsilon_{ij}^{-1}) \tag{1.58}$$

while, as $\left|\frac{\lambda_j}{\lambda_i} z - 2a_{ij}\lambda_j\right| \geq \frac{\lambda_j}{\lambda_i} \frac{|z|}{2}$ if $|z| \geq 4C\ \lambda_i/\lambda_j$,

$$F \leq C_1(\lambda_j/\lambda_i)^{n/2} \left(\frac{\lambda_i}{\lambda_j}\right)^{n/2} \int_{|z| \geq 4C\lambda_i/\lambda_j} \frac{dz}{|z|^{2n}} = \frac{C_1}{n+1} \frac{\operatorname{Vol}(S^{n-1})}{(4C)^n} \left(\frac{\lambda_j}{\lambda_i}\right)^n \tag{1.59}$$

Hence

$$F \leq 0(\varepsilon_{ij}^{2n/n-2}) \tag{1.60}$$

(1.58) and (1.60) prove (E2) under (1.51).

This completes the proof of (E1).

2 First estimates relevant to the theory

C is a constant; ε_{ij} is assumed to be small for $i \neq j$

2.1 Gradient type estimates

Estimate F1:

$$\left|\int_{\mathbb{R}^n} \nabla \frac{\partial \delta_i}{\partial x_i} \nabla \delta_j\right| \le C \lambda_i \varepsilon_{ij} \quad \text{if } i \neq j; \quad \int_{\mathbb{R}^n} \nabla \frac{\partial \delta_i}{\partial x_i} \nabla \delta_i = 0$$

Estimate F2:

$$\left|\int_{\mathbb{R}^n} \nabla \frac{\partial \delta_i}{\partial \lambda_i} \nabla \delta_j\right| \le \frac{C}{\lambda_i} \varepsilon_{ij} \quad \text{if } i \neq j; \quad \int_{\mathbb{R}^n} \nabla \frac{\partial \delta_i}{\partial \lambda_i} \nabla \delta_i = 0$$

Estimate F3:

$$\int_{\mathbb{R}^n} \nabla \delta_i \nabla \delta_j = \varepsilon_{ij} + 0(\varepsilon_{ij}^{n/n-2}) \quad \text{if } i \neq j; \quad \int_{\mathbb{R}^n} |\nabla \delta_i|^2 = C_0 > 0$$

Estimate F4:

$$\left|\int_{\mathbb{R}^n} \nabla \frac{\partial \delta_i}{\partial x_i} \nabla \frac{\partial \delta_j}{\partial x_j}\right| \le C\lambda_i\lambda_j\varepsilon_{ij} \quad \text{if } i \neq j; \quad \int_{\mathbb{R}^n} \left|\nabla \frac{\partial \delta_i}{\partial x_i}\right|^2 = \lambda_i^2 a_0; \quad a_0 > 0$$

Estimate F5:

$$\left|\int_{\mathbb{R}^n} \nabla \frac{\partial \delta_i}{\partial \lambda_i} \nabla \frac{\partial \delta_j}{\partial x_j}\right| \le C \frac{\lambda_j}{\lambda_i} \varepsilon_{ij} \quad \text{if } i \neq j; \quad \int_{\mathbb{R}^n} \nabla \frac{\partial \delta_i}{\partial \lambda_i} \nabla \frac{\partial \delta_i}{\partial x_i} = 0$$

Estimate F6:

$$\left|\int_{\mathbb{R}^n} \nabla \frac{\partial \delta_i}{\partial \lambda_i} \nabla \frac{\partial \delta_j}{\partial \lambda_j}\right| \le \frac{C\varepsilon_{ij}}{\lambda_i\lambda_j} \quad \text{if } i \neq j; \quad \int_{\mathbb{R}^n} \left|\nabla \frac{\partial \delta_i}{\partial \lambda_i}\right|^2 = \frac{\gamma_1}{\lambda_i^2}; \quad \gamma_1 > 0$$

Proof of Estimates F1-F6:

It is easy to check that:

$$\Delta \frac{\partial \delta_i}{\partial \lambda_i} + \frac{n+2}{n-2} \delta_i^{4/n-2} \frac{\partial \delta_i}{\partial \lambda_i} = 0$$

$$\Delta \frac{\partial \delta_i}{\partial x_i} + \frac{n+2}{n-2} \delta_i^{4/n-2} \frac{\partial \delta_i}{\partial x_i} = 0 \tag{2.2}$$

We also recall that:

$$\Delta \delta_i + \delta_i^{\frac{n+2}{n-2}} = 0 \tag{1.6}$$

$$\left| \frac{\partial \delta_i}{\partial \lambda_i} \right| \leq \frac{C}{\lambda_i} \quad ; \quad \left| \frac{\partial \delta_i}{\partial x_i} \right| \leq C\lambda_i \tag{E5}$$

$$\int_{\mathbb{R}^n} \delta_i^{\frac{n+2}{n-2}} \delta_j = \varepsilon_{ij} + 0\ (\varepsilon_{ij}^{n/n-2}) \tag{E1}$$

Then, for $i \neq j$, F1-F6 follow through integration by parts and application of either (2.1), (2.2) or (1.6), together with the bounds provided by (E5) and the estimate (E1) finally.

For $i = j$, F1-F6 are obtained by rescaling and translation invariance.

Remark: These estimates will be improved later on in this work.

2.2 Interaction estimates

We will need:

$$\theta_j \text{ is the solution of } \begin{cases} \Delta \theta_j = 0 \\ \theta_j = \delta_j \big|_{\partial \Omega} \end{cases} ; \quad \theta_j = 0 \text{ if } \Omega = \mathbb{R}^n \tag{2.3}$$

$$\frac{\partial \theta_j}{\partial x_j} \tag{2.4}$$

$$\frac{\partial \theta_j}{\partial \lambda_j}; \tag{2.5}$$

As can be easily checked, we have, denoting $H(x,y)$ the regular part of the Green's function on Ω and $d_j = d(x_j, \partial\Omega)$

$$\left|\theta_j(x) - \frac{1}{\lambda_j^{\frac{n-2}{2}}} C_0 H(x,x_j)\right| = 0\left(\frac{1}{\lambda_j^{\frac{n-2}{2}} d_j^n}\right) \quad \forall x \in \Omega \tag{2.5)'}$$

$$\left|\frac{\partial\theta_j}{\partial x_j}(x) - \frac{1}{\lambda_j^{\frac{n-2}{2}}} C_0 \frac{\partial H}{\partial y}(x,x_j)\right| = 0\left(\frac{1}{\lambda_j^{\frac{n+2}{2}}} \frac{1}{d_j^{n+1}}\right) \quad \forall x \in \Omega$$

$$\left|\frac{\partial\theta_j}{\partial\lambda_j}(x) + \frac{n-2}{2}\frac{1}{\lambda_j^{n/2}} H(x,x_j)\right| = 0\left(\frac{1}{\lambda_j^{\frac{n+4}{2}}} \frac{1}{d_j^n}\right) \quad \forall x \in \Omega$$

Indeed, these estimates hold on the boundary of Ω; hence in Ω by the maximum principle.

[$H(x,y)$ is the solution of:

$$\begin{cases} \Delta_y H(x,y) = 0 \\ H(x,y) = \dfrac{1}{|x-y|^{n-2}} \quad \text{on } \partial\Omega \;] \end{cases} \tag{2.5)''}$$

We want to estimate

$$\int_\Omega \nabla(\Delta_\Omega^{-1} K(x)\ (\Sigma\,\alpha_i\delta_i)^{\frac{n+2}{n-2}})\,\nabla\delta_j = A \tag{2.6}$$

$$\int_\Omega \nabla(\Delta_\Omega^{-1} K(x)\ (\Sigma\,\alpha_i\delta_i)^{\frac{n+2}{n-2}})\,\nabla\frac{\partial\delta_j}{\partial x_j} = B \tag{2.7}$$

$$\int_\Omega \nabla(\Delta_\Omega^{-1} K(x)\ (\Sigma\,\alpha_i\delta_i)^{\frac{n+2}{n-2}})\,\nabla\frac{\partial\delta_j}{\partial\lambda_j} = C \tag{2.8}$$

where K is a smooth and bounded function on $\overline{\Omega}$. For this, we will use, for $n \geq 4$

$$\left|(\Sigma\,\alpha_i\delta_i)^{\frac{n+2}{n-2}} - \Sigma\,\alpha_i^{\frac{n+2}{n-2}}\delta_i^{\frac{n+2}{n-2}} - \frac{n+2}{n-2}(\alpha_j\delta_j)^{4/n-2}\Big(\sum_{i\neq j}\alpha_i\delta_i\Big)\right| \leq \tag{2.9}$$

$$\leq C\Bigg[\sum_{\substack{i\neq j\\ i\neq k}}(\alpha_i\delta_i)^{4/n-2}\,\mathrm{Inf}(\alpha_i\delta_i,\alpha_k\delta_k) + \sum_{i\neq j}(\alpha_j\delta_j)^{4/n-2^{-1}}\,\mathrm{Inf}((\alpha_j\delta_j)^2,(\alpha_i\delta_i)^2)+$$

$$+ \sum_{i \neq j} \operatorname{Inf}((\alpha_i \delta_i)^{4/n-2}, (\alpha_j \delta_j)^{4/n-2}) (\sum_{i \neq j} \alpha_i \delta_i) \Big]$$

(2.9) is just (1.4) with $j = i_0$ and $\alpha_i \delta_i = a_i$.

For $n = 3$, we will just make a direct computation.

Through integration by parts in (2.6), (2.7), (2.8) and using (2.3), (2.4) and (2.5), we have:

$$A = \int_\Omega \nabla(\Delta_\Omega^{-1} K(x) (\Sigma \alpha_i \delta_i)^{\frac{n+2}{n-2}}) \nabla \delta_j = - \int_\Omega K(x) (\Sigma \alpha_i \delta_i)^{\frac{n+2}{n-2}} (\delta_j - \theta_j) \quad (2.10)$$

$$B = \int_\Omega \nabla(\Delta_\Omega^{-1} K(x) (\Sigma \alpha_i \delta_i)^{\frac{n+2}{n-2}}) \nabla \frac{\partial \delta_j}{\partial x_j} = - \int_\Omega K(x) (\Sigma \alpha_i \delta_i)^{\frac{n+2}{n-2}} (\frac{\partial \delta_j}{\partial x_j} - \frac{\partial \theta_j}{\partial x_j}) \quad (2.11)$$

$$C = \int_\Omega \nabla(\Delta_\Omega^{-1} K(x) (\Sigma \alpha_i \delta_i)^{\frac{n+2}{n-2}}) \nabla \frac{\partial \delta_j}{\partial \lambda_j} = - \int_\Omega K(x) (\Sigma \alpha_i \delta_i)^{\frac{n+2}{n-2}} (\frac{\partial \delta_j}{\partial \lambda_j} - \frac{\partial \theta_j}{\partial \lambda_j}) \quad (2.12)$$

We then have:

<u>Lemma 2.1</u>:

$$\Big| A + \Sigma \alpha_i^{\frac{n+2}{n-2}} \int_\Omega K(x) \delta_i^{\frac{n+2}{n-2}} (\delta_j - \theta_j) + \frac{n+2}{n-2} \alpha_j^{4/n-2} \int_\Omega K(x) \delta_j^{\frac{n+2}{n-2}} (\sum_{i \neq j} \alpha_i \delta_i) - \frac{n+2}{n-2} \alpha_j^{4/n-2} \times$$

$$\times \int_\Omega K(x) \delta_j^{4/n-2} \theta_j (\sum_{i \neq j} \alpha_i \delta_i) \Big|$$

$$\leq 0 (\sum_{i \neq k} \varepsilon_{ik}^{n/n-2} \operatorname{Log} \varepsilon_{ik}^{-1}) \qquad \text{for } n \geq 4$$

$$\leq 0 (\sum_{\substack{i \neq j \\ i \neq k}} \varepsilon_{ik}^{3} \operatorname{Log} \varepsilon_{ik}^{-1} + \sum_{i \neq k} \varepsilon_{ij}^{2} (\operatorname{Log} \varepsilon_{ij}^{-1})^{2/3}) \quad \text{for } n = 3$$

$$\Big| B + \Sigma \alpha_i^{\frac{n+2}{n-2}} \int_\Omega K(x) \delta_i^{\frac{n+2}{n-2}} (\frac{\partial \delta_j}{\partial x_j} - \frac{\partial \theta_j}{\partial x_j}) + \frac{n+2}{n-2} \alpha_j^{4/n-2} \int_\Omega K(x) \delta_j^{4/n-2} (\sum_{i \neq j} \alpha_i \delta_i) \times$$

$$\times (\frac{\partial \delta_j}{\partial x_j} - \frac{\partial \theta_j}{\partial x_j}) \Big|$$

$$\leq 0(\lambda_j \sum_{i\neq k} \varepsilon_{ik}^{n/n-2} \operatorname{Log} \varepsilon_{ik}^{-1}) \qquad \text{for } n \geq 4$$

$$\leq 0(\lambda_j \left| \sum_{\substack{i\neq j \\ i\neq k}} \varepsilon_{ik}^3 \operatorname{Log} \varepsilon_{ik}^{-1} + \sum_{i\neq j} \varepsilon_{ij}^2 (\operatorname{Log} \varepsilon_{ij}^{-1})^{2/3} \right|) \qquad \text{for } n = 3$$

$$\Big| C + \Sigma \alpha_i^{\frac{n+2}{n-2}} \int_\Omega K(x)\, \delta_i^{\frac{n+2}{n-2}} \left(\frac{\partial \delta_j}{\partial \lambda_j} - \frac{\partial \theta_j}{\partial \lambda_j}\right)$$

$$+ \frac{(n+2)\, \alpha_j^{4/n-2}}{n-2} \int_\Omega K(x)\, \delta_j^{4/n-2} \left(\sum_{i\neq j} \alpha_i \delta_i\right) \times$$

$$\times \left(\frac{\partial \delta_j}{\partial \lambda_j} - \frac{\partial \theta_j}{\partial \lambda_j}\right) \Big|$$

$$\leq 0(\frac{1}{\lambda_j} \sum_{i\neq k} \varepsilon_{ik}^{n/n-2} \operatorname{Log} \varepsilon_{ik}^{-1}) \qquad \text{for } n \geq 4$$

$$\leq 0 \left(\frac{1}{\lambda_j} \left| \sum_{\substack{i\neq j \\ i\neq k}} \varepsilon_{ik}^3 \operatorname{Log} \varepsilon_{ik}^{-1} + \sum_{i\neq j} \varepsilon_{ij}^2 (\operatorname{Log} \varepsilon_{ij}^{-1})^{2/3} \right| \right) \qquad \text{for } n = 3$$

<u>Proof of Lemma 2.1</u>:

We first note that:

$$\theta_j \leq \delta_j \; ; \quad \left| \frac{\partial \theta_j}{\partial x_j} \right| \leq C\, \lambda_j \delta_j \; ; \quad \left| \frac{\partial \theta_j}{\partial \lambda_j} \right| \leq \frac{C \delta_j}{\lambda_j} \tag{2.13}$$

This follows from the fact that these bounds hold on $\partial\Omega$ and from the fact that:

$$-\Delta \delta_j > 0 \text{ in } \Omega, \text{ while } \Delta \theta_j = \Delta \frac{\partial \theta_j}{\partial x_j} = \Delta \frac{\partial \theta_j}{\partial \lambda_j} = 0. \tag{2.14}$$

For $n \geq 4$, the inequalities of Lemma 2.1 follow from the following estimates which we will prove:

$$\int \delta_j \sum_{\substack{i\neq j \\ i\neq k}} (\alpha_i \delta_i)^{4/n-2} \operatorname{Inf}(\alpha_i \delta_i, \alpha_k \delta_k) \leq C \left(\sum_{\substack{i\neq k \\ i\neq j}} \varepsilon_{ik}^{n/n-2} \operatorname{Log} \varepsilon_{ik}^{-1} + \varepsilon_{ij}^{n/n-2} \operatorname{Log} \varepsilon_{ij}^{-1} \right) \tag{2.15}$$

$$\int \delta_j \sum_{i\neq j} (\alpha_j\delta_j)^{\frac{4}{n-2}-1} \operatorname{Inf}((\alpha_j\delta_j)^2, (\alpha_i\delta_i)^2) \leq C \sum_{i\neq j} \varepsilon_{ij}^{n/n-2} \operatorname{Log} \varepsilon_{ij}^{-1} \tag{2.16}$$

$$\int \delta_j \sum_{i\neq j} \operatorname{Inf}((\alpha_i\delta_i)^{4/n-2}, (\alpha_j\delta_j)^{4/n-2}) \sum_{i\neq j} \alpha_i\delta_i \leq \sum_{i\neq j} \varepsilon_{ij}^{n/n-2} \operatorname{Log} \varepsilon_{ij}^{-1} \tag{2.17}$$

Using (2.13) and (E5), (2.15), (2.16) and (2.17) yield, through (2.9), the estimates of Lemma 2.1 in the case $n \geq 4$.

We now prove (2.15), (2.16) and (2.17).

We have

$$A_1 = \int \delta_j \sum_{\substack{i\neq j\\ i\neq k}} (\alpha_i\delta_i)^{4/n-2} \operatorname{Inf}(\alpha_i\delta_i, \alpha_k\delta_k) \tag{2.18}$$

$$\leq C \sum_{\substack{i\neq j\\ i\neq k}} \int \alpha_i^{n/n-2} \delta_i^{n/n-2} (\alpha_k\delta_k)^{4/n-2} \delta_j \leq$$

$$\leq C \sum_{\substack{i\neq j\\ i\neq k}} \alpha_i^{n/n-2} \alpha_k^{4/n-2} \int \delta_i^{n/n-2} (\delta_k^{n/n-2} + \delta_j^{n/n-2})$$

Hence, by (E2),

$$A_1 \leq C \sum_{\substack{i\neq j\\ i\neq k}} (\varepsilon_{ik}^{n/n-2} \operatorname{Log} \varepsilon_{ik}^{-1} + \varepsilon_{ij}^{n/n-2} \operatorname{Log} \varepsilon_{ij}^{-1}) \tag{2.19}$$

Next, we have:

$$A_2 = \int \delta_j \sum_{i\neq j} (\alpha_j\delta_j)^{\frac{4}{n-2}-1} \operatorname{Inf}((\alpha_j\delta_j)^2, (\alpha_i\delta_i)^2) \leq \tag{2.20}$$

$$\underset{n\geq 4}{\leq} \int \delta_j \sum_{i\neq j} (\alpha_j\delta_j)^{2/n-2} (\alpha_i\delta_i)^{n/n-2}$$

as,

$$\frac{n}{n-2} \leq 2 \quad \text{for } n \geq 4 \tag{2.21}$$

Hence, by (E2)

$$A_2 \leq C(\sum_{i\neq j} \varepsilon_{ij}^{n/n-2} \operatorname{Log} \varepsilon_{ij}^{-1}) \tag{2.22}$$

Finally, we have:

$$A_3 = \int \delta_j \sum_{i\neq j} \operatorname{Inf}((\alpha_i \delta_i)^{4/n-2}, (\alpha_j \delta_j)^{4/n-2})(\sum_{i\neq j} \alpha_i \delta_i) \leq \tag{2.23}$$

$$\leq \sum_{i\neq j} \int \delta_j (\alpha_i \delta_i)^{2/n-2} (\alpha_j \delta_j)^{2/n-2} (\alpha_i \delta_i)$$

Hence, by (E2)

$$A_3 \leq C \sum_{i\neq j} \varepsilon_{ij}^{n/n-2} \operatorname{Log} \varepsilon_{ij}^{-1} \tag{2.24}$$

(2.19), (2.22), (2.24) yield the inequalities (2.15), (2.16) and (2.17). In the case $n = 3$, (2.20) does not hold. We then have:

$$A_2 \leq C \int \delta_j^4 \sum_{i\neq j} \delta_i^2 \tag{2.25}$$

Thus, by (E3),

$$A_2 \leq C \sum_{i\neq j} \varepsilon_{ij}^2 (\operatorname{Log} \varepsilon_{ij}^{-1})^{2/3} \tag{2.26}$$

(2.22) and (2.24) hold. Hence the proof of Lemma 2.1 in the case $n = 3$.

Next, we turn to the following quantities:

Estimate F7: $\int K(x) \delta_i^{\frac{n+2}{n-2}} \delta_j = C_0^{2n/n-2} K(x_i) C_1 \varepsilon_{ij} + o(\varepsilon_{ij}) + 0\left(\frac{1}{\lambda_i^{\frac{n+2}{2}} \lambda_j^{\frac{n-2}{2}}}\right)$;

where $o(\varepsilon_{ij}) \leq \frac{C}{\lambda_i} 0([|DK(x_i)| + \frac{1}{\lambda_i}] \varepsilon_{ij}((\operatorname{Log} \varepsilon_{ij}^{-1})^{\frac{n-2}{2}} \left(\frac{\operatorname{Log} \lambda_i}{\lambda_i}\right)^{2/n} + 1))$; if K is constant, $o(\varepsilon_{ij}) + 0\left(\frac{1}{\lambda_i^{\frac{n+2}{2}} \lambda_j^{\frac{n-2}{2}}}\right)$ is, by E1, $0(\varepsilon_{ij}^{n/n-2})$.

Estimate F8: $\int K(x)\,\delta_i^{\frac{n+2}{n-2}}\,\theta_j = C_0^{\frac{n+2}{n-2}}\,\dfrac{K(x_i)\,\theta_j(x_i)}{\lambda_i^{\frac{n-2}{2}}}\,C_1 + 0\left(\dfrac{1}{\lambda_i^{\frac{n+2}{2}}}\,\dfrac{1}{\lambda_j^{\frac{n+2}{2}}}\,\dfrac{\log\lambda_i}{d_j^n}\right)$

where $C_1 = \int \dfrac{r^{n-1}\,dr\,d\sigma}{(1+r^2)^{\frac{n+2}{2}}}$ and $d_j = \operatorname{Inf} d(x_j, \partial\Omega)$

Estimate F9: $\int K(x)\,\delta_j^{4/n-2}\,\delta_i\theta_j \le \dfrac{C}{\lambda_j^{\frac{n-2}{2}}\,d_j^{n-2}}\,\dfrac{1}{\lambda_i^{\frac{(n-2)^2}{2(n+2)}}}\left(\varepsilon_{ij}^{4/n+2} + \dfrac{1}{(\lambda_i\lambda_j)^{2\frac{n-2}{n+2}}}\right)$

Estimate F10:

$$\frac{1}{C_0^{2n/n-2}}\int K(x)\left(\Sigma\,\alpha_i\delta_i\right)^{2n/n-2} = \Sigma\left(\alpha_i.\,C_0\right)^{2n/n-2}\left\{K(x_i)\int\frac{r^{n-1}\,dr\,d\sigma}{(1+r^2)^n} + \frac{\Delta K(x_i)}{\lambda_i^2}\times\right.$$

$$\left.\times\int\frac{|x|^2\,dx}{(1+|x|^2)^n} + o_K\left(\frac{1}{\lambda_i^2}\right)\right\} + \frac{2n}{n-2}\,C_0^{2n/n-2}\sum_{i\neq j}\alpha_i^{\frac{n+2}{n-2}}\,\alpha_j.\,C_1\,\varepsilon_{ij}K(x_i) + o\Big(\sum_{i\neq j}\varepsilon_{ij}\Big) +$$

$$+\ 0\left(\Sigma\,\frac{1}{\lambda_i^{\frac{n+2}{2}}\,\lambda_j^{\frac{n-2}{2}}}\right) + \begin{cases} 0\Big(\sum\limits_{i\neq j}\varepsilon_{ij}^{n/n-2}\,\mathrm{Log}\,\varepsilon_{ij}^{-1}\Big) & n\ge 4\\[2mm] 0\Big(\sum\limits_{i\neq j}\varepsilon_{ij}^{2}\,(\mathrm{Log}\,\varepsilon_{ij}^{-1})^{2/3}\Big) & n = 3\end{cases}$$

Here $o(\varepsilon_{ij}) + 0\left(\Sigma\,\dfrac{1}{\lambda_i^{\frac{n+2}{2}}\,\lambda_j^{\frac{n-2}{2}}}\right)$ are the same as in F7 and, if K is constant, there is no $o_K(1/\lambda_i^2)$ term in the expansion.

Estimate F11: $\int K(x)\,\delta_i^{\frac{n+2}{n-2}}\,\dfrac{\partial\delta_j}{\partial x_j} = C_0^{\frac{2n}{n-2}}\Big[DK(x_i)\,\varepsilon_{ij} - K(x_i)\,\dfrac{n+2}{n}\,\lambda_i\lambda_j C_7(x_j - x_i)\times$

$$\times\,\varepsilon_{ij}^{n/n-2}\Big] + 0(\varepsilon_{ij}) + 0\left(\frac{1}{\lambda_j^{n/2}\,\lambda_i^{\frac{n-2}{2}}}\right) + o(\varepsilon_{ij}) + 0\Big(\varepsilon_{ij}^{\frac{n+1}{n-2}}\,\lambda_i\lambda_j\,|x_i - x_j|\Big);$$

with $0(\varepsilon_{ij}) = \frac{n+2}{n-2} \int_{B_i} (K(x) - K(x_i))\, \delta_i^{4/n-2}\, \delta_j \frac{\partial \delta_i}{\partial x_i}$; $o(\varepsilon_{ij})$ as in F7.

$$\int \delta_j^{\frac{n+2}{n-2}} \frac{\partial \delta_i}{\partial x_i} = \int_{\mathbb{R}^n} \frac{\partial}{\partial x_i} (\delta_i^{\frac{n+2}{n-2}})\, \delta_j = (n-2)\, \lambda_i \lambda_j C_0^{2n/n-2} C_1 (x_j - x_i)\, \varepsilon_{ij}^{n/n-2} +$$

$$+ 0(\varepsilon_{ij}^{\frac{n+1}{n-2}} \lambda_i \lambda_j |x_i - x_j|) = \frac{\partial \varepsilon_{ij}}{\partial x_i} C_0^{2n/n-2} C_1 + 0(\varepsilon_{ij}^{\frac{n+1}{n-2}} \lambda_i \lambda_j |x_i - x_j|)$$

Estimate F12: $\int K(x)\, \delta_i^{\frac{n+2}{n-2}} \frac{\partial \theta_j}{\partial x_j} = C_0^{\frac{n+2}{n-2}} \frac{K(x_i)}{\lambda_i^{\frac{n-2}{2}}} \frac{\partial \theta_j}{\partial x_j}(x_i)\, C_4 +$

$+ 0\left(\frac{1}{\lambda_i^{\frac{n+2}{2}}} \frac{1}{\lambda_j^{\frac{n-2}{2}}} \frac{\operatorname{Log} \lambda_i}{d_j^{n+1}} \right)$ where C_4 and d_j are defined in F8.

Estimate F14: $\int K(x)\, \delta_j^{4/n-2} \delta_i \frac{\partial \delta_j}{\partial x_j} = K(x_j) \frac{n-2}{n+2} \int \delta_i^{\frac{n+2}{n-2}} \frac{\partial \delta_j}{\partial x_j} + 0\left(\varepsilon_{ij} + \frac{1}{\lambda_j^{n/2} \lambda_i^{\frac{n-2}{2}}} \right)$

$$\int K(x)\, \delta_j^{\frac{n+2}{n-2}} \frac{\partial \delta_j}{\partial x_j} = C_0^{\frac{2n}{n-2}} \left[(DK(x_j)) \int \frac{dy}{(1+|y|^2)^n} + \frac{1}{\lambda_j^2} \int (D^3 K(x_j) . y . y) \frac{dy}{(1+|y|^2)^n} \right] +$$

$+ o\left(\frac{1}{\lambda_j^2} \right) + 0 \frac{1}{\lambda_j^n}$; $0(\varepsilon_{ij}) = \int_{B_j} (K(x) - K(x_j))\, \delta_j^{4/n-2} \delta_i \frac{\partial \delta_j}{\partial x_j}$

Estimate F15: $\left| \int K(x)\, \delta_j^{4/n-2} \delta_i \frac{\partial \theta_j}{\partial x_j} \right| \leq \frac{C}{\lambda_j^{\frac{n-2}{2}} d_j^{n-1}} \times \frac{1}{\lambda_i^{\frac{(n-2)^2}{2(n+2)}}} \left(\varepsilon_{ij}^{4/n+2} + \right.$

$\left. + \frac{1}{(\lambda_i \lambda_j)^{2 . \frac{n-2}{n+2}}} \right)$

Estimate F16: $\int K(x)\, \delta_i^{\frac{n+2}{n-2}} \frac{\partial \delta_j}{\partial \lambda_j} = K(x_i) \int \delta_i^{\frac{n+2}{n-2}} \frac{\partial \delta_j}{\partial \lambda_j} + o\left(\frac{\varepsilon_{ij}}{\lambda_j}\right) + 0\left(\frac{1}{\lambda_i^{\frac{n+2}{2}} \lambda_j^{n/2}} \right)$;

$o(\varepsilon_{ij})$ as in F7.

$$\int \delta_i^{\frac{n+2}{n-2}} \frac{\partial \delta_j}{\partial \lambda_j} = -\frac{(n-2)}{2} C_1 C_0^{\frac{2n}{n-2}} \left(\frac{1}{\lambda_i} - \frac{\lambda_i}{\lambda_j^2} + \lambda_i |x_i - x_j|^2 \right) \varepsilon_{ij} + \frac{1}{\lambda_j} 0(\varepsilon_{ij}^{n/n-2} \operatorname{Log} \varepsilon_{ij}^{-1}) =$$

$$= C_1 C_0^{2n/n-2} \frac{\partial \varepsilon_{ij}}{\partial \lambda_j} + \frac{1}{\lambda_j} 0(\varepsilon_{ij}^{n/n-2} \operatorname{Log} \varepsilon_{ij}^{-1})$$

Estimate F17: $\int K(x)\, \delta_i^{\frac{n+2}{n-2}} \frac{\partial \theta_j}{\partial \lambda_j} = \frac{K(x_i)}{\lambda_i^{\frac{n-2}{2}}} \frac{\partial \theta_j}{\partial \lambda_j}(x_i)\, C_4 + 0\left(\frac{1}{\lambda_i^{\frac{n+2}{2}}} \frac{1}{\lambda_j^{\frac{n}{2}}} \frac{\operatorname{Log} \lambda_i}{d_j^n} \right)$

where C_4 and d_j are defined in F8.

Estimate F18: $\int K(x)\, \delta_j^{4/n-2} \delta_i \frac{\partial \delta_j}{\partial \lambda_j} = K(x_j) \frac{n-2}{n+2} \int \delta_i^{\frac{n+2}{n-2}} \frac{\partial \delta_j}{\partial \lambda_j} + \frac{1}{\lambda_j} o(\varepsilon_{ij}) +$

$$+ 0\left(\frac{1}{\lambda_j^{\frac{n+4}{2}} \lambda_i^{\frac{n-2}{2}}} \right) \quad \text{for } i \neq j$$

$$\int K(x)\, \delta_j^{\frac{n+2}{n-2}} \frac{\partial \delta_j}{\partial \lambda_j} = \left(-\frac{n-2}{2n} \frac{\Delta K(x_j)}{\lambda_j^3} C_6 + 0\left(\frac{1}{\lambda_j^{n+1}}\right) + o\left(\frac{1}{\lambda_j^3}\right) \right) C_0^{\frac{2n}{n-2}}$$

Estimate F19: $\left| \int K(x)\, \delta_j^{4/n-2} \delta_i \frac{\partial \theta_j}{\partial \lambda_j} \right| \leq \frac{C}{\lambda_j^{n/2} d_j^{n-2}} \times$

$$\times \frac{1}{\lambda_i^{\frac{(n-2)^2}{2(n+2)}}} \left(\varepsilon_{ij}^{4/n+2} + \frac{1}{(\lambda_i \lambda_j)^{2 \cdot \frac{n-2}{n+2}}} \right)$$

Estimate F20: $\int K(x)\, \delta_i^{\frac{n+2}{n-2}} \frac{\partial \delta_j}{\partial x_j} = -C_1 K(x_i) \frac{n+2}{n} \lambda_i \lambda_j K(x_i)(x_j - x_i) \varepsilon_{ij}^{n/n-2} +$

$$+ \lambda_j\, o(\varepsilon_{ij}) + 0\left(\frac{1}{\lambda_i^{\frac{n+2}{2}} \lambda_j^{\frac{n-4}{2}}} \right); \quad o(\varepsilon_{ij}) \text{ as in F7.}$$

Estimate F21: $\int K(x)\,\delta_j^{4/n-2}\delta_i \frac{\partial\delta_j}{\partial x_j} = K(x_j)\frac{n-2}{n+2}\int \delta_i^{\frac{n+2}{n-2}} \frac{\partial \delta_j}{\partial x_j} + \lambda_j\, o(\varepsilon_{ij}) +$

$$+\, 0\left(\frac{1}{\lambda_j^{n/2}\lambda_i^{\frac{n-2}{2}}}\right); \quad o(\varepsilon_{ij}) \text{ as in F7.}$$

Proof of Estimate F7: We consider a ball B_i centred at x_i, of radius ε:

$$\int K(x)\,\delta_i^{\frac{n+2}{n-2}}\delta_j = K(x_i)\int_{B_i}\delta_i^{\frac{n+2}{n-2}}\delta_j + \int_{B_i}(K(x)-K(x_i))\,\delta_i^{\frac{n+2}{n-2}}\delta_j + \tag{2.27}$$

$$+ \int_{B_i^c} K(x)\,\delta_i^{\frac{n+2}{n-2}}\delta_j$$

$$= K(x_i)\int \delta_i^{\frac{n+2}{n-2}}\delta_j + \int_{B_i}(K(x)-K(x_i))\,\delta_i^{\frac{n+2}{n-2}}\delta_j + \int_{B_i^c}(K(x)+K(x_i))\,\delta_i^{\frac{n+2}{n-2}}\delta_j$$

We first estimate:

$$\int_{B_i^c} \delta_i^{\frac{n+2}{n-2}}\delta_j \tag{2.28}$$

Let B_j be a ball centred at x_j, of radius ε.
We have

$$\int_{B_i^c}\delta_i^{\frac{n+2}{n-2}}\delta_j \le C\left[\frac{1}{\lambda_i^{\frac{n+2}{2}}}\int_{B_j}\delta_j + \left(\int_{B_i^c}\delta_i^{2n/n-2}\right)^{\frac{n+2}{2n}}\left(\int_{B_j^c}\delta_j^{2n/n-2}\right)^{\frac{n-2}{2n}}\right] \tag{2.29}$$

where we upperbounded, outside B_i, δ_i by $\dfrac{1}{\lambda_i^{\frac{n-2}{2}}\varepsilon^{n-2}}$.

Hence

$$\int_{B_i^c}\delta_i^{\frac{n+2}{n-2}}\delta_j \le C\left[\frac{1}{\lambda_i^{\frac{n-2}{2}}}\int_{B_j}\delta_j + \frac{1}{\lambda_i^{\frac{n+2}{2}}}\cdot\frac{1}{\lambda_j^{\frac{n-2}{2}}}\right] \tag{2.30}$$

On the other hand:

$$\int_{B_j} \delta_j = \frac{1}{\lambda_j^{\frac{n+2}{2}}} \int_{r\leq\lambda_j\varepsilon} \frac{r^{n-1} dr d\sigma}{(1+r^2)^{\frac{n-2}{2}}} \leq \frac{C}{\lambda_j^{\frac{n-2}{2}}} \text{ ; where } d\sigma \text{ is the} \tag{2.31}$$

volume element on S^{n-1}.

(2.30) and (2.31) yield:

$$\int_{B_i^c} \delta_i^{\frac{n+2}{n-2}} \delta_j (K(x) + K(x_i)) \leq C\left[\frac{1}{\lambda_i^{\frac{n+2}{2}}} \frac{1}{\lambda_j^{\frac{n-2}{2}}}\right] \tag{2.32}$$

We are now left with:

$$\int_{B_i} (K(x) - K(x_i)) \delta_i^{\frac{n+2}{n-2}} \delta_j \tag{2.33}$$

K can be expanded around x_i; and we have

$$K(x) - K(x_i) = DK(x_i).(x-x_i) + 0(|x-x_i|^2) \tag{2.34}$$

Hence

$$\int_{B_i} (K(x) - K(x_i)) \delta_i^{\frac{n+2}{n-2}} \delta_j = \int_{B_i} DK(x_i).(x-x_i) \delta_i^{\frac{n+2}{n-2}} \delta_j + 0(\int_{B_i} |x-x_i|^2 \delta_i^{\frac{n+2}{n-2}} \delta_j). \tag{2.35}$$

We first estimate:

$$\int_{B_i} |x-x_i|^2 \delta_i^{\frac{n+2}{n-2}} \delta_j \tag{2.36}$$

we have:

$$\int_{B_i} |x-x_i|^2 \delta_i^{\frac{n+2}{n-2}} \delta_j = \frac{1}{\lambda_i^2} \int_{B_i} \delta_i^{n/n-2} \delta_j - \frac{1}{\lambda_i^2} \int_{B_i} \delta_i^{\frac{n+2}{n-2}} \delta_j \tag{2.37}$$

we have

$$\int_{B_i} \delta_i^{n/n-2} \delta_j \leq \left(\int \delta_i^{n/n-2} \delta_j^{n/n-2}\right)^{\frac{n-2}{n}} \left(\int_{B_i} \delta_i^{n/n-2 \; 2/n}\right) \tag{2.38}$$

Now

$$\left(\int_{B_i} \delta_i^{n/n-2}\right)^{2/n} \leq C \cdot \frac{1}{\lambda_i}\left(\int_{1\leq r\leq\lambda_i\varepsilon} \frac{r^{n-1}dr}{r^n}\right)^{2/n} = C_1 \frac{(\operatorname{Log}\lambda_i)^{2/n}}{\lambda_i} \tag{2.39}$$

while by E2:

$$\left(\int \delta_i^{n/n-2}\delta_j^{n/n-2}\right)^{\frac{n-2}{n}} = 0\left(\varepsilon_{ij}(\operatorname{Log}\varepsilon_{ij}^{-1})^{\frac{n-2}{2}}\right) \tag{2.40}$$

Hence

$$\int_{B_i} |x-x_i|^2 \delta_i^{\frac{n+2}{n-2}}\delta_j = \frac{1}{\lambda_i^2}\, 0\left(\varepsilon_{ij}\left[\left(\operatorname{Log}\varepsilon_{ij}^{-1}\right)^{\frac{n-2}{2}} \frac{(\operatorname{Log}\lambda_i)^{2/n}}{\lambda_i} + 1\right]\right) \tag{2.41}$$

We now estimate

$$\int_{B_i} DK(x_i).(x-x_i)\, \delta_i^{\frac{n+2}{n-2}}\delta_j \tag{2.42}$$

which we upperbound by:

$$|DK(x_i)| \int_{B_i} |x-x_i|\, \delta_i^{\frac{n+2}{n-2}}\delta_j \leq \frac{|DK(x_i)|}{\lambda_i}\int_{B_i} \delta_i^{n/n-2}\delta_j \leq \tag{2.43}$$

$$\leq \frac{|DK(x_i)|}{\lambda_i}\, 0\left(\varepsilon_{ij}\left(\operatorname{Log}\varepsilon_{ij}^{-1}\right)^{\frac{n-2}{2}} \frac{(\operatorname{Log}\lambda_i)^{2/n}}{\lambda_i}\right)$$

using (2.39) and (2.40).

On the other hand, we have:

$$\int_{B_i} (|x-x_i| + |x-x_i|^2)\, \delta_i^{\frac{n+2}{n-2}}\delta_j \leq C\varepsilon \int \delta_i^{\frac{n+2}{n-2}}\delta_j = o(\varepsilon_{ij}) \tag{2.44}$$

Thus, using (2.43) and (2.41), we have:

$$\left|\int_{B_i} (K(x)-K(x_i))\, \delta_i^{\frac{n+2}{n-2}}\delta_j\right| = o(\varepsilon_{ij}) \leq \frac{C}{\lambda_i}\, 0(\varepsilon_{ij}\left[(\operatorname{Log}\varepsilon_{ij}^{-1})^{\frac{n-2}{2}} \frac{(\operatorname{Log}\lambda_i)^{2/n}}{\lambda_i} + 1\right]$$
$$\times\left[|DK(x_i)| + \frac{1}{\lambda_i}\right] \tag{2.45}$$

This, together with (2. 32) and (E1) yield the desired estimate.

Proof of Estimate F8: We want to estimate

$$\int K(x)\ \delta_i^{\frac{n+2}{n-2}}\ \theta_j \tag{2.44}$$

which is equal to:

$$\int K(x)\,\delta_i^{\frac{n+2}{n-2}}\theta_j = K(x_i)\,\theta_j(x_i)\int \delta_i^{\frac{n+2}{n-2}} + \int_{B_i}\left[\theta_j K(x) - K(x_i)\,\theta_j(x_i)\right]\delta_i^{\frac{n+2}{n-2}} +$$

$$+ \int_{B_i^c}(K\theta_j + K(x_i)\,\theta_j(x_i))\,\delta_i^{\frac{n+2}{n-2}} = \frac{K(x_i)\,\theta_j(x_i)}{\lambda_i^{\frac{n-2}{2}}}\int\frac{r^{n-1}\,dr\,d\sigma}{(1+r^2)^{\frac{n+2}{2}}} + 0\left(\frac{1}{\lambda_i^{\frac{n+2}{2}}}\ \operatorname*{Sup}_{\partial\Omega}\theta_j\right) +$$

$$+ \int_{B_i}\left[K\theta_j(x) - K(x_i)\,\theta_j(x_i)\right]\delta_i^{\frac{n+2}{n-2}} \tag{2.45}$$

We consider

$$\int_{B_i}\left[K\theta_j(x) - K(x_i)\,\theta_j(x_i)\right]\delta_i^{\frac{n+2}{n-2}} \tag{2.46}$$

and we expand $K\theta_j$ around x_i:

$$K\theta_j(x) - K(x_i)\,\theta_j(x_i) = D(K\theta_j)(x_i)\,(x-x_i) + 0(|x-x_i|^2 \operatorname*{Sup}_{\Omega} D^2(K\theta_j)) \tag{2.47}$$

By oddness, we have:

$$\int_{B_i} D(K\theta_j)(x_i)\,.\,(x-x_i)\ \delta_i^{\frac{n+2}{n-2}} = 0 \tag{2.48}$$

Hence

$$\int_{B_i}\left[K\theta_j(x) - K(x_i)\,\theta_j(x_i)\right]\delta_i^{\frac{n+2}{n-2}} = \operatorname*{Sup}_{\Omega}|D^2(K\theta_j)|\,0\left(\int_{B_i}|x-x_i|^2\,\delta_i^{\frac{n+2}{n-2}}\right) \tag{2.49}$$

Now

$$\int_{B_i} |x-x_i|^2 \delta_i^{\frac{n+2}{n-2}} = \frac{1}{\lambda_i^{\frac{n+2}{2}}} \int_{r \le \lambda_i \varepsilon} C_0^{\frac{n+2}{n-2}} \frac{r^{n+1} drd\sigma}{(1+r^2)^{\frac{n+2}{2}}} = 0\left(\frac{\operatorname{Log} \lambda_i}{\lambda_i^{\frac{n+2}{2}}}\right) \tag{2.50}$$

It remains thus to estimate:

$$\operatorname*{Sup}_{\partial\Omega} \theta_j \; ; \quad \operatorname*{Sup}_{\Omega} |D^2(K\theta_j)|$$

The two quantities are, as can be checked, upperbounded by $\frac{C}{d_j^2} \operatorname*{Sup}_{\partial\Omega} \theta_j$ where

$$d_j = \inf_{x \in \partial\Omega} |x-x_j| \tag{2.52}$$

$$\operatorname*{Sup}_{\partial\Omega} \theta_j \le \frac{1}{\lambda_j^{\frac{n-2}{2}} d_j^{n-2}} \tag{2.53}$$

Thus:

$$\int K(x) \delta_i^{\frac{n+2}{n-2}} \theta_j = C_0^{\frac{n+2}{n-2}} \frac{K(x_i)\theta_j(x_i)}{\lambda_i^{\frac{n-2}{2}}} C_1 + 0\left(\frac{1}{\lambda_i^{\frac{n+2}{2}}} \frac{1}{\lambda_j^{\frac{n-2}{2}}} \frac{\operatorname{Log} \lambda_i}{d_j^n}\right) \tag{2.54}$$

where

$$C_1 = \int \frac{r^{n-1} drd\sigma}{(1+r^2)^{\frac{n+2}{2}}} \tag{2.55}$$

(2.54) is the desired estimate.

<u>Proof of Estimate F9:</u> We want to estimate

$$\int K(x) \delta_j^{4/n-2} \delta_i \theta_j \tag{2.56}$$

This, we upperbound by:

$$\operatorname{Sup} \theta_j \left| \int K(x) \delta_j^{4/n-2} \delta_i \right| \le C \operatorname*{Sup}_{\partial\Omega} \theta_j \int \delta_j^{4/n-2} \delta_i \tag{2.57}$$

we have

$$\int \delta_j^{4/n-2} \delta_i \leq \left(\int \delta_j^{\frac{n+2}{n-2}} \delta_i \right)^{4/n+2} \times \left(\int_{B_i} \delta_i \right)^{\frac{n-2}{n+2}} + \int_{B_i^c} \delta_j^{4/n-2} \delta_i \tag{2.58}$$

where B_i is a ball centred at x_i of radius ε.

Thus, using (2.31), we derive

$$\int \delta_j^{4/n-2} \delta_i \leq C \varepsilon_{ij}^{4/n+2} \cdot \frac{1}{\lambda_i^{\frac{(n-2)^2}{2(n+2)}}} + \int_{B_i^c} \delta_j^{4/n-2} \delta_i \tag{2.59}$$

Now

$$\int_{B_i^c} \delta_j^{4/n-2} \delta_i \leq \left(\int_{B_i^c} \delta_i^{\frac{n+2}{n-2}} \right)^{\frac{n-2}{n+2}} \times \left(\int \delta_j^{\frac{n+2}{n-2}} \right)^{4/n+2} \leq \tag{2.60}$$

$$\leq \frac{1}{\lambda_i^{\frac{2(n-2)}{n+2}}} \frac{C}{\lambda_i^{\frac{(n-2)^2}{2(n+2)}}} \cdot \frac{1}{\lambda_j^{\frac{2(n-2)}{n+2}}}$$

Thus

$$\int K(x) \delta_j^{4/n-2} \delta_i \theta_j \leq \frac{1}{\lambda_j^{\frac{n-2}{2}} d_j^{n-2}} \left[\frac{1}{\lambda_i^{\frac{(n-2)^2}{2(n+2)}}} \varepsilon_{ij}^{4/n+2} + \frac{1}{\lambda_i^{\frac{(n-2)^2}{2(n+2)}}} \cdot \frac{1}{(\lambda_i \lambda_j)^{2\frac{n-2}{n+2}}} \right] \tag{2.61}$$

which is the desired estimate.

Here again $d_j = \inf d(x_j, \partial\Omega)$.

Proof of Estimate F10: We want to estimate:

$$\int K(x) \left(\Sigma \alpha_i \delta_i\right)^{2n/n-2} \tag{2.62}$$

For this, we use (1.3), and we have

$$\left| \int K(x) \left(\Sigma \alpha_i \delta_i\right)^{2n/n-2} - \Sigma \alpha_i^{2n/n-2} \int K(x) \delta_i^{2n/n-2} - \frac{2n}{n-2} \sum_{i \neq j} \alpha_i^{\frac{n+2}{n-2}} \alpha_j \times \right.$$

$$\times \int K(x)\, \delta_i^{\frac{n+2}{n-2}} \delta_j \Big| \leq M_1 \sum_{i \neq j} \int (\alpha_j \delta_j)^{4/n-2} \inf(\alpha_i \delta_i, \alpha_j \delta_j)^2 \tag{2.63}$$

The right-hand side of inequality (2.63) is the A_2 defined in (2.20) multiplied by α_j and summed on all j's.

Hence, we have by (2.22) if $n \geq 4$ and (2.26) for $n = 3$

$$\sum_{i \neq j} \int (\alpha_j \delta_j)^{4/n-2} \operatorname{Inf}(\alpha_i \delta_i, \alpha_j \delta_j)^2 \leq C \sum_{i \neq j} \varepsilon_{ij}^{n/n-2} \operatorname{Log} \varepsilon_{ij}^{-1} \quad \text{if } n \geq 4 \tag{2.64}$$

$$\leq C \sum_{i \neq j} \varepsilon_{ij}^2 (\operatorname{Log} \varepsilon_{ij}^{-1})^{2/3} \quad \text{if } n = 3$$

By Estimate F7 we know that:

$$\int K(x)\, \delta_i^{\frac{n+2}{n-2}} \delta_j = K(x_i) \varepsilon_{ij} + o(\varepsilon_{ij}) + 0 \left(\frac{1}{\lambda_i^{\frac{n+2}{2}} \lambda_j^{\frac{n-2}{2}}} \right) \tag{2.65}$$

where

$$o(\varepsilon_{ij}) \leq \frac{C}{\lambda_i}\, 0 \left(\varepsilon_{ij} \left(\left(\operatorname{Log} \varepsilon_{ij}^{-1} \right)^{\frac{n-2}{2}} \left(\frac{\operatorname{Log} \lambda_i}{\lambda_i} \right)^{2/n} + 1 \right) \right) \left[|DK(x_i)| + \frac{1}{\lambda_i} \right] \tag{2.66}$$

$\int K(x)\, \delta_i^{2n/n-2}$ is estimated now; and we have:

$$\int K(x)\, \delta_i^{2n/n-2} = K(x_i) \int \delta_i^{2n/n-2} + \int (K(x) - K(x_i))\, \delta_i^{2n/n-2} \tag{2.67}$$

As usual, we consider a ball B_i centred at x_i of radius ε.

Outside B_i we have:

$$\left| \int_{B_i^c} (K(x) - K(x_i))\, \delta_i^{2n/n-2} \right| \leq C \int_{B_i^c} \delta_i^{2n/n-2} \leq \frac{C}{\lambda_i^n} \tag{2.68}$$

Inside B_i, we expand K:

$$\int_{B_i} (K(x) - K(x_i))\, \delta_i^{2n/n-2} = \int_{B_i} DK(x_i) \cdot (x - x_i)\, \delta_i^{2n/n-2} + \tag{2.69}$$

$$+ \int_{B_i} D^2 K(x_i) \cdot (x - x_i) \cdot (x - x_i)\, \delta_i^{2n/n-2} + o \left(\int_{B_i} |x - x_i|^2 \delta_i^{\frac{2n}{n-2}} \right)$$

By oddness, $\int_{B_i} DK(x_i).(x-x_i)\,\delta_i^{2n/n-2}$ is zero; while

$$\int_{B_i} D^2K(x_i).(x-x_i).(x-x_i)\,\delta_i^{2n/n-2} = \frac{\Delta K(x_i)}{\lambda_i^2}\int_{\mathbb{R}^n}\frac{|x|^2dx}{(1+|x|^2)^n} + 0\left(\frac{1}{\lambda_i^n}\right) \quad (2.70)$$

Finally:

$$\int_{B_i} |x-x_i|^2\,\delta_i^{2n/n-2} = \frac{C}{\lambda_i^2} \quad (2.71)$$

Thus, summing up (2.68), (2.69), (2.70) and (2.71) we have:

$$\int K(x)\,\delta_i^{2n/n-2} = K(x_i)\int \delta_i^{2n/n-2} + \quad (2.72)$$

$$+\frac{\Delta K(x_i)}{\lambda_i^2}\int_{\mathbb{R}^n}\frac{|x|^2dx}{(1+|x|^2)^n} + o\left(\frac{1}{\lambda_i^2}\right) + 0\left(\frac{1}{\lambda_i^n}\right)$$

We point out here that, when K is constant, there is obviously no $o\left(\frac{1}{\lambda_i^2}\right)$ in (2.72).

Thus

$$\int K(x)\,(\Sigma\,\alpha_i\delta_i)^{2n/n-2} = \Sigma\,\alpha_i^{2n/n-2}\left\{K(x_i)\int\frac{r^{n-1}drd\sigma}{(1+r^2)^n} + \right. \quad (2.73)$$

$$\left. + \frac{\Delta K(x_i)}{\lambda_i^2}\int_{\mathbb{R}^n}\frac{|x|^2dx}{(1+|x|^2)^n} + o\left(\frac{1}{\lambda_i^2}\right)\right\} + \frac{2n}{n-2}\sum_{i\neq j}\alpha_i^{\frac{n+2}{n-2}}\alpha_j\varepsilon_{ij} + o\left(\sum_{i\neq j}\varepsilon_{ij}\right)$$

$$+ 0\left(\Sigma\frac{1}{\lambda_i^{\frac{n+2}{2}}\lambda_j^{\frac{n-2}{2}}}\right) + 0\left(\sum_{i\neq j}\varepsilon_{ij}^{n/n-2}\,\mathrm{Log}\,\varepsilon_{ij}^{-1}\right)$$

or if $n = 3$ $\qquad 0\left(\sum_{i\neq j}\varepsilon_{ij}^2\,(\mathrm{Log}\,\varepsilon_{ij}^{-1})^{2/3}\right)$

with $o(\Sigma_{ij})$ satisfying (2.66).

This is the desired estimate.

We now turn to estimates F12, F15, F17 and F19, before studying F14 and F18 and we will end with F11 and F16.

Proof of Estimate F12: This estimate is very similar to F8; the only modification lies in the fact that $\sup_{\partial\Omega} \left| \frac{\partial \theta_j}{\partial x_j} \right|$ is now of the order $\frac{1}{d_j} \sup_{\partial\Omega} \theta_j$, due to the fact that the first expression involves a derivative. With this, F12 is identical to F8.

Proof of Estimate F17: This estimate is also very similar to F8; the modification now lies in the fact that $\sup_{\partial\Omega} \left| \frac{\partial \theta_j}{\partial \lambda_j} \right|$ is of the order $\frac{1}{\lambda_j} \sup_{\partial\Omega} \theta_j$. Hence F17.

Proof of Estimate F15: As for F12 with respect to F8, F15 is very similar to F9, taking into account that it involves a derivative in x_j. This yields a division by $\frac{1}{d_j}$ in the estimate.

Proof of Estimate F19: very similar to F9. We divide the estimate by $\frac{1}{\lambda_j}$.

Proof of Estimate F14: We write:

$$\int K(x)\, \delta_j^{4/n-2} \delta_i \frac{\partial \delta_j}{\partial x_j} = \int_{B_j} (K(x) - K(x_j))\, \delta_j^{4/n-2} \delta_i \frac{\partial \delta_j}{\partial x_j} + \tag{2.74}$$

$$+ K(x_j) \int_{B_j} \delta_j^{4/n-2} \delta_i \frac{\partial \delta_j}{\partial x_j} + \int_{B_j^c} K(x)\, \delta_j^{4/n-2} \delta_i \frac{\partial \delta_j}{\partial x_j}$$

$$= K(x_j) \int \frac{n-2}{n+2} \frac{\partial}{\partial x_j} \delta_j^{\frac{n+2}{n-2}} \delta_i + \int_{B_j} (K(x) - K(x_j))\, \delta_j^{4/n-2} \delta_i \frac{\partial \delta_j}{\partial x_j} +$$

$$+ \int_{B_j^c} (K(x) - K(x_j))\, \delta_j^{4/n-2} \delta_i \frac{\partial \delta_j}{\partial x_j}$$

$$= K(x_j) \frac{n-2}{n+2} \int \delta_i^{\frac{n+2}{n-2}} \frac{\partial \delta_j}{\partial x_j} + \text{remainder term.}$$

The estimate on $\int \delta_i^{\frac{n+2}{n-2}} \frac{\partial \delta_j}{\partial x_j}$ will be obtained in the proof of F11.

We estimate presently the remainder term.

We know that

$$\left| \frac{\partial \delta_j}{\partial x_j} \right| \leq C \lambda_j \delta_j \tag{2.75}$$

Thus

$$\left| \int_{B_j^c} (K(x) - K(x_j)) \delta_j^{4/n-2} \delta_i \frac{\partial \delta_j}{\partial x_j} \right| \leq C \lambda_j \int_{B_j^c} \delta_j^{\frac{n+2}{n-2}} \delta_i \tag{2.76}$$

Now:

$$\int_{B_j^c} \delta_j^{\frac{n+2}{n-2}} \delta_i = \int_{B_j^c \cap B_i^c} \delta_j^{\frac{n+2}{n-2}} \delta_i + \int_{B_j^c \cap B_i} \delta_j^{\frac{n+2}{n-2}} \delta_i \leq \frac{C}{\lambda_j^{\frac{n+2}{2}}} \int_{B_i} \delta_i + \tag{2.77}$$

$$+ \left(\int_{B_j^c} \delta_j^{2n/n-2} \right)^{\frac{n+2}{2n}} \left(\int_{B_i^c} \delta_i^{2n/n-2} \right)^{\frac{n-2}{2n}} \leq \frac{C}{\lambda_j^{\frac{n+2}{2}} \lambda_i^{\frac{n-2}{2}}}$$

Thus

$$\left| \int_{B_j^c} (K(x) - K(x_j)) \delta_j^{4/n-2} \delta_i \frac{\partial \delta_j}{\partial x_j} \right| \leq \frac{C}{\lambda_j^{n/2} \lambda_i^{n-2/2}} \tag{2.78}$$

We consider now

$$\int_{B_j} (K(x) - K(x_j)) \delta_j^{4/n-2} \delta_i \frac{\partial \delta_j}{\partial x_j} \tag{2.79}$$

This is upperbounded by:

$$\lambda_j \int_{B_j} |x - x_j|^2 \delta_j^{\frac{n+4}{n-2}} \delta_i \leq \int_{B_j} \delta_j^{\frac{n+2}{n-2}} \delta_i \leq \varepsilon_{ij} \tag{2.80}$$

With (2.78) and (2.80), the remainder term is upperbounded by $\varepsilon_{ij} + \frac{C}{\lambda_j^{n/2} \lambda_i^{n-2/2}}$. This for $i \neq j$ and K non constant.

If K is constant, there is no remainder term and estimating (2.74) amounts to estimating $\int \delta_i^{\frac{n+2}{n-2}} \frac{\partial \delta_j}{\partial x_j}$, which we will do soon after this.

If $i = j$ and K is non constant, we can obtain a better estimate of (2.74).

Indeed:

$$\int K(x)\, \delta_j^{\frac{n+2}{n-2}} \frac{\partial \delta_j}{\partial x_j} = \frac{n-2}{2n} \frac{\partial}{\partial x_j} \int K(x)\, \delta_j^{2n/n-2} = \tag{2.81}$$

$$= \frac{n-2}{2n} \frac{\partial}{\partial x_j} \int K(\frac{y}{\lambda_j} + x_j) \frac{1}{(1+|y|^2)^n} dy = \frac{n-2}{2n} \int (DK(\frac{y}{\lambda_j} + x_j)\,.) \frac{1}{(1+|y|^2)^n} dy$$

$$= \frac{n-2}{2n} \left[\int_{|y| \geq \varepsilon\lambda_j} (DK(\frac{y}{\lambda_j} + x_j)\,.) \frac{1}{(1+|y|^2)^n} dy + \right.$$

$$\left. + \int_{|y|/\lambda_j \leq \varepsilon} (DK(\frac{y}{\lambda_j} + x_j)\,.) \frac{1}{(1+|y|^2)^n} dy \right]$$

We have:

$$\left| \int_{|y| \geq \varepsilon\lambda_j} (DK(\frac{y}{\lambda_j} + x_j)\,.) \frac{1}{(1+|y|^2)^n} dy \right| \leq \frac{C}{\lambda_j^n} \tag{2.82}$$

while, expanding $DK(x_j + y/\lambda_j)$ around x_j:

$$\int_{|y|/\lambda_j \leq \varepsilon} (DK(y/\lambda_j + x_j)\,.) \frac{1}{(1+|y|^2)^n} dy = (DK(x_j)\,.) \int \frac{dy}{(1+|y|^2)^n} +$$

$$+ 0(\frac{1}{\lambda_j^n}) + \frac{1}{\lambda_j^2}\, 0 \left(\int_{|y|/\lambda_j \leq \varepsilon} \frac{|y|^2 dy}{(1+|y|^2)^n} \right) \tag{2.83}$$

where

$$\frac{1}{\lambda_j^2}\, 0 \left(\int_{|y|/\lambda_j \leq \varepsilon} \frac{|y|^2 dy}{(1+|y|^2)^n} \right) = \frac{1}{\lambda_j^2} \int_{|y|/\lambda_j \leq \varepsilon} (D^3 K(x_j)\,.\,y\,.\,y) \frac{dy}{(1+|y|^2)^n} +$$

$$+ o\left(\frac{1}{\lambda_j^2} \int_{|y|/\lambda_j \leq \varepsilon} \frac{|y|^2 dy}{(1+|y|^2)^n} \right)$$

$$= \frac{1}{\lambda_j^2} \int (D^3K(x_j).y.y) \frac{dy}{(1+|y|^2)^n} + o(\frac{1}{\lambda_j^2}) + 0(\frac{1}{\lambda_j^n}). \tag{2.84}$$

Thus

$$\int K(x) \delta_j^{\frac{n+2}{n-2}} \frac{\partial \delta_j}{\partial x_j} = (DK(x_j).) \int \frac{dy}{(1+|y|^2)^n} + \frac{1}{\lambda_j^2} \int (D^3K(x_j).y.y) \frac{dy}{(1+|y|^2)^n} +$$

$$+ o(\frac{1}{\lambda_j^2}) + 0(\frac{1}{\lambda_j^n}) \tag{2.85}$$

Summing up, we have estimate (2.85) if $i = j$; we have

$$\int \delta_j^{\frac{n+2}{n-2}} \frac{\partial \delta_j}{\partial x_j} = 0 \tag{2.86}$$

and we have, for $i \neq j$, following a procedure similar to the one used in the proof of F7:

$$\int K(x) \delta_j^{4/n-2} \delta_i \frac{\partial \delta_j}{\partial x_j} = K(x_j) \frac{n-2}{n+2} \int \delta_i^{\frac{n+2}{n-2}} \frac{\partial \delta_j}{\partial x_j} + o(\varepsilon_{ij}) + 0\left(\frac{1}{\lambda_j^{n/2} \lambda_i^{\frac{n-2}{2}}}\right) \tag{2.87}$$

(2.85), (2.86), (2.87) together with F11 yield F14.

<u>Proof of F18:</u> We have, following (2.74):

$$\int K(x) \delta_j^{4/n-2} \delta_i \frac{\partial \delta_j}{\partial \lambda_j} = K(x_j) \frac{n-2}{n+2} \int \delta_i^{\frac{n+2}{n-2}} \frac{\partial \delta_j}{\partial \lambda_j} + \text{remainder term.} \tag{2.88}$$

The remainder term splits into two parts:

$$\int_{B_i^c} (K(x) - K(x_j)) \delta_j^{4/n-2} \delta_i \frac{\partial \delta_j}{\partial \lambda_j} \tag{2.89}$$

and

$$\int_{B_j} (K(x) - K(x_j)) \delta_j^{4/n-2} \delta_i \frac{\partial \delta_j}{\partial \lambda_j} \tag{2.90}$$

Using the fact that:

$$\left| \frac{\partial \delta_j}{\partial \lambda_j} \right| \leq \frac{C\delta_j}{\lambda_j} \tag{2.91}$$

we upperbound (2.89), as in (2.78), by:

$$\left| \int_{B_j^c} (K(x) - K(x_j))\, \delta_j^{4/n-2} \delta_i \frac{\partial \delta_j}{\partial \lambda_j} \right| \leq \frac{C}{\lambda_j^{\frac{n+4}{2}} \lambda_i^{\frac{n-2}{2}}} \tag{2.92}$$

We treat now $\int_{B_j} (K(x) - K(x_j))\, \delta_j^{4/n-2} \delta_i \frac{\partial \delta_j}{\partial \lambda_j}$

This is upperbounded by

$$\frac{C}{\lambda_j} \int_{B_j} |x - x_j|\, \delta_j^{\frac{n+2}{n-2}} \delta_i \tag{2.93}$$

Thus, by:

$$C \frac{\varepsilon}{\lambda_j} \int_{B_j} \delta_j^{\frac{n+2}{n-2}} \delta_i = C\varepsilon \frac{\varepsilon_{ij}}{\lambda_j} \tag{2.94}$$

Therefore, we have:

$$|\text{remainder term}| \leq \frac{1}{\lambda_j} o(\varepsilon_{ij}) + 0\left(\frac{1}{\lambda_j^{\frac{n+4}{2}} \lambda_i^{\frac{n-2}{2}}} \right) \tag{2.95}$$

We are left with estimating $\int \delta_i^{\frac{n+2}{n-2}} \frac{\partial \delta_j}{\partial \lambda_j}$ which is carried out in the proof of F16.

Meanwhile, we study the case $i = j$.

We then have,

$$\int K(x)\, \delta_j^{\frac{n+2}{n-2}} \frac{\partial \delta_j}{\partial \lambda_j} = \frac{n-2}{2n} \frac{\partial}{\partial \lambda_j} \int K(x)\, \delta_j^{2n/n-2} \tag{2.96}$$

$$= \frac{n-2}{2n} \frac{\partial}{\partial \lambda_j} \int K\left(\frac{y}{\lambda_j} + x_j\right) \frac{dy}{(1+|y|^2)^n} = -\frac{n-2}{2n} \frac{1}{\lambda_j} \int DK\left(\frac{y}{\lambda_j} + x_j\right) \cdot \left(\frac{y}{\lambda_j}\right) \frac{dy}{(1+|y|^2)^n}$$

$$= -\frac{n-2}{2n}\frac{1}{\lambda_j}\left\{\int_{|y|/\lambda_j\le\varepsilon} D^2K(x_j).\left(\frac{y}{\lambda_j}\right).\left(\frac{y}{\lambda_j}\right)\frac{dy}{(1+|y|^2)^n} + \right.$$

$$\left. + o\left(\int_{|y|/\lambda_j\le\varepsilon}\frac{|y|^2}{\lambda_j^2}\frac{dy}{(1+|y|^2)^n}\right) + 0\left(\int_{|y|/\lambda_j\varepsilon}\frac{|y|}{\lambda_j}\frac{dy}{(1+|y|^2)^n}\right)\right\}$$

$$= -\frac{n-2}{2n}\frac{K(x_j)}{\lambda_j^3}c_6 + 0\left(\frac{1}{\lambda_j^{n+1}}\right) + o\left(\frac{1}{\lambda_j^3}\right)$$

where

$$c_6 = \frac{1}{n}\int_{\mathbb{R}^n}\frac{|y|^2 dy}{(1+|y|^2)^n} \tag{2.97}$$

(2.95), (2.96) and (2.97), together with F16, yield F18.

Proof of Estimates F11 and F16: We first estimate:

$$\int_{\mathbb{R}^n}\frac{\partial}{\partial\lambda_j}\left(\delta_j^{\frac{n+2}{n-2}}\right)\delta_i = \frac{n+2}{n-2}\int_{\mathbb{R}^n}\delta_j^{4/n-2}\frac{\partial\delta_j}{\partial\lambda_j}\delta_i = \int_{\mathbb{R}^n}\frac{\partial\delta_j}{\partial\lambda_j}\delta_i^{\frac{n+2}{n-2}} \tag{2.98}$$

we have:

$$\frac{\partial}{\partial\lambda_j}\delta_j^{\frac{n+2}{n-2}} = \frac{n+2}{2}\left(\frac{\lambda_j}{1+\lambda_j^2|x-x_j|^2}\right)^{n/2}\left\{\frac{1}{1+\lambda_j^2|x-x_j|^2} - \frac{2\lambda_j^2|x-x_j|^2}{(1+\lambda_j^2|x-x_j|^2)^2}\right\}$$

$$= \frac{n+2}{2\lambda_j}\delta_j^{\frac{n+2}{n-2}}\left(1 - \frac{2\lambda_j^2|x-x_j|^2}{1+\lambda_j^2|x-x_j|^2}\right) \tag{2.99}$$

Thus

$$\int_{\mathbb{R}^n}\frac{\partial}{\partial\lambda_j}\left(\delta_j^{\frac{n+2}{n-2}}\right)\delta_i = \frac{n+2}{2\lambda_j}\int_{\mathbb{R}^n}\delta_j^{\frac{n+2}{n-2}}\delta_i - (n+2)\lambda_j\int_{\mathbb{R}^n}\delta_j^{\frac{n+2}{n-2}}\frac{|x-x_j|^2}{1+\lambda_j^2|x-x_j|^2}\delta_i \tag{2.100}$$

hence by (E1)

$$= \frac{n+2}{2\lambda_j} c_1 \varepsilon_{ij} - (n+2)\lambda_j \int_{\mathbb{R}^n} \delta_j^{\frac{n+2}{n-2}} \frac{|x-x_j|^2}{1+\lambda^2 |x-x_j|^2} \delta_i + 0\left(\frac{\varepsilon_{ij}^{n/n-2}}{\lambda_j}\right)$$

Let

$$I = \int_{\mathbb{R}^n} \frac{|x-x_j|^2}{1+\lambda_j^2 |x-x_j|^2} \delta_j^{\frac{n+2}{n-2}} \delta_i \tag{2.101}$$

Setting $y = x - x_j$, we have:

$$I = \int_{\mathbb{R}^n} \frac{|y|^2}{1+\lambda_j^2 |y|^2} \frac{\lambda_j^{\frac{n+2}{2}}}{(1+\lambda_j^2 |y|^2)^{\frac{n+2}{2}}} \frac{\lambda_i^{\frac{n-2}{2}}}{(1+\lambda_i^2 |y+x_j-x_i|^2)^{\frac{n-2}{2}}} dx \tag{2.102}$$

Let

$$\theta = \mathrm{Max}\left\{\frac{\lambda_i}{\lambda_j} ; \frac{\lambda_j}{\lambda_i} ; \lambda_i\lambda_j |x_i - x_j|^2\right\} \tag{2.103}$$

Assume first:

$$\theta = \lambda_j/\lambda_i \tag{2.104}$$

Let

$$z = \lambda_j y \tag{2.105}$$

Then

$$I = \frac{\lambda_i^{n-2/2}}{\lambda_j^{n+2/2}} \int \frac{|z|^2 dz}{(1+|z|^2)^{\frac{n+4}{2}} \left(1+\left|\frac{\lambda_i}{\lambda_j} z - \lambda_i(x_i-x_j)\right|^2\right)^{\frac{n-2}{2}}} \tag{2.106}$$

We then split I into two parts:

$$I = I_1 + I_2 = \frac{1}{\lambda_j^2}\left(\frac{\lambda_i}{\lambda_j}\right)^{n-2/2}\left(\int_{|z|\le 1/4\,\lambda_j/\lambda_i} \frac{|z|^2 dz}{(1+|z|^2)^{\frac{n+4}{2}} \left(1+\left|\frac{\lambda_i}{\lambda_j} z - \lambda_i(x_i-x_j)\right|^2\right)^{\frac{n-2}{2}}}\right.$$

$$+ \int_{|z| \geq 1/4 \lambda_j/\lambda_i} \frac{|z|^2 dz}{(1+|z|^2)^{\frac{n+4}{2}} (1+\frac{\lambda_i}{\lambda_j} z - \lambda_i (x_i - x_j) |^2)^{\frac{n-2}{2}}} \Bigg) \tag{2.107}$$

First we estimate I_2:

$$I_2 \leq \frac{1}{\lambda_j^2} \left(\frac{\lambda_i}{\lambda_j}\right)^{\frac{n-2}{2}} \int_{|z| \geq \frac{1}{4}\lambda_j/\lambda_i} \frac{|z|^2 dz}{(1+|z|^2)^{\frac{n+4}{2}}} \leq \frac{C}{\lambda_j^2} \left(\frac{\lambda_i}{\lambda_j}\right)^{\frac{n+2}{2}} \tag{2.108}$$

Next, we have:

$$1+ \left| \frac{\lambda_i}{\lambda_j} z - \lambda_i (x_i - x_j) \right|^2 = (1+\lambda_i^2 |x_i - x_j|^2) \left\{ 1 + \left(\frac{\lambda_i}{\lambda_j}\right)^2 \frac{|z|^2}{1+\lambda_i^2 |x_i - x_j|^2} - \right.$$

$$\left. - 2 \frac{\lambda_i (x_i - x_j)}{1+\lambda_i^2 |x_i - x_j|^2} \frac{\lambda_i}{\lambda_j} z \right\} \tag{2.109}$$

Thus, if $|z| \leq \frac{1}{4} \lambda_j/\lambda_i$, we have:

$$\left| \frac{1}{(1+\left|\frac{\lambda_i}{\lambda_j} z - \lambda_i (x_i - x_j)\right|^2)^{\frac{n-2}{2}}} - \frac{1}{(1+\lambda_i^2 |x_i - x_j|^2)^{\frac{n-2}{2}}} \right|$$

$$\leq C \left\{ \frac{\lambda_i}{\lambda_j} |z| + \left(\frac{\lambda_i}{\lambda_j}\right)^2 |z|^2 \right\} \tag{2.110}$$

Hence

$$\frac{\lambda_j^{\frac{n+2}{2}}}{\lambda_i^{\frac{n-2}{2}}} I_1 = \frac{1}{(1+\lambda_i^2 |x_i - x_j|^2)^{\frac{n-2}{2}}} \int_{|z| \leq \frac{1}{4}\lambda_j/\lambda_i} \frac{|z|^2 dz}{(1+|z|^2)^{\frac{n+4}{2}}} + \tag{2.111}$$

$$+ 0 \left(\int_{|z| \leq \frac{1}{4}\lambda_j/\lambda_i} \left(\frac{\lambda_i}{\lambda_j} |z| + \left(\frac{\lambda_i}{\lambda_j}\right)^2 |z|^2 \right) \frac{|z|^2}{(1+|z|^2)^{\frac{n+4}{2}}} dz \right)$$

thus

$$I_1 = \frac{1}{\lambda_j^2}\left(\frac{\lambda_i}{\lambda_j}\right)^{\frac{n-2}{2}}\left(\int \frac{r^{n+1}\,dr\,d\sigma}{(1+r^2)^{\frac{n+4}{2}}} + 0\left(\frac{\lambda_i}{\lambda_j}\right)\right) \cdot \frac{1}{(1+\lambda_i^2|x_i-x_j|^2)^{\frac{n-2}{2}}} \tag{2.112}$$

Now

$$\int \frac{r^{n+1}\,dr\,d\sigma}{(1+r^2)^{\frac{n+4}{2}}} = \left[-\frac{1}{n+2}\,\frac{r^n}{(1+r^2)^{\frac{n+2}{2}}}\right]_0^\infty + \frac{n}{n+2}\int \frac{r^{n-1}\,dr\,d\sigma}{(1+r^2)^{\frac{n+2}{2}}} = \frac{n}{n+2}\,c_1 \tag{2.113}$$

Thus, summing up (2.112), (2.113), (2.108) and (2.100) we derive:

$$\int_{\mathbb{R}^n} \frac{\partial}{\partial \lambda_j}\left(\delta_j^{\frac{n+2}{n-2}}\right)\delta_i = \frac{n+2}{2\lambda_j}c_1\,\varepsilon_{ij} - n\frac{1}{\lambda_j}\left(\frac{\lambda_i}{\lambda_j}\right)^{\frac{n-2}{2}}\frac{1}{(1+\lambda_i^2|x_i-x_j|^2)^{\frac{n-2}{2}}}\,c_1 +$$

$$+ 0\left(\frac{1}{\lambda_j}\left(\frac{\lambda_i}{\lambda_j}\right)^{n/2} + \frac{1}{\lambda_j}\varepsilon_{ij}^{n/n-2}\right) \tag{2.114}$$

We have:

$$\varepsilon_{ij} = \frac{1}{\left(\frac{\lambda_i}{\lambda_j}+\frac{\lambda_j}{\lambda_i}+\lambda_i\lambda_j|x_i-x_j|^2\right)^{\frac{n-2}{2}}}\ ; \quad \varepsilon_{ij} \text{ small; thus } \lambda_j/\lambda_i \text{ large}$$

and λ_i/λ_j small. (2.115)

Hence

$$\frac{1}{(\varepsilon_{ij})^{2/n-2}} = \lambda_i/\lambda_j + \lambda_j/\lambda_i + \lambda_i\lambda_j|x_i-x_j|^2 = \lambda_j/\lambda_i\,(1+o(1)) \tag{2.116}$$

and

$$1+\lambda_i^2|x_i-x_j|^2 = \frac{\lambda_i}{\lambda_j}\left(\frac{\lambda_i}{\lambda_j}+\frac{\lambda_j}{\lambda_i}+\lambda_i\lambda_j|x_i-x_j|^2\right) = \frac{\lambda_i}{\lambda_j}\left(\frac{1}{\varepsilon_{ij}^{2/n-2}} - \lambda_i/\lambda_j\right) =$$

$$= \lambda_i/\lambda_j\,\frac{1}{\varepsilon_{ij}^{2/n-2}} + 0(\varepsilon_{ij}^{4/n-2}) \tag{2.117}$$

Thus

$$\frac{1}{\lambda_j}\left(\frac{\lambda_i}{\lambda_j}\right)^{\frac{n-2}{2}} \frac{1}{(1+\lambda_i^2|x_i-x_j|^2)^{\frac{n-2}{2}}} = \frac{1}{\lambda_j} \frac{\varepsilon_{ij}}{1+0(\varepsilon_{ij}^{4/n-2})} = \frac{1}{\lambda_j}\varepsilon_{ij} + \frac{1}{\lambda_j} 0\left(\varepsilon_{ij}^{\frac{n+2}{n-2}}\right) \tag{2.118}$$

We thus finally derive:

$$\int_{\mathbb{R}^n} \frac{\partial}{\partial\lambda_j}\left(\delta_j^{\frac{n+2}{n-2}}\right)\delta_i = -\frac{n-2}{2\lambda_j} c_1 \varepsilon_{ij} + \frac{1}{\lambda_j} 0(\varepsilon_{ij}^{n/n-2}) \tag{2.119}$$

which is (F16) under (2.104).

We now assume:

$$\theta = \lambda_i/\lambda_j \tag{2.120}$$

This time, we consider $\int_{\mathbb{R}^n} \frac{\partial\delta_j}{\partial\lambda_j} \delta_i^{\frac{n+2}{n-2}}$

we have:

$$\int_{\mathbb{R}^n} \delta_i^{\frac{n+2}{n-2}} \frac{\partial\delta_j}{\partial\lambda_j} = \frac{n-2}{2}\int_{\mathbb{R}^n} \frac{\lambda_i^{\frac{n+2}{2}}}{(1+\lambda_i^2|x-x_i|^2)^{\frac{n+2}{2}}} \cdot \left(\frac{\lambda_j}{1+\lambda_j^2|x-x_j|^2}\right)^{\frac{n-4}{2}} \times$$

$$\times\left(\frac{1}{1+\lambda_j^2|x-x_j|^2} - \frac{2\lambda_j^2|x-x_j|^2}{(1+\lambda_j^2|x-x_j|^2)^2}\right) dx \tag{2.121}$$

$$= \frac{n-2}{2\lambda_j}\int_{\mathbb{R}^n} \delta_i^{\frac{n+2}{n-2}}\delta_j - (n-2)\int_{\mathbb{R}^n} \frac{\lambda_i^{\frac{n+2}{2}}\lambda_j^{n/2}}{(1+\lambda_i^2|x-x_i|^2)^{\frac{n+2}{2}}} \frac{|x-x_j|^2}{(1+\lambda_j^2|x-x_j|^2)^{n/2}} dx$$

$$= \frac{n-2}{2\lambda_j} c_1 \varepsilon_{ij} + 0\left(\frac{1}{\lambda_j}\varepsilon_{ij}^{n/n-2}\right) - (n-2)L$$

$$L = \int_{\mathbb{R}^n} \frac{\lambda_i^{\frac{n+2}{2}} \lambda_j^{n/2}}{(1+\lambda_i^2 |x-x_i|^2)^{\frac{n+2}{2}}} \frac{|x-x_j|^2}{(1+\lambda_j^2 |x-x_j|^2)^{n/2}} dx \tag{2.122}$$

Let

$$y = \lambda_i (x - x_i) \tag{2.123}$$

we have

$$L = \int_{\mathbb{R}^n} \frac{\lambda_j^{n/2} \lambda_i^{\frac{2-n}{2}}}{(1+|y|^2)^{\frac{n+2}{2}}} \frac{\left|\frac{y}{\lambda_i} + x_i - x_j\right|^2}{\left(1+\left|\frac{\lambda_j}{\lambda_i} y + \lambda_j (x_i - x_j)\right|^2\right)^{n/2}} dy = \int_{\mathbb{R}^n} K_{ij}(y)\, dy \tag{2.124}$$

We split L into two summands:

$$L = L_1 + L_2 \tag{2.125}$$

where

$$L_1 = \int_{|y| \le \frac{1}{4}\lambda_i / \lambda_j} K_{ij}(y)\, dy; \quad L_2 = \int_{|y| \ge \frac{1}{4}\lambda_i / \lambda_j} K_{ij}(y)\, dy \tag{2.126}$$

We have

$$\left| \frac{1}{(1+|\lambda_j/\lambda_i y + \lambda_i (x_i - x_j)|^2)^{n/2}} - \frac{1}{(1+\lambda_j^2 |x_i - x_j|^2)^{n/2}} \right| \le \tag{2.127}$$

$$\le C(\lambda_j/\lambda_i |y| + (\lambda_j/\lambda_i)^2 |y|^2) \frac{1}{(1+\lambda_j^2 |x_i - x_j|^2)^{n/2}}$$

as soon as

$$|y| \le 1/4 \lambda_i / \lambda_j \tag{2.128}$$

We also have, under (2.128)

$$|y/\lambda_i + x_i - x_j|^2 = |x_i - x_j|^2 + 2\frac{(x_i - x_j, y)}{\lambda_i} + \frac{|y|^2}{\lambda_i^2} = |x_i - x_j|^2 + 0(\frac{1}{\lambda_j^2}) \tag{2.129}$$

We notice that

$$\int_{|y| \leq 1/4\lambda_i/\lambda_j} \frac{(y, x_i - x_j)\,dy}{(1+|y|^2)^{\frac{n+2}{2}}} = 0; \quad \int_{|y| \leq 1/4\lambda_i/\lambda_j} \frac{|y|\,dy}{(1+|y|^2)^{\frac{n+2}{2}}} = 0(\lambda_j/\lambda_i);$$

$$\int_{|y| \leq 1/4\lambda_i/\lambda_j} \frac{|y|^2 dy}{(1+|y|^2)^{\frac{n+2}{2}}} = 0(\operatorname{Log} \lambda_i/\lambda_j) \tag{2.130}$$

Thus, by (2.127), (2.129) and (2.130):

$$\begin{aligned} L_1 = {} & \frac{1}{\lambda_j}\left(\frac{\lambda_j}{\lambda_i}\right)^{\frac{n-2}{2}} \frac{\lambda_j^2 |x_i - x_j|^2}{(1+\lambda_j^2 |x_i - x_j|^2)^{n/2}} \int_{\mathbb{R}^n} \frac{dy}{(1+|y|^2)^{\frac{n+2}{2}}} + \\ & + 0\left(\left(\frac{\lambda_j}{\lambda_i}\right)^{n+4/2} \frac{\lambda_j^2 |x_i - x_j|^2}{(1+\lambda_j^2 |x_i - x_j|^2)^{n/2}}\right) \\ & + \left(\frac{\lambda_j}{\lambda_i}\right)^{\frac{n+2}{2}} \frac{1}{\lambda_j} \frac{1}{(1+\lambda_j^2 |x_i - x_j|^2)^{n/2}} 0(\operatorname{Log} \lambda_i/\lambda_j) + \\ & + 0\left(\lambda_j \left(\frac{\lambda_j}{\lambda_i}\right)^{\frac{n+2}{2}} \frac{1}{(1+\lambda_j^2 |x_i - x_j|^2)^{n/2}}\right) \times (|x_i - x_j|^2 + \frac{1}{\lambda_j^2}) \\ & + 0\left(\lambda_j \left(\frac{\lambda_j}{\lambda_i}\right)^{\frac{n+2}{2}} \operatorname{Log} \frac{\lambda_i}{\lambda_j} \frac{1}{(1+\lambda_j^2 |x_i - x_j|^2)^{n/2}} \left(|x_i - x_j|^2 + \frac{1}{\lambda_j^2}\right)\right) \end{aligned} \tag{2.131}$$

We thus have:

$$L_1 = \frac{1}{\lambda_j}\left(\frac{\lambda_j}{\lambda_i}\right)^{\frac{n-2}{2}} \frac{\lambda_j^2 |x_i - x_j|^2}{(1+\lambda_j^2 |x_i - x_j|^2)^{n/2}} \int_{\mathbb{R}^n} \frac{dy}{(1+|y|^2)^{\frac{n+2}{2}}} + \tag{2.132}$$

$$+ 0\left(\frac{1}{\lambda_j} \varepsilon_{ij}^{\frac{n+2}{n-2}} \operatorname{Log} \varepsilon_{ij}^{-1}\right)$$

We now estimate L_2

$$L_2 \leq \int_{|y| \geq \frac{1}{4}\lambda_i/\lambda_j} 1/\lambda_j^2 \times \frac{\lambda_j^{n/2}\lambda_i^{\frac{2-n}{2}}}{(1+|y|^2)^{\frac{n+2}{2}}} dy = 0\left(\left(\lambda_j/\lambda_i\right)^{\frac{n+2}{2}} \cdot \frac{1}{\lambda_j}\right)$$

$$= 0\left(\frac{1}{\lambda_j} \varepsilon_{ij}^{n+2/n-2}\right) \tag{2.133}$$

Then

$$L = \frac{1}{\lambda_j}\left(\frac{\lambda_j}{\lambda_i}\right)^{\frac{n-2}{2}} \frac{\lambda_j^2 |x_i-x_j|^2}{(1+\lambda_j^2|x_i-x_j|^2)^{n/2}} c_1 + 0\left(\frac{1}{\lambda_j}\varepsilon_{ij}^{\frac{n+2}{n-2}} \operatorname{Log} \varepsilon_{ij}^{-1}\right) \tag{2.134}$$

Hence, we have:

$$\int_{\mathbb{R}^n} \delta_i^{\frac{n+2}{n-2}} \frac{\partial \delta_j}{\partial \lambda_j} = \frac{n-2}{2\lambda_j} c_1 \varepsilon_{ij} - \frac{(n-2)}{\lambda_j}\left(\frac{\lambda_j}{\lambda_i}\right)^{\frac{n-2}{2}} \frac{\lambda_j^2|x_i-x_j|^2}{(1+\lambda_j^2|x_i-x_j|^2)^{n/2}} c_1 +$$

$$+ 0\left(\frac{1}{\lambda_j}\varepsilon_{ij}^{\frac{n+2}{n-2}} \operatorname{Log} \varepsilon_{ij}\right) \tag{2.135}$$

$$= -\frac{(n-2)}{2\lambda_j} c_1 \varepsilon_{ij}^{n/n-2}\left(-\lambda_i/\lambda_j - \lambda_j/\lambda_i - \lambda_i/\lambda_j |x_i-x_j|^2 + 2 \frac{\lambda_j}{\lambda_i}^{\frac{n-2}{2}} \frac{\lambda_j^2|x_i-x_j|^2}{(1+\lambda_j^2|x_i-x_j|^2)^{n/2}} \cdot \right.$$

$$\left. \cdot\ \varepsilon_{ij}^{-n/n-2}\right) + 0\left(\frac{1}{\lambda_j} \varepsilon_{ij}^{\frac{n+2}{n-2}} \operatorname{Log} \varepsilon_{ij}^{-1}\right)$$

Now

$$(1 + \lambda_j^2|x_i-x_j|^2)^{n/2} = (\lambda_j/\lambda_i)^{n/2}(\lambda_i/\lambda_j + \lambda_i\lambda_j|x_i-x_j|^2)^{n/2} \tag{2.136}$$

$$= (\lambda_j/\lambda_i)^{n/2} \varepsilon_{ij}^{-n/n-2} + 0((\lambda_j/\lambda_i)^{n/2})$$

Thus

$$\int_{\mathbb{R}^n} \delta_i^{\frac{n+2}{n-2}} \frac{\partial \delta_j}{\partial \lambda_j} = -\frac{(n-2)\, c_1}{2\lambda_j} \left[-\frac{\lambda_i}{\lambda_j} + \frac{\lambda_j}{\lambda_i} + \lambda_i \lambda_j |x_i - x_j|^2 \right] \varepsilon_{ij}^{n/n-2} +$$

$$+ \; 0(\varepsilon_{ij/\lambda_j}^{n/n-2}) \qquad (2.137)$$

(2.137) provides Estimate F16 under (2.120).

We now assume, to complete the proof of F16:

$$\theta = \lambda_i \lambda_j |x_i - x_j|^2 \qquad (2.138)$$

We then have two possible cases.

First

$$\lambda_i \leq \lambda_j \qquad (2.139)$$

We then have

$$\int_{\mathbb{R}^n} \frac{\partial}{\partial \lambda_j} \delta_j^{\frac{n+2}{n-2}} \delta_i = \frac{n+2}{2} \frac{c_1}{\lambda_j} \varepsilon_{ij} - (n+2)\lambda_j \int_{\mathbb{R}^n} \delta_j^{\frac{n+2}{n-2}} \frac{|x-x_j|^2}{1+\lambda_j^2 |x-x_j|^2} \delta_i +$$

$$+ \; 0\left(\frac{\varepsilon_{ij}^{n/n-2}}{\lambda_j}\right) \qquad (2.140)$$

We thus have to estimate

$$N = \int_{\mathbb{R}^n} \frac{|x-x_j|^2 \lambda_j^{\frac{n+2}{2}} \lambda_i^{\frac{n-2}{2}}}{(1+\lambda_j^2 |x-x_j|^2)^{\frac{n+4}{2}} (1+\lambda_i^2 |x-x_i|^2)^{n-2/2}} dx \qquad (2.141)$$

$$= \frac{\lambda_i^{\frac{n-2}{2}}}{\lambda_j^{\frac{n+2}{2}}} \int_{\mathbb{R}^n} \frac{|y|^2 dy}{(1+|y|^2)^{\frac{n+4}{2}} (1+|\frac{\lambda_i}{\lambda_j} y - \lambda_i (x_i - x_j)|^2)^{n-2/2}}$$

$$= \frac{1}{\lambda_j^2} \int_{\mathbb{R}^n} \frac{|y|^2 dy}{(1+|y|^2)^{\frac{n+4}{2}} \left(\frac{\lambda_j}{\lambda_i} + \left|\sqrt{\frac{\lambda_i}{\lambda_j}}\, y - \sqrt{\lambda_i\lambda_j}(x_i - x_j)\right|^2\right)^{\frac{n-2}{2}}}$$

To estimate N, we proceed as in the proof of (E1) under (1.27) and (1.29) (Notice that the indices i and j are reversed here with respect to (E1) under (27) and (29)).

As in the proof of (E1), we have:

$$\lambda_j/\lambda_i + \left|\sqrt{\lambda_i/\lambda_j}\, y - \lambda_i\lambda_j[x_i - x_j]\right|^2 = (\lambda_j/\lambda_i + \lambda_i\lambda_j|x_i - x_j|^2) \times \tag{2.142}$$

$$\times \left(1 + \frac{\lambda_i/\lambda_j |y|^2 + 2\lambda_i y.(x_j - x_i)}{\lambda_j/\lambda_i + \lambda_i\lambda_j |x_i - x_j|^2}\right)$$

Thus

$$\frac{1}{\lambda_j^2} \int_{|y| \le \frac{\sqrt{\theta}}{10}} \frac{|y|^2 dy}{(1+|y|^2)^{\frac{n+4}{2}} \left(\frac{\lambda_j}{\lambda_i} + \left|\sqrt{\frac{\lambda_i}{\lambda_j}}\, y - \sqrt{\lambda_i\lambda_j}[x_i - x_j]\right|^2\right)^{\frac{n-2}{2}}} = \tag{2.143}$$

$$= \frac{1}{\lambda_j^2} \frac{1}{(\lambda_j/\lambda_i + \lambda_i\lambda_j|x_i - x_j|^2)^{\frac{n-2}{2}}} \times B$$

where

$$B = \int_{|y| \le \frac{\sqrt{\theta}}{10}} \frac{|y|^2}{(1+|y|^2)^{\frac{n+4}{2}}} \left(1 - \frac{(n-2)\lambda_i y.(x_j - x_i)}{\lambda_j/\lambda_i + \lambda_i\lambda_j|x_i - x_j|^2} + \right. \tag{2.144}$$

$$\left. + \frac{\lambda_i/\lambda_j\, 0(|y|^2)}{\lambda_j/\lambda_i + \lambda_i\lambda_j|x_i - x_j|^2}\right) dy$$

$$= \int_{\mathbb{R}^n} \frac{|y|^2 dy}{(1+|y|^2)^{n+4/2}} + \frac{\lambda_i}{\lambda_j} \frac{1}{\lambda_j/\lambda_i + \lambda_i\lambda_j|x_i - x_j|^2}\, 0(\mathrm{Log}\, \theta) -$$

$$- \int_{|y| \geq \frac{\sqrt{\theta}}{10}} \frac{|y|^2 dy}{(1+|y|^2)^{\frac{n+4}{2}}}$$

$$= \int_{\mathbb{R}^n} \frac{|y|^2 dy}{(1+|y|^2)^{\frac{n+4}{2}}} + \frac{\lambda_i}{\lambda_j} \frac{1}{\lambda_j/\lambda_i + \lambda_i\lambda_j |x_i - x_j|^2} 0(\text{Log } \varepsilon_{ij}^{-1}) + 0(\varepsilon_{ij}^{2/n-2})$$

Now, under (2.138) and (2.139) :

$$\lambda_j/\lambda_i + \lambda_i\lambda_j |x_i - x_j|^2 = \varepsilon_{ij}^{-2/n-2} - \lambda_i/\lambda_j = \varepsilon_{ij}^{-2/n-2}(1+0(\varepsilon_{ij}^{2/n-2})) \tag{2.145}$$

Thus

$$\varepsilon_{ij}^{-2/n-2} = (\lambda_j/\lambda_i + \lambda_i\lambda_j |x_i - x_j|^2)(1 + 0(\varepsilon_{ij}^{2/n-2})) \tag{2.146}$$

or equivalently

$$\frac{1}{\lambda_j/\lambda_i + \lambda_i\lambda_j |x_i - x_j|^2} = \varepsilon_{ij}^{2/n-2}(1 + 0(\varepsilon_{ij}^{2/n-2})) \tag{2.147}$$

From (2.143) and (2.144), we then derive:

$$\frac{1}{\lambda_j^2} \int_{|y| \leq \frac{\sqrt{\theta}}{10}} \frac{|y|^2 dy}{(1+|y|^2)^{\frac{n+4}{2}} \left(\frac{\lambda_j}{\lambda_i} + \left| \sqrt{\frac{\lambda_i}{\lambda_j}} y - \sqrt{\lambda_i\lambda_j}(x_i - x_j) \right|^2 \right)^{\frac{n-2}{2}}} = \tag{2.148}$$

$$= \frac{1}{\lambda_j^2} \left\{ \varepsilon_{ij} \int_{\mathbb{R}^n} \frac{|y|^2 dy}{(1+|y|^2)^{\frac{n+4}{2}}} + 0\left(\varepsilon_{ij}^{n/n-2} \text{Log } \varepsilon_{ij}^{-1} \right) \right\}$$

On the other hand:

$$\int_{|y| \geq \frac{\sqrt{\theta}}{10}} \frac{|y|^2 dy}{(1+|y|^2)^{\frac{n+4}{2}} \left(\frac{\lambda_j}{\lambda_i} + \sqrt{\frac{\lambda_i}{\lambda_j}} y - \sqrt{\lambda_i\lambda_j}(x_i - x_j) |^2 \right)^{\frac{n-2}{2}}} = \int_{|y| \geq \frac{\sqrt{\theta}}{10}} L_{ij}(y) dy$$

$$\int_{\substack{|y| \geq \frac{\sqrt{\theta}}{10} \\ |y-\lambda_j(x_i-x_j)| \geq \frac{1}{10}\lambda_j|x_i-x_j|}} L_{ij}(y)\,dy \qquad + \int_{\substack{|y| \geq \frac{\sqrt{\theta}}{10} \\ |y-\lambda_j(x_i-x_j)| \leq \frac{1}{10}\lambda_j|x_i-x_j|}} L_{ij}(y)\,dy \tag{2.149}$$

Thus

$$\int_{\substack{|y| \geq \frac{\sqrt{\theta}}{10} \\ |y-\lambda_i(x_i-x_j)| \geq \frac{1}{10}\lambda_j|x_i-x_j|}} L_{ij}(y)\,dy \leq \frac{C}{\left(\frac{\lambda_j}{\lambda_i}+\lambda_i\lambda_j|x_i-x_j|^2\right)^{\frac{n-2}{2}}} \int_{|y| \geq \frac{\sqrt{\theta}}{10}} \frac{|y|^2 dy}{(1+|y|^2)^{\frac{n+4}{2}}} =$$

$$= 0(\varepsilon_{ij}^{n/n-2}) \tag{2.150}$$

while, as $|y| \geq \frac{9}{10}\lambda_j|x_i-x_j|$ as soon as $|y-\lambda_j(x_i-x_j)| \leq \frac{1}{10}\lambda_j|x_i-x_j|$,

$$\int_{\substack{|y| \geq \frac{\sqrt{\theta}}{10} \\ |y-\lambda_j(x_i-x_j)| \leq \frac{1}{10}|x_i-x_j|}} L_{ij}(y)\,dy \leq C \,.\, \frac{1}{\lambda_j^{n+2}|x_i-x_j|^{n+2}} \left(\frac{\lambda_j}{\lambda_i}\right)^{\frac{n+2}{2}} \times \tag{2.151}$$

$$\times \int_0^{\lambda_i|x_i-x_j|} \frac{dz}{(1+|z|^2)^{\frac{n-2}{2}}}$$

$$\leq C\frac{1}{\lambda_j^{n+2}|x_i-x_j|^{n+2}}\left(\frac{\lambda_j}{\lambda_i}\right)^{\frac{n+2}{2}} .\, \lambda_i^2|x_i-x_j|^2 = \frac{C}{\lambda_j^{\frac{n+2}{2}}\lambda_i^{\frac{n-2}{2}}} \,.\, \frac{1}{|x_i-x_j|^n} = 0(\varepsilon_{ij}^{n/n-2})$$

Thus, by (2.148) and (2.151),

$$N = \frac{1}{\lambda_j^2}\left\{\varepsilon_{ij}\int_{\mathbb{R}^n} \frac{|y|^2 dy}{(1+|y|^2)^{\frac{n+4}{2}}} + 0(\varepsilon_{ij}^{n/n-2}\operatorname{Log}\varepsilon_{ij}^{-1})\right\} \tag{2.152}$$

Hence by (2.140)

$$\int_{\mathbb{R}^n} \frac{\partial}{\partial\lambda_j} \delta_j^{\frac{n+2}{n-2}} \delta_i = \frac{n+2}{2}\frac{c_1}{\lambda_j}\varepsilon_{ij} - \frac{n+2}{\lambda_j}\varepsilon_{ij}\int_{\mathbb{R}^n}\frac{|y|^2 dy}{(1+|y|^2)^{\frac{n+4}{2}}} + \tag{2.153}$$

$$+ 0\left(\frac{\varepsilon_{ij}^{n/n-2}\ \mathrm{Log}\ \varepsilon_{ij}^{-1}}{\lambda_j}\right)$$

Hence, by (2.113)

$$\int_{\mathbb{R}^n} \frac{\partial}{\partial\lambda_j}\left(\delta_j^{\frac{n+2}{n-2}}\right)\delta_i = -\frac{1}{\lambda_j}\ \frac{n-2}{2}\ c_1\ \varepsilon_{ij} + \frac{1}{\lambda_j}\ 0(\varepsilon_{ij}^{n/n-2}\mathrm{Log}\ \varepsilon_{ij}^{-1}) \tag{2.154}$$

which is (F16) under (2.138) and (2.139), as in this case we have:

$$\frac{\lambda_i}{\lambda_j}\ \varepsilon_{ij}^{n/2} = 0(\varepsilon_{ij}^{n/n-2}). \tag{2.155}$$

The remaining case is now:

$$\theta = \lambda_i\lambda_j\left|x_i - x_j\right|^2 \tag{2.138}$$

and

$$\lambda_j \leq \lambda_i \tag{2.156}$$

The proof is very similar to the preceding cases, especially to the previous proof using, this time, $\int_{\mathbb{R}^n} \delta_i^{\frac{n+2}{n-2}}\ \frac{\partial\delta_j}{\partial\lambda_j}$. As there are no new facts, the proof is not carried out here.

We now turn to

$$\int_{\mathbb{R}^n} \frac{\partial}{\partial x_i}\left(\delta_i^{\frac{n+2}{n-2}}\right)\delta_j \tag{2.157}$$

It is easy to see that:

$$\int_{\mathbb{R}^n} \frac{\partial}{\partial x_i}\left(\delta_i^{\frac{n+2}{n-2}}\right)\delta_j = -\int_{\mathbb{R}^n} \delta_i^{\frac{n+2}{n-2}}\ \frac{\partial\delta_j}{\partial x_j} = -\int_{\mathbb{R}^n} \delta_i\ \frac{\partial}{\partial x_j}\left(\delta_j^{\frac{n+2}{n-2}}\right) \tag{2.158}$$

Thus we can always assume

$$\lambda_i \geq \lambda_j \tag{2.159}$$

We first consider the case

$$\theta = \lambda_i/\lambda_j \tag{2.160}$$

Let

$$y = \lambda_i(x-x_i) \tag{2.161}$$

We have

$$\int_{\mathbb{R}^n} \frac{\partial}{\partial x_i}\left(\delta_i^{\frac{n+2}{n-2}}\right)\delta_j = \frac{n+2}{n-2}\lambda_j^{\frac{n-2}{2}} \frac{1}{\lambda_i^{\frac{n-4}{2}}} \int_{\mathbb{R}^n} \frac{y}{(1+|y|^2)^{\frac{n+4}{2}}\left(1+\left|\frac{\lambda_j}{\lambda_i}y-\lambda_j(x_j-x_i)\right|^2\right)^{\frac{n-2}{2}}}dy$$

(2.162)

which we split as usual into two parts; one of which corresponds to integration on $\{y/|y| \leq 1/4\lambda_i/\lambda_j\}$ and the other one to the complement to this set. Again, we have:

$$1 + \left|\frac{\lambda_j}{\lambda_i}y-\lambda_j(x_j-x_i)\right|^2 = (1+\lambda_j^2|x_j-x_i|^2)\left\{1-2\frac{\lambda_j(x_j-x_i)}{1+\lambda_j^2|x_j-x_i|^2}\cdot\frac{\lambda_j}{\lambda_i}y + \right.$$

$$\left. + \frac{\lambda_j^2}{\lambda_i^2}\frac{|y|^2}{1+\lambda_j^2|x_j-x_i|^2}\right\} \tag{2.163}$$

Thus assuming

$$|y| \leq 1/4\,\lambda_i/\lambda_j \tag{2.164}$$

and expanding in power series, we have:

$$\frac{1}{\left(1+\left|\frac{\lambda_j}{\lambda_i}y-\lambda_j(x_j-x_i)\right|^2\right)^{\frac{n-2}{2}}} = \frac{1}{(1+\lambda_j^2|x_j-x_i|^2)^{\frac{n-2}{2}}} + \tag{2.165}$$

$$+ (n-2)\frac{\lambda_j(x_j-x_i)}{(1+\lambda_j^2|x_j-x_i|^2)^{n/2}}\cdot\frac{\lambda_j}{\lambda_i}y + g(|y|) + R(y)$$

with

$$|R(y)| \leq C \left\{ \frac{\lambda_j^2 |x_j - x_i|^2}{(1+\lambda_j^2 |x_j - x_i|^2)^{\frac{n+2}{2}}} \left(\frac{\lambda_j}{\lambda_i}\right)^2 |y|^2 + \right. \tag{2.166}$$

$$\left. + \frac{\lambda_j |x_j - x_i|}{(1+\lambda_j^2 |x_j - x_i|^2)^{\frac{n+2}{2}}} \left(\frac{\lambda_j}{\lambda_i}\right)^3 |y|^3 \right\}$$

Now, by oddness:

$$\int_{|y| \leq \frac{1}{4}\lambda_i/\lambda_j} \frac{y\, g(|y|)}{(1+|y|^2)^{\frac{n+4}{2}}} dy = 0 \tag{2.167}$$

while

$$\int_{|y| \leq \frac{1}{4}\lambda_i/\lambda_j} \frac{|y|^3 dy}{(1+|y|^2)^{\frac{n+4}{2}}} \leq C \tag{2.168}$$

and

$$\int_{|y| \leq \frac{1}{4}\lambda_i/\lambda_j} \frac{|y|^4}{(1+|y|^2)^{\frac{n+4}{2}}} dy \leq C \operatorname{Log}(\lambda_i/\lambda_j) \tag{2.169}$$

Thus

$$\frac{n+2}{n-2} \frac{\lambda_j^{\frac{n-2}{2}}}{\lambda_i^{\frac{n-4}{2}}} \int_{|y| \leq \frac{1}{4}\lambda_i/\lambda_j} \frac{|y|}{(1+|y|^2)^{\frac{n+4}{2}}} |R(y)|\, dy \leq \tag{2.170}$$

$$\leq C \frac{\lambda_j^{\frac{n-2}{2}}}{\lambda_i^{\frac{n-4}{2}}} \left\{ \frac{\lambda_j^2 |x_j - x_i|^2}{(1+\lambda_j^2 |x_j - x_i|^2)^{\frac{n+2}{2}}} \times \varepsilon_{ij}^{4/n-2} + \frac{\lambda_j |x_j - x_i|}{(1+\lambda_j^2 |x_j - x_i|^2)^{\frac{n+2}{2}}} \varepsilon_{ij}^{6/n-2} \operatorname{Log} \varepsilon_{ij}^{-1} \right\}$$

Now

$$\frac{\lambda_j^{\frac{n-2}{2}}}{\lambda_i^{\frac{n-4}{2}}} \frac{\lambda_j^2 |x_j - x_i|^2}{(1+\lambda_j^2 |x_j-x_i|^2)^{\frac{n+2}{2}}} \leq \left(\frac{\lambda_j}{\lambda_i}\right)^{\frac{n-2}{2}} \lambda_i \lambda_j^2 |x_j - x_i|^2 \tag{2.171}$$

$$\leq \lambda_i \lambda_j \left(\frac{\lambda_j}{\lambda_i}\right)^{\frac{n-1}{2}} \sqrt{\lambda_j \lambda_i} |x_j - x_i| \ |x_j - x_i| \leq C \lambda_i \lambda_j |x_i - x_j| \varepsilon_{ij} \quad \text{as } \theta = \lambda_i / \lambda_j$$

Thus

$$\frac{n+2}{n-2} \frac{\lambda_j^{\frac{n-2}{2}}}{\lambda_i^{\frac{n-4}{2}}} \int_{|y| \leq \frac{1}{4}\lambda_i/\lambda_j} \frac{|y|}{(1+|y|^2)^{\frac{n+4}{2}}} |R(y)| dy = 0 \ (\lambda_i \lambda_j) \varepsilon_{ij}^{\frac{n+2}{n-2}} |x_i - x_j| \tag{2.172}$$

It remains thus to estimate

$$(n+2) \frac{\lambda_j^{n-2/2} / \lambda_i^{n-4/2}}{(1+\lambda_j^2 |x_j - x_i|^2)^{n/2}} \cdot \frac{\lambda_j^2}{\lambda_i} \int_{|y| \leq \frac{1}{4}\lambda_i/\lambda_j} \frac{y(x_j - x_i)}{(1+|y|^2)^{\frac{n+4}{2}}} y \, dy \tag{2.173}$$

which is equal to:

$$\frac{n+2}{n} \frac{\lambda_j^{n-2/2} / \lambda_i^{n-4/2}}{(1+\lambda_j^2 |x_j - x_i|^2)^{n/2}} \cdot \frac{\lambda_j^2}{\lambda_i} \int_{|y| \leq \frac{1}{4}\lambda_i/\lambda_j} (x_j - x_i) \frac{|y|^2}{(1+|y|^2)^{\frac{n+4}{2}}} dy \tag{2.174}$$

$$= \frac{n+2}{n} \frac{\lambda_j^{\frac{n-2}{2}} / \lambda_i^{\frac{n-4}{2}}}{(1+\lambda_j^2 |x_j - x_i|^2)^{n/2}} \frac{\lambda_j^2}{\lambda_i} c_7 (x_j - x_i) + 0 \left(\frac{|x_i - x_j| \lambda_j^{n-2/2} / (\lambda_i)^{n-4/2}}{(1+\lambda_j^2 |x_j - x_i|^2)^{n/2}} \cdot \frac{\lambda_j^4}{\lambda_i^2} \right)$$

Finally

$$\frac{\lambda_j^{n-2/2} / \lambda_i^{n-4/2}}{(1+\lambda_j^2 |x_j - x_i|^2)^{n/2}} \frac{\lambda_j^2}{\lambda_i} = \frac{\lambda_j \lambda_i}{\left(\frac{\lambda_i}{\lambda_j} + \lambda_i \lambda_j |x_j - x_i|^2\right)^{n/2}} = \tag{2.175}$$

$$= \frac{\lambda_j \lambda_i}{\left(\frac{\lambda_j}{\lambda_i} + \frac{\lambda_i}{\lambda_j} + \lambda_i \lambda_j |x_j - x_i|^2 - \frac{\lambda_j}{\lambda_i}\right)^{n/2}} = \lambda_i \lambda_j \varepsilon_{ij}^{n/n-2}(1 + 0(\varepsilon_{ij}^{4/n-2}))$$

Hence

$$\frac{n+2}{n-2} \frac{\lambda_j^{n-2/2}}{\lambda_i^{n-4/2}} \int_{|y| \leq \frac{1}{4}\lambda_i/\lambda_j} \frac{y}{(1+|y|^2)^{\frac{n+4}{2}} (1+|\frac{\lambda_j}{\lambda_i} y - \lambda_j(x_j - x_i)|^2)^{\frac{n-2}{2}}} dy =$$

$$= \frac{n+2}{n} \lambda_i \lambda_j \varepsilon_{ij}^{n/n-2} c_7(x_j - x_i) + 0(\lambda_i \lambda_j \varepsilon_{ij}^{\frac{n+2}{n-2}} |x_i - x_j|) \qquad (2.176)$$

We consider now

$$\frac{n+2}{n-2} \frac{\lambda_j^{\frac{n-2}{2}}}{\lambda_i^{\frac{n-4}{2}}} \int_{|y| \geq \frac{1}{4}\lambda_i/\lambda_j} \frac{y}{(1+|y|^2)^{\frac{n+4}{2}} (1+|\frac{\lambda_j}{\lambda_i} y - \lambda_j(x_j - x_i)|^2)^{\frac{n-2}{2}}} dy \qquad (2.177)$$

which we upperbound by

$$C \cdot \frac{\lambda_j^{\frac{n-2}{2}}}{\lambda_i^{\frac{n-4}{2}}} \varepsilon_{ij}^{3/n-2} = C \lambda_i \lambda_j |x_i - x_j| \varepsilon_{ij}^{n/n-2} \frac{1}{(\lambda_i \lambda_j)^{1/2} |x_i - x_j|} \qquad (2.178)$$

In case,

$$\lambda_j^2 |x_i - x_j|^2 \geq \varepsilon_0 > 0; \quad \varepsilon_0 \text{ arbitrary fixed constant.} \qquad (2.179)$$

(2.178) is satisfactory.

Indeed, we then have:

$$\lambda_i \lambda_j |x_i - x_j|^2 \geq \varepsilon_0 \frac{\lambda_i}{\lambda_j} \geq C \varepsilon_{ij}^{-2/n-2} \qquad (2.180)$$

Thus

$$\frac{n+2}{n-2}\frac{\lambda_j^{\frac{n-2}{2}}}{\lambda_i^{\frac{n-4}{2}}}\int_{|y|\geq\frac{1}{4}\lambda_i/\lambda_j}\frac{|y|}{(1+|y|^2)^{\frac{n+4}{2}}\left(1+\left|\frac{\lambda_j}{\lambda_i}y-\lambda_j(x_j-x_i)\right|^2\right)^{\frac{n-2}{2}}}dy =$$

$$= 0\left(\lambda_i\lambda_j|x_i-x_j|\varepsilon_{ij}^{\frac{n+2}{n-2}}\right) \tag{2.181}$$

Otherwise, we have:

$$|x_i-x_j| \leq \frac{\varepsilon_0}{\lambda_j} \tag{2.182}$$

Hence

$$1 + \left|\frac{\lambda_j}{\lambda_i}y-\lambda_j(x_j-x_i)\right|^2 = 1 + \frac{\lambda_j^2}{\lambda_i^2}|y|^2 - 2\frac{\lambda_j^2}{\lambda_i}y(x_j-x_i)+\lambda_j^2|x_i-x_j|^2 \tag{2.183}$$

$$= \left(1+\frac{\lambda_j^2}{\lambda_i^2}|y|^2\right)\left(1-\frac{2\lambda_j^2/\lambda_i y(x_j-x_i)}{1+\lambda_j^2/\lambda_i^2|y|^2}+\frac{\lambda_j^2|x_i-x_j|^2}{1+\lambda_j^2/\lambda_i^2|y|^2}\right)$$

Thus, if

$$|y| \geq 1/4\lambda_i/\lambda_j \tag{2.184}$$

we have:

$$\frac{|\lambda_j^2/\lambda_i y.(x_j-x_i)|}{1+\lambda_j^2/\lambda_i^2|y|^2} \leq \lambda_i\frac{|x_j-x_i|}{|y|} \leq 4\varepsilon_0 \tag{2.188}$$

$$\frac{\lambda_j^2|x_i-x_j|^2}{1+\lambda_j^2/\lambda_i^2|y|^2} \leq \frac{\lambda_i^2|x_i-x_j|^2}{\lambda_j^2} \leq 16\varepsilon_0^2 \tag{2.186}$$

We then expand $(1+|\lambda_j/\lambda_i y - \lambda_j(x_j-x_i)|^2)^{\frac{n-2}{2}}$ in power series

$$\frac{1}{(1+|\lambda_j/\lambda_i y-\lambda_j(x_j-x_i)|^2)^{\frac{n-2}{2}}} = \frac{1}{(1+\frac{\lambda_j^2}{\lambda_i^2}|y|^2)^{\frac{n-2}{2}}}(1+g(|y|)+R(y)) \qquad (2.187)$$

with

$$|R(y)| \leq C \cdot \frac{\lambda_i|x_i-x_j|}{|y|} \qquad (2.188)$$

Thus

$$\left|\frac{\lambda_j^{\frac{n-2}{2}}}{\lambda_i^{\frac{n-4}{2}}}\int_{|y|\geq\frac{1}{4}\lambda_i/\lambda_j}\frac{|y|}{(1+|y|^2)^{\frac{n+4}{2}}}\cdot\frac{1}{(1+\frac{\lambda_j^2}{\lambda_i^2}|y|^2)^{\frac{n-2}{2}}}R(r)\,dy\right| \leq \qquad (2.189)$$

$$\leq C|x_i-x_j|\lambda_i\lambda_j\left(\frac{\lambda_j}{\lambda_i}\right)^{\frac{n-4}{2}}\int_{|y|\geq\frac{1}{4}\lambda_i/\lambda_j}\frac{dy}{(1+|y|^2)^{\frac{n+4}{2}}} =$$

$$0(\lambda_i\lambda_j|x_i-x_j|\varepsilon_{ij}^{n+4/n-2})$$

The contribution in (2.177) of $1 + g(|y|)$ is zero.

Finally, we thus have under (2.160), summing up (2.176), (2.181) and (2.189):

$$\int_{\mathbb{R}^n}\frac{\partial}{\partial x_i}\left(\delta_i^{\frac{n+2}{n-2}}\right)\delta_j = \frac{n+2}{n}\lambda_i\lambda_j\varepsilon_{ij}^{n/n-2}c_7(x_j-x_i) + \qquad (2.190)$$

$$+ 0(\lambda_i\lambda_j\varepsilon_{ij}^{\frac{n+2}{n-2}}|x_i-x_j|)$$

We are left now with the case:

$$\theta = \lambda_i\lambda_j|x_i-x_j|^2 \; ; \; \lambda_j \leq \lambda_i \qquad (2.191)$$

Let

$$y = \lambda_i(x-x_i) \qquad (2.192)$$

We have:

$$\int_{\mathbb{R}^n} \frac{\partial}{\partial x_i}\left(\delta_i^{\frac{n+2}{n-2}}\right)\delta_j = \frac{n+2}{n-2}\,\frac{\lambda_j^{\frac{n-2}{2}}}{\lambda_i^{\frac{n-4}{2}}}\int_{\mathbb{R}^n}\frac{y\,dy}{(1+|y|^2)^{\frac{n+4}{2}}\left(1+\left|\frac{\lambda_j}{\lambda_i}y-\lambda_j(x_j-x_i)\right|^2\right)^{\frac{n-2}{2}}}$$

$$= \frac{n+2}{n-2}\lambda_i\int_{\mathbb{R}^n}\frac{y\,dy}{(1+|y|^2)^{\frac{n+4}{2}}\left(\frac{\lambda_i}{\lambda_j}+\left|\sqrt{\frac{\lambda_i}{\lambda_j}}y-\sqrt{\lambda_i\lambda_j}(x_j-x_i)\right|^2\right)^{\frac{n-2}{2}}} \tag{2.193}$$

We have:

$$\frac{\lambda_i}{\lambda_j}+\left|\sqrt{\frac{\lambda_i}{\lambda_j}}y-\sqrt{\lambda_i\lambda_j}(x_j-x_i)\right|^2 = \left(\frac{\lambda_i}{\lambda_j}+\lambda_i\lambda_j|x_i-x_j|^2\right)\times \tag{2.194}$$

$$\times\left(1+\frac{\lambda_j/\lambda_i|y|^2-2\lambda_j y(x_j-x_i)}{\lambda_i/\lambda_j+\lambda_i\lambda_j|x_i-x_j|^2}\right)$$

We first assume:

$$|y| \leq \varepsilon_0|x_i-x_j|\sqrt{\lambda_i\lambda_j} \tag{2.195}$$

Then

$$\frac{\frac{\lambda_j}{\lambda_i}|y|^2}{\frac{\lambda_i}{\lambda_j}+|x_i-x_j|^2\lambda_i\lambda_j} \leq \varepsilon_0^2\,\frac{|x_i-x_j|^2\lambda_j^2}{\frac{\lambda_i}{\lambda_j}+|x_i-x_j|^2\lambda_i\lambda_j} \leq \varepsilon_0^2 \tag{2.196}$$

$$\frac{\lambda_j|x_i-x_j||y|}{\frac{\lambda_i}{\lambda_j}+|x_i-x_j|^2\lambda_i\lambda_j} \leq \varepsilon_0\sqrt{\frac{\lambda_j}{\lambda_i}}\;\frac{\lambda_i\lambda_j|x_i-x_j|^2}{\frac{\lambda_i}{\lambda_j}+|x_i-x_j|^2\lambda_i\lambda_j} \leq \varepsilon_0 \tag{2.197}$$

$$\left(\frac{\lambda_i}{\lambda_j}+\left|\sqrt{\frac{\lambda_j}{\lambda_i}}y-(x_j-x_i)\sqrt{\lambda_i\lambda_j}\right|^2\right)^{-\frac{n-2}{2}} = \tag{2.198}$$

$$= g(|y|)+(n-2)\frac{\lambda_j y.(x_j-x_i)}{\left(\frac{\lambda_i}{\lambda_j}+\lambda_i\lambda_j|x_i-x_j|^2\right)^{n/2}} + R(y)$$

with

$$|R(y)| \leq C\left\{\frac{\lambda_j^2|y|^2|x_i-x_j|^2}{\left(\frac{\lambda_i}{\lambda_j}+\lambda_i\lambda_j|x_i-x_j|^2\right)^{n+2/2}} + \frac{\lambda_j/\lambda_i|y|^3|x_i-x_j|\lambda_j}{\left(\frac{\lambda_i}{\lambda_j}+|x_i-x_j|^2\lambda_i\lambda_j\right)^{n+2/2}}\right\} \tag{2.199}$$

Now

$$\int_{|y|\leq\varepsilon_0|x_i-x_j|\sqrt{\lambda_i\lambda_j}} \frac{|y|^3}{(1+|y|^2)^{\frac{n+4}{2}}}dy \leq C \tag{2.200}$$

while

$$\int_{|y|\leq\varepsilon_0|x_i-x_j|\sqrt{\lambda_i\lambda_j}} \frac{|y|^4}{(1+|y|^2)^{\frac{n+4}{2}}}dy \leq C \operatorname{Log} \varepsilon_{ij}^{-1} \tag{2.201}$$

$$(n+2)\int_{|y|\leq\varepsilon_0|x_i-x_j|\sqrt{\lambda_i\lambda_j}} \frac{y(y.(x_j-x_i))}{(1+|y|^2)^{\frac{n+4}{2}}}dy = \tag{2.202}$$

$$= \frac{n+2}{n}\int_{|y|\leq\varepsilon_0|x_i-x_j|\sqrt{\lambda_i\lambda_j}} \frac{|y|^2}{(1+|y|^2)^{\frac{n+4}{2}}}dy\quad(x_j-x_i) =$$

$$= \frac{n+2}{n}c_7(x_j-x_i) + 0(\varepsilon_{ij}^{2/n-2}|x_i-x_j|)$$

Finally

$$\frac{1}{(\lambda_i/\lambda_j+\lambda_i\lambda_j|x_i-x_j|^2)^{n/2}} = \varepsilon_{ij}^{n/n-2} + 0(\varepsilon_{ij}^{\frac{n+2}{n-2}}) \tag{2.203}$$

Thus

$$\frac{\lambda_j^{\frac{n-2}{2}}}{\lambda_i^{\frac{n-4}{2}}} \frac{n+2}{n-2} \int_{|y|\le\varepsilon_0|x_i-x_j|\sqrt{\lambda_i\lambda_j}} \frac{y\,dy}{(1+|y|^2)^{\frac{n+4}{2}}(1+|\frac{\lambda_j}{\lambda_i}y-\lambda_j(x_j-x_i)|^2)^{\frac{n-2}{2}}} =$$

$$= \frac{n+2}{n}\lambda_i\lambda_j\, c_7(x_j-x_i)\,\varepsilon_{ij}^{n/n-2} + 0(\varepsilon_{ij}^{\frac{n+2}{n-2}}|x_i-x_j|) \tag{2.204}$$

We finally estimate:

$$\frac{\lambda_j^{\frac{n-2}{2}}}{\lambda_i^{\frac{n-4}{2}}} \int_{|y|\ge\varepsilon_0|x_i-x_j|\sqrt{\lambda_i\lambda_j}} \frac{|y|\,dy}{(1+|y|^2)^{\frac{n+4}{2}}(1+|\frac{\lambda_j}{\lambda_i}y-\lambda_j(x_j-x_i)|^2)^{\frac{n-2}{2}}} \le$$

$$\le \frac{\lambda_j^{\frac{n-2}{2}}}{\lambda_i^{\frac{n-4}{2}}} \int_{|y|\ge\varepsilon_0|x_i-x_j|\sqrt{\lambda_i\lambda_j}} \frac{|y|\,dy}{(1+|y|^2)^{\frac{n+4}{2}}} \le C\frac{\lambda_j^{\frac{n-2}{2}}}{\lambda_i^{\frac{n-4}{2}}} \frac{1}{(\lambda_i\lambda_j)^{3/2}|x_i-x_j|^3}$$

$$\le C\frac{\lambda_j^{\frac{n-3}{2}}}{\lambda_i^{\frac{n-3}{2}}} \frac{1}{\lambda_i\lambda_j^2|x_i-x_j|^4}\lambda_i\lambda_j|x_i-x_j| \le C\,\varepsilon_{ij}^{\frac{n+1}{n-2}}\lambda_i\lambda_j|x_i-x_j| \tag{2.205}$$

Summing up all our estimates, we have

$$\int_{\mathbb{R}^n} \frac{\partial}{\partial x_i}\delta_i^{\frac{n+2}{n-2}}\delta_j = \frac{n+2}{n}\lambda_i\lambda_j c_7(x_j-x_i)\,\varepsilon_{ij}^{n/n-2} + 0\left(\varepsilon_{ij}^{\frac{n+1}{n-2}}\lambda_i\lambda_j|x_i-x_j|\right) \tag{2.206}$$

which yields the second estimate of F11.

We now want to estimate:

$$\int K(x)\,\delta_i^{\frac{n+2}{n-2}}\,\frac{\partial\delta_j}{\partial x_j} = \frac{\partial}{\partial x_j}\int K(x)\,\delta_i^{\frac{n+2}{n-2}}\,\delta_j = \tag{2.207}$$

$$= \frac{\partial}{\partial x_j}\int K(x+x_j)\,\delta(x_i-x_j,\lambda_i)^{\frac{n+2}{n-2}}\,\delta(o,\lambda_j)\quad dx$$

$$= \int\Big[DK(x+x_j)\,.\Big]\,\delta(x_i-x_j,\lambda_i)^{\frac{n+2}{n-2}}\,\delta(o,\lambda_j)\,dx\ -$$

$$-\ \int K(x+x_j)\,\frac{\partial}{\partial x_i}\left[\delta(x_i-x_j,\lambda_i)^{\frac{n+2}{n-2}}\right]\delta(o,\lambda_j)\,dx$$

$$= \int\big[DK(x)\,.\big]\,\delta_i^{\frac{n+2}{n-2}}\,\delta_j\,dx - \int K(x)\,\frac{n+2}{n-2}\,\delta_i^{4/n-2}\,\frac{\partial\delta_i}{\partial x_i}\,\delta_j\,dx$$

Thus by (F7) and F14:

$$\int K(x)\,\delta_i^{\frac{n+2}{n-2}}\,\frac{\partial\delta_j}{\partial x_j} = DK(x_i)\,\varepsilon_{ij} + o(\varepsilon_{ij}) + 0\left(\frac{1}{\lambda_i^{\frac{n+2}{2}}\lambda_j^{\frac{n-2}{2}}}\right) - K(x_i)\int\delta_j^{\frac{n+2}{n-2}}\,\frac{\partial\delta_i}{\partial x_i}$$

$$+\ 0\left(\varepsilon_{ij} + \frac{1}{\lambda_j^{n/2}\lambda_i^{\frac{n-2}{2}}}\right) \tag{2.208}$$

Thus, by (2.206)

$$\int K(x)\,\delta_i^{\frac{n+2}{n-2}}\,\frac{\partial\delta_j}{\partial x_j} = DK(x_i)\,\varepsilon_{ij} - K(x_i)\,\frac{n+2}{n}\,\lambda_i\lambda_j c_7(x_j-x_i)\,\varepsilon_{ij}^{n/n-2} \tag{2.209}$$

$$+\ 0(\varepsilon_{ij}) + 0\left(\frac{1}{\lambda_j^{n/2}\lambda_i^{\frac{n-2}{2}}}\right) + o(\varepsilon_{ij}) + 0\left(\varepsilon_{ij}^{\frac{n+1}{n-2}}\,\lambda_i\lambda_j\,|x_i-x_j|\right)$$

where

$$o(\varepsilon_{ij}) \leq \frac{C}{\lambda_i} \, 0 \left(\varepsilon_{ij} \left((\operatorname{Log} \varepsilon_{ij}^{-1})^{\frac{n-2}{2}} \left(\frac{\operatorname{Log} \lambda_i}{\lambda_i} \right)^{2/n} + 1 \right) \left[|DK(x_i)| + \frac{1}{\lambda_i} \right] \right) \tag{2.210}$$

and

$$0(\varepsilon_{ij}) = \frac{n+2}{n-2} \int_{B_i} (K(x) - K(x_i)) \, \delta_i^{4/n-2} \delta_j \frac{\partial \delta_i}{\partial x_i} \tag{2.211}$$

as in (2.79) with i replaced by j.

Indeed, $o(\varepsilon_{ij})$ comes from $\int [DK(x).] \delta_j dx$ which is estimated in F7, where $o(\varepsilon_{ij})$ satisfies the inequality (2.210).

While $0(\varepsilon_{ij})$ comes from $\int K(x) \, \delta_i^{4/n-2} \frac{\partial \delta_i}{\partial x_i} \delta_j dx$ which is estimated in F14, with a $0(\varepsilon_{ij})$ satisfying (2.79), up to changing i in j.

This yields the first estimate in F11.

For the first estimate in F16, one proceeds as in F7.

This completes the proof of Estimates F7-F19.

The proof of F20-F21 follows the proof of F7, using the fact that $\left|\frac{\partial \delta_j}{\partial x_j}\right| \leq C \, \delta_j \lambda_j$.

We end this section with the following estimates, whose proof is very similar to that of F11 and F16.

<u>Estimate F22</u>: $$\int_{\mathbb{R}^n} \nabla \frac{\partial \delta_i}{\partial x_i} \nabla \frac{\partial \delta_j}{\partial x_j} = \lambda_i \lambda_j \, 0(\varepsilon_{ij}^{n/n-2})$$

<u>Estimate F23</u>: $$\int_{\mathbb{R}^n} \nabla \frac{\partial \delta_i}{\partial x_i} \nabla \frac{\partial \delta_j}{\partial \lambda_j} = 0(\lambda_i |x_i - x_j| \varepsilon_{ij}^{n/n-2})$$

With respect to F11-F16, there are no new facts in the proof of F22-F23. The computations are very similar and more precise formulae than the ones stated can be derived. As there are enough estimates, already and to come, in this work, we will not repeat here such very similar proofs. However, we show how one can prove quite rapidly slightly weaker estimates, namely:

$\underline{\text{F22}'}$: $\int_{\mathbb{R}^n} \nabla \frac{\partial \delta_i}{\partial x_i} \nabla \frac{\partial \delta_j}{\partial x_j} = \lambda_i \lambda_j 0(\varepsilon_{ij}^{n/n-2} \operatorname{Log} \varepsilon_{ij}^{-1})$

$\underline{\text{F23}'}$: $\int_{\mathbb{R}^n} \nabla \frac{\partial \delta_i}{\partial x_i} \nabla \frac{\partial \delta_j}{\partial \lambda_j} = 0(\lambda_i |x_i - x_j| \varepsilon_{ij}^{n/n-2} \operatorname{Log} \varepsilon_{ij}^{-1})$.

Proof of F22': We have

$$\left| \nabla \frac{\partial \delta_i}{\partial x_i} \right| = (n-2) \left| \nabla \frac{\lambda_i^{\frac{n+2}{2}} (x - x_i)}{(1+\lambda_i^2 |x-x_i|^2)^{n/2}} \right| \le C \lambda_i \delta_i^{n/n-2} \tag{2.212}$$

Hence, by (E2)

$$\left| \int_{\mathbb{R}^n} \nabla \frac{\partial \delta_i}{\partial x_i} \nabla \frac{\partial \delta_j}{\partial x_j} \right| \le C \lambda_i \lambda_j \int_{\mathbb{R}^n} \delta_i^{n/n-2} \delta_j^{n/n-2} \tag{2.213}$$

$$= \lambda_i \lambda_j 0(\varepsilon_{ij}^{n/n-2} \operatorname{Log} \varepsilon_{ij}^{-1})$$

Proof of F23': We have

$$\nabla \delta_j = (2-n) \frac{\lambda_j^{\frac{n+2}{2}} (x-x_j)}{(1+\lambda_j^2 |x-x_j|^2)^{n/2}} = (2-n) \frac{\lambda_j (x-x_j)}{(1+\lambda_j^2 |x-x_j|^2)^{n/2}} \tag{2.214}$$

Hence

$$\left| \frac{\partial}{\partial \lambda_j} \nabla \delta_j \right| \le C |x-x_j| \delta_j^{n/n-2} \tag{2.215}$$

This, together with (2.212) implies:

$$\left| \int_{\mathbb{R}^n} \nabla \frac{\partial \delta_i}{\partial x_i} \nabla \frac{\partial \delta_j}{\partial \lambda_j} \right| \le C \lambda_i \int_{\mathbb{R}^n} \delta_i^{n/n-2} \delta_j^{n/n-2} |x-x_j| \tag{2.216}$$

Hence

$$\left| \int_{\mathbb{R}^n} \nabla \frac{\partial \delta_i}{\partial x_i} \nabla \frac{\partial \delta_j}{\partial \lambda_j} \right| \le C \lambda_i \varepsilon_{ij}^{n/n-2} \operatorname{Log} \varepsilon_{ij}^{-1} \tag{2.217}$$

Now, when $x_i = x_j$, $\int_{\mathbb{R}^n} \nabla \frac{\partial \delta_i}{\partial x_i} \nabla \frac{\partial \delta_j}{\partial \lambda_j} = 0$.

Furthermore:

$$\left| \nabla \frac{\partial}{\partial x_j} \frac{\partial \delta_j}{\partial \lambda_j} \right| = C \left| \nabla \frac{\partial}{\partial \lambda_j} \left[\frac{\lambda_j^{\frac{n-2}{2}} \lambda_j^2 (x-x_j)}{(1+\lambda_j^2 |x-x_j|^2)^{n/2}} \right] \right| \tag{2.218}$$

$$= C \left| \frac{n+2}{2} \nabla \left[\frac{\lambda_i^{n/2} (x-x_j)}{(1+\lambda_j^2 |x-x_j|^2)^{n/2}} \right] - n \nabla \left[\frac{\lambda_j^{n/2} (x-x_j) \lambda_j^2 |x-x_j|^2}{(1+\lambda_j^2 |x-x_j|^2)^{n/2+1}} \right] \right|$$

$$\leq C \, \delta_j^{n/n-2}$$

Thus

$$\left| \frac{\partial}{\partial x_j} \int_{\mathbb{R}^n} \nabla \frac{\partial \delta_i}{\partial x_i} \nabla \frac{\partial \delta_j}{\partial \lambda_j} \right| \leq C \lambda_i \int \delta_i^{n/n-2} \delta_j^{n/n-2} \tag{2.219}$$

$$= \lambda_i \, 0(\varepsilon_{ij}^{n/n-2} \operatorname{Log} \varepsilon_{ij}^{-1})$$

(2.217), (2.219) and the fact that $\int_{\mathbb{R}^n} \nabla \frac{\partial \delta_i}{\partial x_i} \nabla \frac{\partial \delta_j}{\partial \lambda_j}$ depends only on $x_i - x_j$, λ_i, λ_j imply Estimate F23'.

3 The v-part of u

We introduce, for v such that $\nabla v \in L^2(\mathbb{R}^n)$, $v \in L^{\frac{2n}{n-2}}(\mathbb{R}^n)$

$$v_0 = \text{projection of } v \text{ onto } H_0^1(\Omega) \tag{3.1}$$

i.e. v_0 satisfies

$$\begin{cases} \Delta v_0 = \Delta v \\ v_0 = 0 \big|_{\partial\Omega} \end{cases} \tag{3.2}$$

We have:

$$\int_\Omega |\nabla v_0|^2 = -\int_\Omega \Delta v_0 . v_0 = -\int_\Omega \Delta v . v_0 = -\int_\Omega \nabla v . \nabla v_0 \tag{3.3}$$

hence

$$\int_\Omega |\nabla v_0|^2 \leq \int_\Omega |\nabla v|^2 \tag{3.4}$$

We assume on v:

$$\begin{cases} \int_{\mathbb{R}^n} \nabla v . \nabla \delta_i = 0 & \forall i \in [1,p] \\ \int_{\mathbb{R}^n} \nabla v . \nabla \frac{\partial \delta_i}{\partial x_i} = 0 & \forall i \in [1,p] \\ \int_{\mathbb{R}^n} \nabla v . \nabla \frac{\partial \delta_i}{\partial \lambda_i} = 0 & \forall i \in [1,p] \end{cases} \tag{G0}$$

3.1 The second derivative and v

We first prove on such a v the following proposition:

Proposition 3.1: Assume the ε_{ij}'s are small enough for $i \neq j$; $(i, j) \in [1, p]^2$. Let v satisfy (G0).

Then, there exists a constant $\alpha_0(p) > 0$ such that:

$$\int |\nabla v|^2 - \frac{n+2}{n-2} \sum_i \int \delta_i^{4/n-2} v^2 \geq \alpha_0 \int |\nabla v|^2$$

The proof of Proposition 3.1 requires the following construction.

For δ_i given, $i = 1, \ldots, p$, we introduce:

$$\Omega_i = \{x \in \mathbb{R}^n \text{ s.t } |x - x_i| < \frac{1}{8\lambda_i} \underset{i \neq j}{\operatorname{Min}} \varepsilon_{ij}^{-1/n-2} ; |x - x_j| > \frac{1}{8\lambda_j} \varepsilon_{ij}^{-1/n-2} \tag{3.5}$$

$$\text{for those } j \text{ such that } \lambda_j \geq \lambda_i \}$$

By construction $\Omega_i \cap \Omega_j = \emptyset$ for $i \neq j$.

Indeed, if x belongs to $\Omega_i \cap \Omega_j$, $i \neq j$, we have, assuming $\lambda_j \geq \lambda_i$:

$$\frac{1}{8\lambda_j \varepsilon_{ij}^{1/n-2}} < |x - x_j| < \frac{1}{8\lambda_j \varepsilon_{ij}^{1/n-2}} \tag{3.6}$$

a contradiction.

Now let

$$Q_i = \text{orthogonal projection on } H_0^1(\Omega_i) \tag{3.7}$$

and for φ in $H = \{ \varphi \in L^{\frac{2n}{n-2}}(\mathbb{R}^n) \, ; \nabla \varphi \in L^2(\mathbb{R}^n) \}$

$$\varphi_i = Q_i \varphi \tag{3.8}$$

We then have the following lemma:

Lemma 3.2: if the ε_{ij}'s are small enough for $i \neq j$, then:

$$\int_{\mathbb{R}^n} \delta_j^{\frac{n+2}{n-2}} |\varphi - \varphi_j| \leq C\left(\sum_{i \neq j} \varepsilon_{ij}^{1/2}\right)\left(\int |\nabla \varphi|^2\right)^{1/2}$$

From Lemma 3.2, we now derive the proof of Proposition 3.1.

Proof of Proposition 3.1

Consider

$$v_i = Q_i v \tag{3.9}$$

Let

$$E_i^- = \text{span}\{\delta_i, \frac{\partial \delta_i}{\partial x_i}, \frac{\partial \delta_i}{\partial \lambda_i}\} \tag{3.10}$$

$$E_i^+ = (E_i^-)^{\perp}; \quad \text{the orthogonal being taken in the sense of} \tag{3.11}$$

the scalar product $\int \nabla \varphi . \nabla \psi$ for and ψ in H.

On E_i^+, we have:

$$\int |\nabla \varphi|^2 - \frac{n+2}{n-2} \int \varphi^2 \delta_i^{4/n-2} \geq \alpha_1 \int |\nabla \varphi|^2 ; \alpha_1 > 0; \tag{3.12}$$

$\varphi \in E_i^+$

On E_i^-, we, of course, have:

$$\left| \int |\nabla \varphi|^2 - \frac{n+2}{n-2} \int \varphi^2 \delta_i^{4/n-2} \right| \leq C \int |\nabla \varphi|^2 ; \tag{3.13}$$

C given independent of i; $\varphi \in E_i^-$

We split v:

$$v_i = v_i^- + v_i^+ ; \; v_i^- \in E_i^- ; \; v_i^+ \in E_i^+ \tag{3.14}$$

We have:

$$v_i^- = (\int \nabla v_i \nabla \delta_i) \frac{\delta_i}{\int |\nabla \delta_i|^2} + \int \nabla v_i \nabla \frac{\partial \delta_i}{\partial x_i} \frac{\frac{\partial \delta_i}{\partial x_i}}{\int |\nabla \frac{\partial \delta_i}{\partial x_i}|^2} + \tag{3.15}$$

$$+ \int \nabla v_i \nabla \frac{\partial \delta_i}{\partial \lambda_i} \frac{\frac{\partial \delta_i}{\partial \lambda_i}}{\int |\nabla \frac{\partial \delta_i}{\partial \lambda_i}|^2}$$

The notation $\int \nabla v_i \nabla \frac{\partial \delta_i}{\partial x_i} \frac{\frac{\partial \delta_i}{\partial x_i}}{\int |\nabla \frac{\partial \delta_i}{\partial x_i}|^2}$ should be understood as a summation on

each component of x_i in $\mathbb{R}^n$.

We estimate $\int |\nabla v_i^-|^2$

We have

$$\int \nabla v_i \nabla \delta_i = -\int \delta_i^{\frac{n+2}{n-2}} (v_i - v) \tag{3.16}$$

as v satisfies (G0), i.e. $\int \nabla v \nabla \delta_i = -\int \delta_i^{\frac{n+2}{n-2}} v = 0$

Hence, by Lemma 3.2:

$$\left| \int \nabla v_i . \nabla \delta_i \right| \leq C \sum_{i \neq j} \varepsilon_{ij}^{1/2} \left(\int |\nabla v|^2 \right)^{1/2} \tag{3.17}$$

Next, using again (G0):

$$\int \nabla v_i \nabla \frac{\partial \delta_i}{\partial x_i} = \int \nabla (v_i - v) \nabla \frac{\partial \delta_i}{\partial x_i} = -\int \frac{n+2}{n-2} \delta_i^{4/n-2} \frac{\partial \delta_i}{\partial x_i} (v_i - v) \tag{3.18}$$

Hence, using (E.5), and Lemma 3.2:

$$\left| \int \nabla v_i . \nabla \frac{\partial \delta_i}{\partial x_i} \right| \leq C \lambda_i \int \delta_i^{\frac{n+2}{n-2}} |v_i - v| \leq C \lambda_i \sum_{i \neq j} \varepsilon_{ij}^{\frac{n-2}{2}} \left(\int |\nabla v|^2 \right)^{1/2} \tag{3.19}$$

The same argument yields also with (E6):

$$\left| \int \nabla v_i . \nabla \frac{\partial \delta_i}{\partial \lambda_i} \right| \leq \frac{C}{\lambda_i} \sum_{i \neq j} \varepsilon_{ij}^{1/2} \left(\int |\nabla v|^2 \right)^{1/2} \tag{3.20}$$

Thus, as we know that

$$\int \left| \nabla \frac{\partial \delta_i}{\partial x_i} \right|^2 = \lambda_i^2 \delta_0 \quad \text{(Estimate F4)} \tag{3.21}$$

$$\int \left|\nabla \frac{\partial \delta_i}{\partial \lambda_i}\right|^2 = \delta_1/\lambda_i^2 \qquad \text{(Estimate F6)} \tag{3.22}$$

we derive from (3.17), (3.19), (3.20) and the definition in (3.15) of v_i^-:

$$\left(\int |\nabla v_i^-|^2\right)^{1/2} \leq C \sum_{i\neq j} \varepsilon_{ij}^{1/2} \left(\int |\nabla v|^2\right)^{1/2} \tag{3.23}$$

Then using (3.12) and (3.13), we get

$$\begin{aligned}\int |\nabla v_i|^2 - \frac{n+2}{n-2}\int v_i^2 \delta_i^{4/n-2} &\geq \int |\nabla v_i^+|^2 - \frac{n+2}{n-2}\int v_i^{+2}\delta_i^{4/n-2} - \\ &- C\int |\nabla v_i^-|^2 - C\left(\int |\nabla v_i^+|^2\right)^{1/2}\left(\int |\nabla v_i^-|^2\right)^{1/2} \\ &\geq \alpha_1 \int |\nabla v_i|^2 - C'\left(\sum_{i\neq j}\varepsilon_{ij}^{1/2}\right)^2 \int |\nabla v|^2 - \\ &- C\left(\int |\nabla v_i^+|^2\right)^{1/2}\left(\int |\nabla v_i^-|^2\right)^{1/2}\end{aligned} \tag{3.24}$$

We upperbound, with M a suitable constant:

$$C\left(\int |\nabla v_i^+|^2\right)^{1/2}\left(\int |\nabla v_i^-|^2\right)^{1/2} \leq \alpha_1/2 \int |\nabla v_i^+|^2 + M\int |\nabla v_i^-|^2 \tag{3.25}$$

This, with (3.23) yields:

$$\int |\nabla v_i|^2 - \frac{n+2}{n-2}\int v_i^2 \delta_i^{4/n-2} \geq \alpha_1/2 \int |\nabla v_i|^2 - C''\left(\sum_{i\neq j}\varepsilon_{ij}\right)\int |\nabla v|^2 \tag{3.26}$$

As v_i is the orthogonal projection of v on $H_0^1(\Omega_i)$ and as the Ω_i's are disjoint, we have, by (3.4) applied to Ω_i:

$$\sum_i \int |\nabla v_i|^2 \leq \sum_i \int_{\Omega_i} |\nabla v|^2 \tag{3.27}$$

On another hand, we have:

$$\begin{aligned}\int \delta_i^{4/n-2}|v^2 - v_i^2| &\leq \left(\int |v+v_i|^{2n/n-2}\right)^{\frac{n-2}{2n}} \left(\int \delta_i^{\frac{8n}{n^2-4}} |v-v_i|^{2n/n+2}\right)^{\frac{n+2}{2n}} \\ &\leq C\left(\int |\nabla v|^2\right)^{1/2} \times \left(\int \delta_i^{\frac{n+2}{n-2}}|v-v_i|\right)^{\frac{8n}{(n+2)^2}} \left(\int |v-v_i|^{2n/n-2}\right)^{\frac{(n-2)^2}{(n+2)^2}}\end{aligned} \tag{3.28}$$

Thus, using Lemma 3.2,

$$\int \delta_i^{4/n-2} |v^2 - v_i^2| \leq C \int |\nabla v|^2 \left(\sum_{i \neq j} \varepsilon_{ij}^{1/2} \right)^{\frac{8n}{(n+2)^2}} \tag{3.29}$$

(3.26) and (3.29) yield:

$$\int |\nabla v|^2 - \frac{n+2}{n-2} \sum_i \int \delta_i^{4/n-2} v^2 \geq \int_{(\cup \Omega_i)^c} |\nabla v|^2 + \tag{3.30}$$

$$+ \sum_i \left[\int_{\Omega_i} |\nabla v|^2 - \int_{\Omega_i} |\nabla v_i|^2 \right] + \alpha_1/2 \sum_i \int_{\Omega_i} |\nabla v_i|^2 - o(1) \int |\nabla v|^2$$

Now, by (3.27),

$$\sum_i \left[\int_{\Omega_i} |\nabla v|^2 - \int_{\Omega_i} |\nabla v_i|^2 \right] + \alpha_1/2 \sum_i \int |\nabla v_i|^2 \geq \tag{3.31}$$

$$\geq \operatorname{Inf}(\alpha_1/2, 1) \sum_i \int_{\Omega_i} |\nabla v|^2 = \bar{\alpha}_0 \sum_i \int_{\Omega_i} |\nabla v|^2$$

Thus, with (3.30)

$$\int |\nabla v|^2 - \frac{n+2}{n-2} \sum_i \int \delta_i^{4/n-2} v^2 \geq \bar{\alpha}_0 \int |\nabla v|^2 - \tag{3.32}$$

$$- o(1) \int |\nabla v|^2 \geq \bar{\alpha}_0/2 \int |\nabla v|^2 = \alpha_0 \int |\nabla v|^2$$

hence Proposition 3.1.

Proof of Lemma 3.2: We first prove the following.

Let

$$\delta = \frac{1}{(1+|x|^2)^{\frac{n-2}{2}}} \; ; \quad \Delta \delta = -c_2 \delta^{\frac{n+2}{n-2}} \; ; \quad c_2 > 0 \tag{3.33}$$

Let

$$w \in L^{2n/n-2} \; ; \quad \nabla w \in L^2(\mathbb{R}^n) \tag{3.34}$$

Let

$$\lambda > 0\ ; \quad B_\lambda = \{x \in \mathbb{R}^n / |x| < \lambda\} \tag{3.35}$$

and

$$h \text{ defined by } \begin{cases} \Delta h = 0 \text{ in } B_\lambda \\ h = w\big|_{\partial B_\lambda} \end{cases} \tag{3.36}$$

We then have

$$\int_{B_\lambda} \delta^{n+2/n-2} |h| \leq \frac{C}{\lambda^{\frac{n-2}{2}}} \left(\int_{\mathbb{R}^n} |\nabla w|^2 dx \right)^{1/2} \tag{3.37}$$

Indeed, first notice that we can take w to be positive. Otherwise replacing w by $|w| = \tilde{w}$ and h by $\tilde{h}$, we have:

$$|h| \leq \tilde{h} \tag{3.38}$$

and

$$\int_{\mathbb{R}^n} |\nabla w|^2 dx = \int_{\mathbb{R}^n} |\nabla \tilde{w}|^2 dx \tag{3.39}$$

Hence if (3.37) holds with $w \geq 0$, it will hold for all w.

If we assume w to be positive, (3.37) becomes:

$$\int_{B_\lambda} \delta^{n+2/n-2} h \leq \frac{C}{\lambda^{n/2}} \left(\int_{\mathbb{R}^n} |\nabla w|^2 dx \right)^{1/2} \tag{3.40}$$

We prove now (3.40)

We have:

$$\int_{B_\lambda} \delta^{n+2/n-2} h = -c_2^{-1} \int_{B_\lambda} \Delta\delta\, h = -c_2^{-1} \int_{\delta B_\lambda} \frac{\partial \delta}{\partial n} h + \tag{3.41}$$

$$c_2 \int \nabla \left(\delta - \frac{1}{(1+\lambda^2)^{\frac{n-2}{2}}} \right) \nabla h = -c_2^{-1} \int_{\partial B_\lambda} \frac{\partial \delta}{\partial n} h \leq \frac{c_3}{\lambda^{n-1}} \int_{\partial B_\lambda} h$$

Let

$$\bar{h}(x) = \lambda^{\frac{n-2}{2}} h(\lambda x) \ ; \quad \bar{w}(x) = \lambda^{\frac{n-2}{2}} w(\lambda x) \tag{3.42}$$

We have

$$\begin{cases} \Delta \bar{h} = 0 & \text{in } B_1 = \{x / |x| < 1\} \\ \bar{h} = \bar{w} \big|_{\partial B_1} \end{cases} \tag{3.43}$$

and

$$\int_{\partial B_1} \bar{h} \leq C \left[\left(\int_{B_1} |\nabla \bar{w}|^2 dx \right)^{1/2} + \left(\int_{B_1} \bar{w}^{2n/n-2} \right)^{\frac{n-2}{2n}} \right] \tag{3.44}$$

$$\leq C \left(\int_{\mathbb{R}^n} |\nabla \bar{w}|^2 \right)^{1/2}$$

Now

$$\int_{\mathbb{R}^n} |\nabla \bar{w}|^2 = \int_{\mathbb{R}^n} |\nabla w|^2 \tag{3.45}$$

and

$$\int_{\partial B_1} \bar{h} = \frac{1}{\lambda^{\frac{n-2}{2}}} \int_{\partial B_\lambda} h \tag{3.46}$$

(3.45), (3.46) and (3.41) yield (3.37).

We turn now to the proof of Lemma 3.2.

First assume we are dealing with two sets only, Ω_1 and Ω_2, as in (3.5).

The general case will be easily derived from this situation.

Take $j = 1$ in Lemma 3.2.

We also make a translation and a dilation, so that $\lambda_1 = 1$ and $x_1 = 0$.

Thus:

$$\bar{\varphi} = \lambda_1^{\frac{n-2}{2}} \varphi(\lambda_1 x + x_1) \; ; \quad \bar{\delta}_i = \lambda_1^{\frac{n-2}{2}} \delta_i(\lambda_1 x + x_1) \tag{3.47}$$

Notice that

$$\bar{\varepsilon}_{12} = \varepsilon_{12} \; ; \quad \bar{\lambda}_1 = 1 \; ; \quad \bar{\lambda}_2 = \lambda_2/\lambda_1 \; ; \quad \bar{x}_1 = 0; \quad \bar{x}_2 = \lambda_1(x_1 - x_2) \tag{3.48}$$

Assume first that:

$$\lambda_1 \geq \lambda_2 \; ; \quad \text{hence } \bar{\lambda}_2 \leq 1 \tag{3.49}$$

Then

$$\bar{\Omega}_1 = \{x \mid |x| < \frac{1}{8\varepsilon_{12}^{\frac{1}{n-2}}}\} \tag{3.50}$$

$$\bar{\varphi}_1 = \bar{\varphi} - \bar{h} \; ; \quad \text{with } \Delta\bar{h} = 0 \text{ in } \bar{\Omega}_1; \quad \bar{h} = \bar{\varphi} \text{ on } \partial\bar{\Omega}_1 \tag{3.51}$$

we have:

$$\int_{\mathbb{R}^n} \bar{\delta}_1^{\frac{n+2}{n-2}} |\bar{\varphi} - \bar{\varphi}_1| \leq \int_{\bar{\Omega}_1} \bar{\delta}_1^{\frac{n+2}{n-2}} |\bar{\varphi} - \bar{\varphi}_1| + \tag{3.52}$$

$$+ \int_{|x| \geq \frac{1}{8\varepsilon_{12}^{\frac{1}{n-2}}}} \frac{1}{(1+|x|^2)^{\frac{n+2}{2}}} |\bar{\varphi}|$$

By (3.37), we have:

$$\int_{\bar{\Omega}_1} \bar{\delta}_1^{\frac{n+2}{n-2}} |\bar{\varphi} - \bar{\varphi}_1| \leq C\,\varepsilon_{12}^{1/2} \left(\int_{\mathbb{R}^n} |\nabla\bar{\varphi}|^2 dx \right)^{1/2} \tag{3.53}$$

$$= C\,\varepsilon_{12}^{1/2} \left(\int_{\mathbb{R}^n} |\nabla\varphi|^2 \right)^{1/2}$$

On the other hand:

$$\int_{|x| \geq \frac{1}{8\varepsilon_{12}^{1/n-2}}} \frac{1}{(1+|x|^2)^{\frac{n+2}{2}}} |\bar{\varphi}| \tag{3.54}$$

$$\leq C \left(\int_{\mathbb{R}^n} |\nabla \bar{\varphi}|^2 \right)^{1/2} \cdot \varepsilon_{12}^{\frac{n+2}{2(n-2)}} = C\,\varepsilon_{12}^{\frac{n+2}{2(n-2)}} \left(\int_{\mathbb{R}^n} |\nabla \varphi|^2 \right)^{1/2}$$

Now

$$\int_{\mathbb{R}^n} \bar{\delta}_1^{\frac{n+2}{n-2}} |\bar{\varphi} - \bar{\varphi}_1| = \int_{\mathbb{R}^n} \delta_1^{\frac{n+2}{n-2}} |\varphi - \varphi_1| \tag{3.55}$$

(3.55), (3.54) and (3.53) yield Lemma 3.2 under 3.49.

We now consider the other case, namely:

$$\lambda_2 \geq \lambda_1 \; ; \quad \lambda_2 \geq 1 \tag{3.56}$$

Then

$$\bar{\Omega}_1 = \{x / |x| < \frac{1}{\varepsilon_{12}^{1/n-2}} \text{ and } |x - \bar{x}_2| > \frac{1}{8\bar{\lambda}_2 \varepsilon_{12}^{1/n-2}} \} \tag{3.57}$$

Let then

$$\tilde{\Omega} = \{x / |x| < \frac{1}{8\varepsilon_{12}^{1/n-2}} \} \; ; \quad \tilde{W} = \{x / |x - \bar{x}_2| > \frac{1}{8\bar{\lambda}_2 \varepsilon_{12}^{1/n-2}} \} \tag{3.58}$$

Let

$$\tilde{\varphi}_1 = \text{orthogonal projection of } \bar{\varphi} \text{ on } H_0^1(\tilde{\Omega}) \tag{3.59}$$

$$\tilde{\psi}_1 = \text{orthogonal projection of } \bar{\varphi} \text{ on } H_0^1(\tilde{W}) \tag{3.60}$$

As already noticed, we may always assume

$$\bar{\varphi} \geq 0 \tag{3.61}$$

Then, $\partial\bar{\Omega}_1 \subset \partial\tilde{\Omega} \cup \partial\tilde{W}$

Hence:

$$|\bar{\varphi} - \bar{\varphi}_1| \leq -[\tilde{\varphi}_1 + \tilde{\psi}_1] + 2\bar{\varphi} \quad \text{in } \mathbb{R}^n \tag{3.61'}$$

and

$$\int_{\mathbb{R}^n} \bar{\delta}_1^{\frac{n+2}{n-2}} |\bar{\varphi} - \bar{\varphi}_1| \leq \int_{\tilde{\Omega}} \bar{\delta}_1^{\frac{n+2}{n-2}} (\bar{\varphi} - \tilde{\varphi}_1) + \int_{\tilde{W}} \bar{\delta}_1^{\frac{n+2}{n-2}} (\bar{\varphi} - \tilde{\psi}_1) + \tag{3.62}$$

$$+ \int_{\mathbb{R}^n - \tilde{\Omega}} \bar{\delta}_1^{\frac{n+2}{n-2}} \bar{\varphi} + \int_{\mathbb{R}^n / \tilde{W}} \bar{\delta}_1^{\frac{n+2}{n-2}} \bar{\varphi}$$

We already have an estimate on $\int_{\mathbb{R}^n / \tilde{\Omega}} \bar{\delta}_1^{\frac{n+2}{n-2}} |\bar{\varphi}|$ from (3.54), which is suitable for the inequality we wish to prove.

We estimate now

$$\int_{\mathbb{R}^n / \tilde{W}} \bar{\delta}_1^{\frac{n+2}{n-2}} |\bar{\varphi}| = \int_{\bar{\lambda}_2 |x - \bar{x}_2| \leq \frac{1}{8\varepsilon_{12}^{1/n-2}}} \bar{\delta}_1^{\frac{n+2}{n-2}} |\bar{\varphi}| \tag{3.63}$$

We claim that:

$$\bar{\delta}_1 = \frac{1}{(1+|x|^2)^{\frac{n-2}{2}}} \leq \frac{2^{\frac{n-2}{2}} \bar{\lambda}_2^{\frac{n-2}{2}}}{(1+\bar{\lambda}_2^2 |x-\bar{x}_2|^2)^{\frac{n-2}{2}}} \quad \text{if } \bar{\lambda}_2 |x - \bar{x}_2| \leq \frac{1}{8\varepsilon_{12}^{1/n-2}} \tag{3.64}$$

and ε_{12} is small enough.

Indeed setting $y = \bar{\lambda}_2 [x - \bar{x}_2]$, (3.64) is equivalent to:

$$1 + |y|^2 \leq 2\bar{\lambda}_2 (1 + |\frac{y}{\bar{\lambda}_2} + \bar{x}_2|^2) \quad \text{if } |y| \leq \frac{1}{8\varepsilon_{12}^{1/n-2}} \tag{3.65}$$

Thus

$$|y|^2 (1 - \frac{2}{\bar{\lambda}_2}) - 4y.\bar{x}_2 + 1 \leq 2\bar{\lambda}_2 + 2\bar{\lambda}_2 |\bar{x}_2|^2 = 2(\varepsilon_{12}^{-\frac{2}{n-2}} - 1) \tag{3.66}$$

if $|y| \leq \frac{1}{8\varepsilon_{12}^{1/n-2}}$

(3.66) holds if:

$$\frac{1}{64\varepsilon_{12}^{2/n-2}}\left(1 - \frac{2}{\bar{\lambda}_2}\right) + \frac{4}{8}\frac{|\bar{x}_2|}{\varepsilon_{12}^{1/n-2}} + 1 \leq 2\,\frac{1}{\varepsilon_{12}^{2/n-2}} - 1 \tag{3.67}$$

Now, we have:

$$\sqrt{\bar{\lambda}_2}\,|\bar{x}_2| \leq \varepsilon_{12}^{-1/n-2} \tag{3.68}$$

Hence (3.66) holds if:

$$\frac{1}{\varepsilon_{12}^{2/n-2}} \geq 3 + \frac{1}{8(\sqrt{\bar{\lambda}_2}\,\varepsilon_{12}^{1/n-2})^2} + \frac{1}{2\varepsilon_{12}^{1/n-2}(\sqrt{\bar{\lambda}_2}\,\varepsilon_{12}^{1/n-2})} \tag{3.69}$$

As $\bar{\lambda}_2 \geq 1$, (3.69) holds for ε_{12} small enough; thus (3.64) holds if ε_{12} is small enough.

We have then

$$\int_{\mathbb{R}^n/\tilde{W}} \bar{\delta}_1^{\frac{n+2}{n-2}} |\bar{\varphi}| \leq C \left(\int_{\mathbb{R}^n/\tilde{W}} \bar{\delta}_1^{2n/n-2}\right)^{\frac{n+2}{2n}} \left(\int |\nabla\bar{\varphi}|^2\right)^{1/2} \tag{3.70}$$

Thus by (3.64):

$$\int_{\mathbb{R}^n/\tilde{W}} \bar{\delta}_1^{\frac{n+2}{n-2}} |\bar{\varphi}| \leq C \left(\int_{\mathbb{R}^n/\tilde{W}} \bar{\delta}_1^{n/n-2}\bar{\delta}_2^{n/n-2}\right)^{\frac{n+2}{2n}} \left(\int|\nabla\bar{\varphi}|^2\right)^{1/2} \leq$$

$$\leq C\left(\int_{\mathbb{R}^n} \bar{\delta}_1^{n/n-2}\bar{\delta}_2^{n/n-2}\right)^{\frac{n+2}{2n}} \left(\int|\nabla\bar{\varphi}|^2\right)^{1/2} \tag{3.70}$$

Using (E2), (3.70) yields:

$$\int_{\mathbb{R}^n/\tilde{W}} \bar{\delta}_1^{\frac{n+2}{n-2}} |\bar{\varphi}| \leq C\,\varepsilon_{ij}^{\frac{n+2}{2(n-2)}} (\mathrm{Log}\,\varepsilon_{ij}^{-1})^{\frac{n+2}{2n}} \left(\int|\nabla\bar{\varphi}|^2\right)^{1/2} \tag{3.71}$$

(3.71) is satisfactory also in view of (3.62) and the desired result.

It remains to estimate two quantities:

$$\int_{\overline{\Omega}} \bar{\delta}_1^{\frac{n+2}{n-2}} (\bar{\varphi} - \tilde{\varphi}_1) \tag{3.72}$$

$$\int_{\tilde{W}} \bar{\delta}_1^{\frac{n+2}{n-2}} (\bar{\varphi} - \tilde{\psi}_1) \tag{3.73}$$

For (3.72), we already worked out a satisfactory estimate in (3.37) and (3.53).

Thus, $\int_{\bar{\lambda}_2 |x-\bar{x}_2| \geq \frac{1}{8\varepsilon_{12}^{1/n-2}}} \bar{\delta}_1^{\frac{n+2}{n-2}} |\bar{\varphi} - \tilde{\psi}_1|$ remains.

This is, with c_3 a constant:

$$\int_{\tilde{W}} \bar{\delta}_1^{\frac{n+2}{n-2}} (\bar{\varphi} - \tilde{\psi}_1) = -c_3 \int_{\tilde{W}} \Delta\bar{\delta}_1 (\bar{\varphi} - \tilde{\psi}_1) = c_3 \int_{\tilde{W}} \nabla(\bar{\delta}_1 - \bar{\theta}_1) \nabla(\bar{\varphi} - \tilde{\psi}_1) -$$

$$- c_3 \int_{\partial\tilde{W}} \frac{\partial}{\partial n} (\bar{\delta}_1 - \bar{\theta}_1)(\bar{\varphi} - \tilde{\psi}_1) = -c_3 \int_{\partial\tilde{W}} \frac{\partial}{\partial n} (\bar{\delta}_1 - \bar{\theta}_1)(\bar{\varphi} - \tilde{\psi}_1) \tag{3.74}$$

where

$$\bar{\theta}_1 \text{ satisfies } \Delta\bar{\theta}_1 = 0 \quad \bar{\theta}_1 = \bar{\delta}_1 \big|_{\partial\tilde{W}} \tag{3.75}$$

(We used in (3.74) the fact that $\Delta(\bar{\varphi} - \tilde{\psi}_1) = 0$ on $\tilde{W}$ by definition of $\tilde{\psi}_1$.)

Thus:

$$\left| \int_{\tilde{W}} \bar{\delta}_1^{\frac{n+2}{n-2}} (\bar{\varphi} - \tilde{\psi}_1) \right| \leq C \operatorname{Sup}_{\partial\tilde{W}} \left| \frac{\partial}{\partial n} (\bar{\delta}_1 - \bar{\theta}_1) \right| \int_{\partial\tilde{W}} \bar{\varphi} = \tag{3.76}$$

$$= C \operatorname{Sup}_{\partial\tilde{W}} \left| \frac{\partial}{\partial n} (\bar{\delta}_1 - \bar{\theta}_1) \right| \int_{\bar{\lambda}_2 |x-\bar{x}_2| = \frac{1}{8\varepsilon_{12}^{1/n-2}}} \bar{\varphi}$$

Thus by (3.40):

$$\left|\int_{\tilde{W}} \bar{\delta}_1^{\frac{n+2}{n-2}} (\bar{\varphi} - \tilde{\psi}_1)\right| \leq C \operatorname{Sup}_{\partial\tilde{W}} \left|\frac{\partial}{\partial n}(\bar{\delta}_1 - \bar{\theta}_1)\right| . \frac{1}{\bar{\lambda}_2^{\frac{n-2}{2}} \varepsilon_{12}^{1/2}} \left(\int |\nabla \varphi|^2\right)^{1/2} \tag{3.77}$$

$\tilde{W}$ is the exterior of a ball around $\bar{x}_2$ of radius $\dfrac{1}{8\varepsilon_{12}^{1/n-2}\bar{\lambda}_2}$

By a homogeneity argument, one checks that:

$$\operatorname{Sup}_{\partial\tilde{W}} \left|\frac{\partial}{\partial n}(\bar{\delta}_1 - \bar{\theta}_1)\right| \leq C \operatorname{Sup}_{\partial\tilde{W}} \left|\frac{\partial}{\partial n}\bar{\delta}_1\right| = C' \operatorname{Sup}_{\partial\tilde{W}} \frac{\left|x . \frac{x-x_2}{|x-x_2|}\right|}{(1+|x|^2)^{n/2}} \tag{3.78}$$

We have:

$$\operatorname{Sup}_{\partial\tilde{W}} \frac{\left|x . \frac{x-x_2}{|x-x_2|}\right|}{(1+|x|^2)^{n/2}} = \operatorname{Sup}_{u\in S^{n-1}} \frac{|(x_2+ru) . u|}{(1+r^2+|x_2|^2+2x_2 . ur)^{n/2}} ; \tag{3.79}$$

$$r = \frac{1}{8\bar{\lambda}_2\varepsilon_{12}^{1/n-2}}$$

Hence

$$\operatorname{Sup}_{\partial\tilde{W}} \frac{\left|x . \frac{x-x_2}{|x-x_2|}\right|}{(1+|x|^2)^{n/2}} = \frac{r + |x_2|}{(1+r^2+|x_2|^2-2r|x_2|)^{n/2}} \tag{3.80}$$

and

$$\left|\int_{\tilde{W}} \bar{\delta}_1^{\frac{n+2}{n-2}} (\bar{\varphi} - \tilde{\psi}_1)\right| \leq C\rho \left(\int |\nabla \varphi|^2\right)^{1/2} \tag{3.81}$$

with

$$\rho = \frac{r+|x_2|}{(1+r^2+|x_2|^2-2r|x_2|)^{n/2}} \times r^{\frac{n-2}{2}} \tag{3.82}$$

In the case $|x_2| \leq 1$, then $\varepsilon_{12}^{2/n-2}$ is of the order $\frac{1}{\bar{\lambda}_2}$; hence r is of the order $\frac{1}{\sqrt{\bar{\lambda}_2}}$; and ρ is then upperbounded by $C\, r^{n-2/2}$; hence by $C\frac{1}{\bar{\lambda}_2^{n-2/4}}$ or either $C\,\varepsilon_{12}^{1/2}$.

This is a satisfactory bound.

In the case $|x_2| \geq 1$, then a first expansion of $\varepsilon_{12}^{2/n-2}$ is $\frac{1}{\bar{\lambda}_2+\bar{\lambda}_2|x_2|^2}$ and r is upperbounded by $\frac{|x_2|}{2}$ and is of the order $\frac{|x_2|}{\sqrt{\bar{\lambda}_2}}$. Then $(r-|x_2|)^2$ is lowerbounded by $\frac{|x_2|^2}{4}$ and ρ is thus upperbounded by

$$C \cdot \frac{|x_2|}{|x_2|^n} \cdot \frac{|x_2|^{n-2/2}}{\bar{\lambda}_2^{n-2/4}} ;$$

hence by

$$\frac{1}{(\bar{\lambda}_2|x_2|^2)^{\frac{n-2}{4}}}$$

or else $\varepsilon_{12}^{1/2}$, yielding a satisfactory bound also in this case.

This completes the proof of Lemma 3.2 in the case where we are dealing with two points.

In the general case, one introduces the sets, assuming $\lambda_1 = 1$; $x_1 = 0$ and $p \geq 0$:

$$\theta_i = \{x \in \mathbb{R}^n / |x| < \frac{1}{\varepsilon_{1i}^{1/n-2}} \text{ and } |x-x_i| > \frac{1}{\lambda_i\,\varepsilon_{1i}^{1/n-2}} \text{ if } \lambda_i > 1\} \quad (3.83)$$

Then $\partial\Omega_1 \subset \cup\, \partial\theta_i$

Let

$$h_i : \Delta h_i = 0 \qquad h_i = \varphi \Big|_{\partial\theta_i} \tag{3.84}$$

$$h : \Delta h = 0 \qquad h = \varphi \Big|_{\partial\Omega_i} \tag{3.85}$$

Then $h = \varphi - \underset{\sim}{\varphi}_1$ where $\underset{\sim}{\varphi}_1$ is the orthogonal projection on $H_0^1(\Omega_1)$ of φ and $h_i = \varphi - \varphi_i$ where φ_i is the orthogonal projection on $H_0^1(\theta_i)$ of φ.

Then the above arguments, in the case of two points, imply:

$$\int_{\theta_i} \delta_1^{\frac{n+2}{n-2}} h_i \leq C\,\varepsilon_{1i}^{1/2} \left(\int |\nabla\varphi|^2\right)^{1/2} \tag{3.86}$$

$$\int_{\mathbb{R}^n/\theta_i} \delta_1^{\frac{n+2}{n-2}} \varphi \leq C\,\varepsilon_{1i}^{1/2} \left(\int |\nabla\varphi|^2\right)^{1/2} \tag{3.87}$$

From (3.86) and (3.87) the general case follows.

3.2 The interaction with v; Estimates

We will need the following estimates on a v satisfying (G0)

<u>Estimate G1:</u> $\displaystyle\int_{\mathbb{R}^n} \nabla v.\nabla\delta_i = \int_{\mathbb{R}^n} \nabla v.\nabla \frac{\partial\delta_i}{\partial x_i} = \int_{\mathbb{R}^n} \nabla v.\nabla \frac{\partial\delta_i}{\partial\lambda_i} = 0$

(by definition of v)

<u>Estimate G2:</u> $\displaystyle\left|\int_{\mathbb{R}^n} \nabla v.\nabla \frac{\partial^2\delta_i}{\partial x_i^2}\right| \leq C\,\lambda_i^2\left(\int |\nabla v|^2\right)^{1/2}$

<u>Estimate G3:</u> $\displaystyle\left|\int_{\mathbb{R}^n} \nabla v.\nabla \frac{\partial^2\delta_i}{\partial x_i\,\partial\lambda_i}\right| \leq C\left(\int |\nabla v|^2\right)^{1/2}$

<u>Estimate G4:</u> $\displaystyle\left|\int_{\mathbb{R}^n} \nabla v.\nabla \frac{\partial^2\delta_i}{\partial\lambda_i^2}\right| \leq \frac{C}{\lambda_i^2}\left(\int |\nabla v|^2\right)^{1/2}$

For the following estimates, we choose ε and ε' such that $0 < \varepsilon < \varepsilon'/4$.

The quantity $o_{\varepsilon'}(\varepsilon_{ij})$ is the same and is in fact:

$$o_{\varepsilon'}(\varepsilon_{ij}) = 0\left(\int_{|v|\leq \varepsilon' \sum_{i\neq j} \alpha_i \delta_i} \delta_j |v| \left(\sum_{k\neq j} (\alpha_k \delta_k)^{4/n-2} + \right.\right.$$

$$\left.\left. + (\alpha_j \delta_j)^{\frac{4}{n-2}-1} \operatorname{Inf}(\alpha_j \delta_j, \sum_{i\neq j} \alpha_j \delta_j)\right)\right)$$

Estimate G5:

$$\left|\int_{\mathbb{R}^n} \nabla \Delta_\Omega^{-1}\left(K(x)\left[(\Sigma \alpha_i \delta_i + v)^{\frac{n+2}{n-2}} - (\Sigma \alpha_i \delta_i)^{\frac{n+2}{n-2}}\right]\right) \nabla \delta_j\right|$$

$$\leq 0_\varepsilon \left(\int_{|v|\geq \varepsilon \Sigma \alpha_i \delta_i} \delta_j |v|^{\frac{n+2}{n-2}}\right) + 0_{\varepsilon'}\left(\int_{|v|\leq \varepsilon \Sigma \alpha_i \delta_i} \delta_j^{4/n-2} v^2\right) +$$

$$+ o_{\varepsilon'}(\varepsilon_{ij}) + 0\left(\frac{1}{\lambda_j^{\frac{n+2}{2}} d_j^{\frac{n+2}{2}}}\left(\int_{\Omega^c} |v|^{2n/n-2}\right)^{\frac{n-2}{2n}} + \right.$$

$$\left. + \frac{(\int |\nabla v|^2)^{1/2}}{\lambda_j^{\frac{n+2}{2}} d_j^{n-2}}\right) + 0_K\left(\left(\int |\nabla v|^2\right)^{1/2}\left[\frac{|DK(x_j)|}{\lambda_j} + \frac{1}{\lambda_j^2}\right]\right);$$

$d_j = d(x_j; \partial\Omega)$; $0_K(\int |\nabla v|^2 \frac{1}{\lambda_j})$

appears when K is non constant; $\theta = \inf(\frac{4}{n-2}, 1)$;

$$\left(\int_{\Omega^c} |v|^{2n/n-2}\right)^{\frac{n-2}{2n}} \leq C\, \Sigma \frac{1}{(\lambda_i d_i)^{\frac{n-2}{2}}} \quad \text{if } v = -\Sigma c_i \delta_i \text{ on } \Omega^c$$

Estimate G6:

$$\left|\int_{\mathbb{R}^n} \nabla\left(\Delta_\Omega^{-1}\left(K(x)\left[(\Sigma \alpha_i \delta_i + v)^{\frac{n+2}{n-2}}\right) - (\Sigma \alpha_i \delta_i)^{\frac{n+2}{n-2}}\right]\right) \nabla \frac{\partial \delta_j}{\partial x_j}\right|$$

$$\le \left[O_\varepsilon\left(\int_{|v|\ge \varepsilon\Sigma\alpha_i\delta_i} \delta_j |v|^{\frac{n+2}{n-2}}\right) + O_{\varepsilon'}\left(\int_{|v|\le\varepsilon\Sigma\alpha_i\delta_i}\delta_j^{4/n-2} v^2\right) + \right.$$

$$\left. + o_{\varepsilon'}(\varepsilon_{ij}) \right]\lambda_j + O\left(\frac{1}{\lambda_j^{\frac{n}{2}} d_j^{\frac{n+2}{2}}}\left(\int_{\Omega^c}|v|^{2n/n-2}\right)^{\frac{n-2}{2n}} + \right.$$

$$\left. + \frac{(\int|\nabla v|^2)^{1/2}}{\lambda_j^n d_j^{n-2}}\right) + O_K\left((\int|\nabla v|^2)^{1/2}\left(|DK(x_j)| + \frac{1}{\lambda_j}\right)\right)$$

$d_j = d(x_j, \partial\Omega)$; $\theta = \inf(\frac{4}{n-2}, 1)$; $O_K\left((\int|\nabla v|^2)^{1/2}\left(|DK(x_j)| + \frac{1}{\lambda_j}\right)\right)$ appears when K is non constant;

$$\left(\int_{\Omega^c}|v|^{2n/n-2}\right)^{\frac{n-2}{2n}} \le C\,\Sigma\,\frac{1}{(\lambda_i d_i)^{\frac{n-2}{2}}} \quad \text{if } v = -\Sigma\alpha_i\delta_i \text{ on } \Omega^c.$$

<u>Estimate G7:</u>

$$\left|\int_{\mathbb{R}^n}\nabla\Delta_\Omega^{-1}\left(K(x)\left[(\Sigma\alpha_i\delta_i+v)^{\frac{n+2}{n-2}} - (\Sigma\alpha_i\delta_i)^{\frac{n+2}{n-2}}\right]\right)\nabla\frac{\partial\delta_j}{\partial\lambda_j}\right| \le$$

$$\le \frac{1}{\lambda_j}\left[O_\varepsilon\left(\int_{|v|\ge\varepsilon\Sigma\alpha_i\delta_i}\delta_j|v|^{\frac{n+2}{n-2}}\right) + O_{\varepsilon'}\left(\int_{|v|\le\varepsilon\Sigma\alpha_i\delta_i}\delta_j^{4/n-2}v^2\right) + \right.$$

$$\left. + o_{\varepsilon'}(\varepsilon_{ij})\right] + O\left(\frac{1}{\lambda_j^{\frac{n+4}{2}}}\frac{1}{d_j^{\frac{n+2}{2}}}\left(\int_{\Omega^c}|v|^{2n/n-2}\right)^{\frac{n-2}{2n}} + \frac{(\int|\nabla v|^2)^{1/2}}{\lambda_j^{\frac{n+4}{2}} d_j^{n-2}}\right) +$$

$$+ O_K\left(\left(\int|\nabla v|^2\right)^{1/2}\left(\frac{|DK(x_j)|}{\lambda_j^2} + \frac{1}{\lambda_j^3}\right)\right)$$

$d_j = d(x_j, \partial\Omega)$; $\theta = \mathrm{Inf}(\frac{4}{n-2}, 1)$; $O_K\left((\int|\nabla v|^2)^{1/2}\left[\frac{|DK(x_j)|}{\lambda_j^2} + \frac{1}{\lambda_j^3}\right]\right)$ appears when K is non constant. $\left(\int_{\Omega^c}|v|^{2n/n-2}\right)^{\frac{n-2}{2n}} \leq C\,\Sigma \frac{1}{(\lambda_i d_i)^{\frac{n-2}{2}}}$ if

$v = -\Sigma\,\alpha_i\delta_i$ on Ω^c.

Estimate G8:

$$\left|\int_{\mathbb{R}^n}\nabla\Delta_\Omega^{-1}\left(K(x)\left[(\Sigma\,\alpha_i\delta_i+v)^{\frac{n+2}{n-2}} - (\Sigma\,\alpha_i\delta_i)^{\frac{n+2}{n-2}} - \frac{n+2}{n-2}(\Sigma\,\alpha_i\delta_i)^{4/n-2}v\right]\right)\nabla v\right|$$

$$\leq \begin{cases} 0(\int|\nabla v|^2)^{n/n-2} & \text{if } n \geq 6 \\ 0((\int|\nabla v|^2)^{3/2}) & \text{if } n < 6 \end{cases}$$

Estimate G9:

$$\left|\int_{\mathbb{R}^n}K(x)(\Sigma\alpha_i\delta_i)^{4/n-2}v(\Sigma\alpha_j\theta_j)\right| \leq \begin{cases} C\left(\Sigma\frac{1}{d_j^{n-2}}\right)\left(\Sigma\frac{1}{\lambda_j^{\frac{n+2}{2}}}\right)\left(\int|\nabla v|^2\right)^{1/2} & \text{if } n \neq 6 \\ C\left(\Sigma\frac{1}{d_j^4}\right)\left(\Sigma\frac{(\mathrm{Log}\,\lambda_j)^{2/3}}{\lambda_j^4}\right)\left(\int|\nabla v|^2\right)^{1/2} & \text{if } n=6 \end{cases}$$

where $d_j = d(x_j;\,\partial\Omega)$; θ_j is defined in (2.3).

Estimate G10:

$$\left|\int_{\mathbb{R}^n}K(x)(\Sigma\,\alpha_i\delta_i)^{\frac{n+2}{n-2}}v\right| = \begin{cases} 0\left(\left(\int|\nabla v|^2\right)^{1/2}\left[\sum_{i\neq j}\varepsilon_{ij}^{\frac{n+2}{2(n-2)}}\left(\mathrm{Log}\,\varepsilon_{ij}^{-1}\right)^{\frac{n+2}{2n}} + \Sigma\frac{|DK(x_i)|}{\lambda_i} + \frac{1}{\lambda_i^2}\right]\right) & \text{if } n \geq 6 \\ 0\left(\left(\int|\nabla v|^2\right)^{1/2}\left[\sum_{i\neq j}\varepsilon_{ij}(\mathrm{Log}\,\varepsilon_{ij}^{-1})^{\frac{n-2}{2}} + \Sigma\frac{|DK(x_i)|}{\lambda_i} + \frac{1}{\lambda_i^2}\right]\right) & \text{if } n < 6 \end{cases}$$

the term $\Sigma \frac{|DK(x_i)|}{\lambda_i} + \frac{1}{\lambda_i^2}$ appearing if K is non constant.

Estimate G11:

$$\int_{\mathbb{R}^n} K(x)\,(\Sigma\,\alpha_i\delta_i)^{4/n-2} v^2 = \Sigma\,\alpha_i^{4/n-2} K(x_i) \int_{\mathbb{R}^n} \delta_i^{4/n-2} v^2 +$$

$$+\,0\left(\int|\nabla v|^2 \left(\sum_{i\neq j} (\alpha_i\alpha_j)^{2/n-2} \varepsilon_{ij}^{2/n-2} (\text{Log}\,\varepsilon_{ij}^{-1})^{2/n}\right)\right) +$$

$$+\,0_K\left(\int|\nabla v|^2 \left(\Sigma\,\frac{|DK(x_i)|}{\lambda_i} + \frac{(\text{Log}\,\lambda_i)^{2/n}}{\lambda_i^2}\right)\right)$$

or if $n = 3 + 0\left(\int|\nabla v|^2 \left(\sum_{i\neq j} \alpha_i^3\alpha_j(\varepsilon_{ij}\ \text{Log}\,\varepsilon_{ij}^{-1})\right)\right)$

The term 0_K appears when K is not constant.

Estimate G12:

$$\int_{\mathbb{R}^n} K(x)\,\delta_i^{4/n-2} v^2 = K(x_i)\int_{\mathbb{R}^n} \delta_i^{4/n-2} v^2 + 0_K\left(\int|\nabla v|^2 \left(|DK(x_i)|\,\frac{1}{\lambda_i} + \frac{1}{\lambda_i^2}(\text{Log}\,\lambda_i)^{2/n}\right)\right)$$

When K is constant, 0_K does not appear.

Estimate G13:

$$\Bigg|\int_{\mathbb{R}^n} K(x)\,(\Sigma\,\alpha_i\delta_i+v)^{2n/n-2} - \int_{\mathbb{R}^n} K(x)\,(\Sigma\,\alpha_i\delta_i)^{2n/n-2} - \frac{2n}{(n-2)}\int_{\mathbb{R}^n} K(x)\,(\Sigma\,\alpha_i\delta_i)^{\frac{n+2}{n-2}} v -$$

$$- \frac{2n(n-2)}{(n-2)^2}\int_{\mathbb{R}^n} K(x)\,(\Sigma\,\alpha_i\delta_i)^{4/n-2} v^2\Bigg|$$

$$\leq \begin{cases} 0((\int|\nabla v|^2)^{n/n-2}) & \text{if } n \geq 6 \\ 0((\int|\nabla v|^2)^{3/2}) & \text{if } n < 6 \end{cases}$$

We start by proving some easy estimates, namely G11 and G12.

Proof of Estimate G12: We have:

$$\int_{\mathbb{R}^n} K(x)\,\delta_i^{4/n-2} v^2 = K(x_i)\int_{\mathbb{R}^n} \delta_i^{4/n-2} v^2 + \int_{B_i} (K(x)-K(x_i))\,\delta_i^{4/n-2} v^2 +$$

$$+ \int_{B_i^c} (K(x)+K(x_i))\,\delta_i^{4/n-2} v^2 \tag{3.88}$$

We first estimate $\int_{B_i^c} (K(x)+K(x_i))\,\delta_i^{4/n-2} v^2$, which is upperbounded by:

$$C\int|\nabla v|^2 \left(\int_{B_i^c} \delta_i^{2n/n-2}\right)^{2/n} = \frac{C}{\lambda_i^2}\int|\nabla v|^2 \tag{3.89}$$

while

$$\left|\int_{B_i} (K(x)-K(x_i))\,\delta_i^{4/n-2} v^2\right| \le |DK(x_i)| \int_{B_i} |x-x_i|\,\delta_i^{4/n-2} v^2 + \tag{3.90}$$

$$+ C\int_{B_i} |x-x_i|^2 \delta_i^{4/n-2} v^2$$

$$\le C\int|\nabla v|^2 \left(|DK(x_i)| \left(\int_{B_i} |x-x_i|^{n/2}\delta_i^{2n/n-2}\right)^{2/n} + \right.$$

$$\left. + \left(\int_{B_i} |x-x_i|^n \delta_i^{2n/n-2}\right)^{2/n}\right) \le C\int|\nabla v|^2 \left(|DK(x_i)|\frac{1}{\lambda_i} + \frac{1}{\lambda_i^2}(\mathrm{Log}\,\lambda_i)^{2/n}\right)$$

Thus

$$\int_{\mathbb{R}^n} K(x)\,\delta_i^{4/n-2} v^2 = K(x_i)\int_{\mathbb{R}^n} \delta_i^{4/n-2} v^2 + \tag{3.91}$$

$$+ 0\left(\int|\nabla v|^2 \left(|DK(x_i)|\frac{1}{\lambda_i} + \frac{1}{\lambda_i^2}(\mathrm{Log}\,\lambda_i)^{2/n}\right)\right).$$

We point out here that the 0_K in (3.91) does not appear if K is constant.

Proof of Estimate G11: We have:

$$\left|\int_{\mathbb{R}^n} K(x)\left(\Sigma\,\alpha_i\delta_i\right)^{4/n-2} v^2 - \Sigma\,\alpha_i^{4/n-2}\int_{\mathbb{R}^n} K(x)\,\delta_i^{4/n-2} v^2\right| \le \tag{3.92}$$

$$\leq C \int_{\mathbb{R}^n} |\nabla v|^2 dx \left\{ \int_{\mathbb{R}^n} \left| (\Sigma \alpha_i \delta_i)^{4/n-2} - \Sigma \alpha_i^{4/n-2} \delta_i^{4/n-2} \right|^{n/2} dx \right\}^{2/n}$$

By inequality (1.2), this is upperbounded by:

$$C \int_{\mathbb{R}^n} |\nabla v|^2 \left\{ \int_{\mathbb{R}^n} \sum_{\substack{i,j \\ i \neq j}} \left[\operatorname{Sup} \left((\alpha_j \delta_i)^{\frac{6-n}{n-2}}, (\alpha_j \delta_j)^{\frac{6-n}{n-2}} \right) \operatorname{Inf}(\alpha_i \delta_i, \alpha_j \delta_j) \right]^{n/2} \right\}^{2/n} \tag{3.93}$$

We now have two cases:

1st case: $n \geq 4$

then

$$n/2 \geq \frac{n}{n-2} \tag{3.94}$$

and by (E2):

$$\int_{\mathbb{R}^n} \sum_{i \neq j} \left\{ \operatorname{Sup} \left((\alpha_i \delta_i)^{\frac{6-n}{n-2}}, (\alpha_j \delta_j)^{\frac{6-n}{n-2}} \right) \operatorname{Inf}(\alpha_i \delta_i, \alpha_j \delta_j) \right\}^{n/2} \leq \tag{3.95}$$

$$\leq C \int_{\mathbb{R}^n} \sum_{i \neq j} (\alpha_i \delta_i)^{n/n-2} (\alpha_j \delta_j)^{n/n-2} = 0 \left(\sum_{\substack{i,j \\ i \neq j}} \alpha_i^{n/n-2} \alpha_j^{n/n-2} \varepsilon_{ij}^{n/n-2} \operatorname{Log} \varepsilon_{ij}^{-1} \right)$$

We thus have, in this case:

$$\left| \int_{\mathbb{R}^n} K(x) (\Sigma \alpha_i \delta_i)^{4/n-2} v^2 - \Sigma \alpha_i^{4/n-2} \int K(x) \delta_i^{4/n-2} v^2 \right| \leq \tag{3.96}$$

$$\leq \int |\nabla v|^2 \left(\sum_{\substack{i,j \\ i \neq j}} \alpha_i^{2/n-2} \alpha_j^{2/n-2} \varepsilon_{ij}^{2/n-2} (\operatorname{Log} \varepsilon_{ij}^{-1})^{2/n} \right)$$

(3.96) with G12 yields:

$$\int_{\mathbb{R}^n} K(x) (\Sigma \alpha_i \delta_i)^{4/n-2} v^2 = \Sigma \alpha_i^{4/n-2} K(x_i) \int_{\mathbb{R}^n} \delta_i^{4/n-2} v^2 + 0 \left(\frac{\int |\nabla v|^2}{\lambda_i} \right) +$$

$$+ 0 \left(\int |\nabla v|^2 \left(\sum_{i \neq j} \alpha_i^{2/n-2} \alpha_j^{2/n-2} \varepsilon_{ij}^{2/n-2} (\operatorname{Log} \varepsilon_{ij}^{-1})^{2/n} \right) \right) \tag{3.97}$$

where again we point out that the term $0\left(\int \frac{|\nabla v|^2}{\lambda_i}\right)$ does not exist if K is constant. For $n \geq 4$, (3.97) is the estimate given by G11.

2nd case: $n = 3$

Then

$$\int_{\mathbb{R}^3} \left|(\Sigma \alpha_i \delta_i)^4 - \Sigma \alpha_i^4 \delta_i^4\right|^{3/2} dx \leq C \sum_{i \neq j} \int_{\mathbb{R}^3} \alpha_i^{9/2} \delta_i^{9/2} \operatorname{Inf}\, (\alpha_i \delta_i)^{3/2}, (\alpha_j \delta_j)^{3/2} \tag{3.98}$$

By (E3), we then have:

$$\int_{\mathbb{R}^3} \left|(\Sigma \alpha_i \delta_i)^4 - \Sigma \alpha_i^4 \delta_i^4\right|^{3/2} dx \leq C \sum_{i \neq j} \alpha_i^{9/2} \alpha_j^{3/2} \{\varepsilon_{ij} (\operatorname{Log} \varepsilon_{ij}^{-1})^{1/3}\}^{3/2} \tag{3.99}$$

Thus

$$\left|\int_{\mathbb{R}^3} K(x) (\Sigma \alpha_i \delta_i)^4 v^2 - \Sigma \alpha_i^4 \int_{\mathbb{R}^3} K(x) \delta_i^4 v^2\right| \leq \tag{3.100}$$

$$\leq C \int_{\mathbb{R}^3} |\nabla v|^2 \times \sum_{i \neq j} \alpha_i^3 \alpha_j (\varepsilon_{ij} \operatorname{Log} \varepsilon_{ij}^{-1})$$

This together with G12 yields:

$$\int_{\mathbb{R}^3} K(x) (\Sigma \alpha_i \delta_i)^4 v^2 = \Sigma \alpha_i^4 K(x_i) \int \delta_i^4 v^2 + 0\left(\int |\nabla v|^2 \left(\sum_{i \neq j} \alpha_i^3 \alpha_j (\varepsilon_{ij} \operatorname{Log} \varepsilon_{ij}^{-1})\right)\right)$$

$$+ 0_K\left(\int |\nabla v|^2 \left(\Sigma |DK(x_i)| \frac{1}{\lambda_i} + \Sigma \frac{(\operatorname{Log} \lambda_i)^{2/n}}{\lambda_i^2}\right)\right) \tag{3.101}$$

where we point out again that there is no $0(\int |\nabla v|^2 / \lambda_i)$ if K is constant. Hence, Estimate G11.

Proof of Estimate G10: We have:

$$\left|\int_{\mathbb{R}^n} K(x) (\Sigma \alpha_i \delta_i)^{\frac{n+2}{n-2}} v - \int_{\mathbb{R}^n} K(x) \Sigma \alpha_i^{\frac{n+2}{n-2}} \delta_i^{\frac{n+2}{n-2}} v\right| \leq \tag{3.102}$$

$$\leq C \left(\int |\nabla v|^2\right)^{1/2} \left\{ \int \sum_{i\neq j} (\alpha_i \delta_i)^{\frac{8n}{n^2-4}} \operatorname{Inf}\left((\alpha_i\delta_i), (\alpha_j\delta_j)\right)^{2n/n+2} \right\}^{\frac{n+2}{2n}}$$

To estimate the right-hand side, we consider two cases:

<u>1st case</u>

$$n \geq 6 \ ; \quad \text{then } \ 2n/n+2 \geq \frac{n}{n-2} \tag{3.103}$$

Thus, by (E2), we derive:

$$\left| \int_{\mathbb{R}^n} K(x) \left(\Sigma \alpha_i \delta_i\right)^{\frac{n+2}{n-2}} v - \int_{\mathbb{R}^n} K(x) \, \Sigma \, \alpha_i^{\frac{n+2}{n-2}} \delta_i^{\frac{n+2}{n-2}} v \right| \leq \tag{3.104}$$

$$\leq C \left(\int |\nabla v|^2\right)^{1/2} \left(\sum_{i\neq j} \varepsilon_{ij}^{\frac{n+2}{2(n-2)}} \left(\operatorname{Log} \varepsilon_{ij}^{-1}\right)^{\frac{n+2}{2n}} \right)$$

<u>2nd case</u>

$$n < 6 \ ; \quad \text{then } \ \frac{2n}{n+2} < \frac{n}{n-2} \tag{3.105}$$

By (E3), we then have:

$$\left| \int_{\mathbb{R}^n} K(x) \left(\Sigma \alpha_i \delta_i\right)^{\frac{n+2}{n-2}} v - \int_{\mathbb{R}^n} K(x) \, \Sigma \, \alpha_i^{\frac{n+2}{n-2}} \delta_i^{\frac{n+2}{n-2}} v \right| \leq \tag{3.106}$$

$$\leq C \left(\int |\nabla v|^2\right)^{1/2} \left(\sum_{i\neq j} \varepsilon_{ij} \left(\operatorname{Log} \varepsilon_{ij}^{-1}\right)^{\frac{n-2}{n}} \right)$$

We now estimate $\int_{\mathbb{R}^n} K(x) \, \alpha_i^{\frac{n+2}{n-2}} \delta_i^{\frac{n+2}{n-2}} v$

We have, by G1

$$\int_{\mathbb{R}^n} K(x) \, \delta_i^{\frac{n+2}{n-2}} v = \int_{\mathbb{R}^n} \left(K(x) - K(x_i) \right) \delta_i^{\frac{n+2}{n-2}} v \tag{3.107}$$

Thus:

$$\left|\int_{\mathbb{R}^n} K(x)\,\delta_i^{\frac{n+2}{n-2}} v\right| \le C\Big[\int_{B_i} |DK(x_i)|\,|x-x_i|\,\delta_i^{\frac{n+2}{n-2}}|v| + \tag{3.108}$$

$$+ |x-x_i|^2\,\delta_i^{\frac{n+2}{n-2}}|v| + \int_{B_i^c}\delta_i^{\frac{n+2}{n-2}}|v|\Big]$$

Now

$$\left(\int_{B_i}|x-x_i|^{\frac{2n}{n+2}}\,\delta_i^{2n/n-2}\right)^{\frac{n+2}{2n}} \le \frac{C}{\lambda_i}\;; \tag{3.109}$$

$$\left(\int_{B_i}|x-x_i|^{\frac{4n}{n+2}}\,\delta_i^{2n/n-2}\right)^{\frac{n+2}{2n}} \le \frac{C}{\lambda_i^2}\;;\quad \left(\int_{B_i^c}\delta_i^{2n/n-2}\right)^{\frac{n+2}{2n}} \le \frac{C}{\lambda_i^{\frac{n+2}{2}}}$$

Thus

$$\left|\int_{\mathbb{R}^n} K(x)\,\delta_i^{\frac{n+2}{n-2}} v\right| \le \left(\int|\nabla v|^2\right)^{1/2} 0_K\left(\frac{|DK(x_i)|}{\lambda_i}+\frac{1}{\lambda_i^2}\right) \tag{3.110}$$

Here again the term $\left(\int|\nabla v|^2\right)^{1/2} 0_K\left(\frac{|DK(x_i)|}{\lambda_i}+\frac{1}{\lambda_i^2}\right)$ appears when K is non constant.

We thus find:

$$\int_{\mathbb{R}^n} K(x)\,(\Sigma\alpha_i\delta_i)^{\frac{n+2}{n-2}} v = 0\left(\left(\int|\nabla v|^2\right)^{1/2}\Big[\sum_{i\ne j}\varepsilon_{ij}^{\frac{n+2}{2(n-2)}}(\operatorname{Log}\varepsilon_{ij}^{-1})^{\frac{n+2}{2n}} + \right.$$

$$\left. + \sum_i \frac{1}{\lambda_i}|DK(x_i)| + \frac{1}{\lambda_i^2}\Big]\right) \quad \text{if } n \ge 6 \tag{3.111}$$

$$= 0\left(\left(\int|\nabla v|^2\right)^{1/2}\Big[\sum_{i\ne j}\varepsilon_{ij}(\operatorname{Log}\varepsilon_{ij}^{-1})^{\frac{n-2}{n}} + \sum_i\frac{1}{\lambda_i}|DK(x_i)| + \frac{1}{\lambda_i^2}\Big]\right)$$

if $n < 6$

the term $\Sigma \frac{1}{\lambda_i^2} + \frac{|DK(x_i)|}{\lambda_i}$ appearing if K is non constant.

(3.111) is the desired estimate.

Proof of Estimate G9: We have:

$$\left| \int_{\mathbb{R}^n} K(x) (\Sigma \alpha_i \delta_i)^{4/n-2} v(\Sigma \alpha_j \boldsymbol{\theta}_j) \right| \leq \tag{3.112}$$

$$\leq C \sup_{\partial \Omega} \Sigma \alpha_j \theta_j \left(\int |\nabla v|^2 \right)^{1/2} \Sigma \left(\int_\Omega \delta_i^{\frac{8n}{n^2-4}} \right)^{\frac{n+2}{2n}}$$

If $\Omega = \mathbb{R}^n$, $\theta_j = 0$ and there is nothing to estimate. If $\Omega \neq \mathbb{R}^n$, then Ω is bounded in our hypotheses and (3.112) makes sense.

Now:

$$\int_\Omega \delta_i^{\frac{8n}{n^2-4}} = \int_\Omega \frac{\lambda_i^{4n/n+2}}{(1+\lambda_i^2 |x-x_i|^2)^{\frac{4n}{n+2}}} dx = \frac{1}{\lambda_i^{\frac{n(n-2)}{n+2}}} \int_{\lambda_i \Omega} \frac{dx}{(1+|x|^2)^{4n/n+2}}$$

$$\leq \begin{cases} \dfrac{C}{\lambda_i^{4n/n+2}} & \text{if } n \neq 6 \\[2ex] \dfrac{C \log \lambda_i}{\lambda_i^3} & \text{if } n = 6 \end{cases} \tag{3.113}$$

On the other hand:

$$\sup_{\partial \Omega} \Sigma \alpha_j \theta_j \leq C \Sigma \frac{1}{d_j^{\frac{n-2}{2}}} \cdot \frac{1}{\lambda_j^{\frac{n-2}{2}}} ; \text{ where } d_j = d(x_j, \partial \Omega). \tag{3.114}$$

Thus

$$\left| \int_{\mathbb{R}^n} K(x) (\Sigma \alpha_i \delta_i)^{4/n-2} v (\Sigma \alpha_j \theta_j) \right| \le \begin{cases} C \left(\Sigma \frac{1}{d_j^{n-2}} \right) \left(\Sigma \frac{1}{\lambda_j^{\frac{n+2}{2}}} \right) (\int |\nabla v|^2)^{1/2} & \text{if } n \neq 6 \\ C \left(\Sigma \frac{(\mathrm{Log}\, \lambda_j)^{2/3}}{\lambda_j^4} \right) \left(\Sigma \frac{1}{d_j^{n-2}} \right) \int |\nabla v|^2 {}^{1/2} & n = 6 \end{cases} \tag{3.115}$$

This is G9.

<u>Proof of Estimate G8:</u> Let

$$v_0 = \text{orthogonal projection of } v \text{ onto } H_0^1(\Omega). \tag{3.116}$$

We have:

$$\int_{\mathbb{R}^n} \nabla \left(\Delta_\Omega^{-1} \quad K(x) \left[(\Sigma \alpha_i \delta_i + v)^{\frac{n+2}{n-2}} - (\Sigma \alpha_i \delta_i)^{\frac{n+2}{n-2}} - \frac{n+2}{n-2} (\Sigma \alpha_i \delta_i)^{4/n-2} v \right] \right) \nabla v =$$

$$= \int_{\mathbb{R}^n} K(x) \left[(\Sigma \alpha_i \delta_i + v)^{\frac{n+2}{n-2}} - (\Sigma \alpha_i \delta_i)^{\frac{n+2}{n-2}} - \frac{n+2}{n-2} (\Sigma \alpha_i \delta_i)^{4/n-2} v \right] v_0 dx \tag{3.117}$$

Using inequality (1.1), we upperbound this quantity by:

$$M \left[\int_\Omega K(x) |v|^{\frac{n+2}{n-2}} |v_0| dx + \int_\Omega K(x) (\Sigma \alpha_i \delta_i)^{\frac{6-n}{n-2}} \mathrm{Inf} \left((\Sigma \alpha_i \delta_i)^2, v^2 \right) |v_0| dx \right] \tag{3.118}$$

which we upperbound by:

$$M_1 \left[\int_\Omega |v|^{\frac{n+2}{n-2}} |v_0| dx + \int_{\{x | \Sigma \alpha_i \delta_i (x) \ge |v(x)| \}} (\Sigma \alpha_i \delta_i)^{\frac{6-n}{n-2}} v^2 (x) |v_0| dx \right] \tag{3.119}$$

hence by:

$$M_1\left[\int_\Omega |v|^{\frac{n+2}{n-2}}|v_0|dx + \right. \tag{3.120}$$

$$+\left(\int_{\{x|\Sigma\alpha_i\delta_i(x)\geq|v(x)|\}}(\Sigma\alpha_i\delta_i)^{\frac{2(6-n)n}{n^2-4}}|v|^{\frac{4n}{n+2}}dx\right)^{\frac{n+2}{2n}}\times$$

$$\times\left(\int|v_0|^{2n/n-2}\right)^{\frac{n-2}{2n}}$$

Using a Sobolev inequality and (3.116), we have:

$$\int_\Omega |v|^{\frac{n+2}{n-2}}|v_0|dx + \int|v_0|^{2n/n-2} \leq C\left(\int|\nabla v|^2\right)^{n/n-2} \tag{3.121}$$

On the other hand:

$$(\Sigma\alpha_i\delta_i)^{\frac{2n(6-n)}{n^2-4}}\chi_{\{x|\Sigma\alpha_i\delta_i(x)\geq|v(x)|\}}|v|^{\frac{4n}{n+2}} \leq |v|^{2n/n-2} \tag{3.122}$$

if $n \geq 6$

and

$$\int_{\{x|\Sigma\alpha_i\delta_i(x)\geq|v(x)|\}}(\Sigma\alpha_i\delta_i)^{\frac{2n(6-n)}{n^2-4}}|v|^{\frac{4n}{n+2}} \leq C\left(\int|v|^{2n/n-2}\right)^{2\frac{n-2}{n+2}}$$

if $n < 6$ (3.123)

Thus

$$\left|\int_{\mathbb{R}^n}\nabla\Delta_\Omega^{-1}K(x)\left[(\Sigma\alpha_i\delta_i+v)^{\frac{n+2}{n-2}} - (\Sigma\alpha_i\delta_i)^{\frac{n+2}{n-2}} - \frac{n+2}{n-2}(\Sigma\alpha_i\delta_i)^{4/n-2}v\right]\nabla v\right| \leq$$

$$\leq \begin{cases} C\left(\int|\nabla v|^2\right)^{n/n-2} & \text{if } n \geq 6 \\ C\left(\left(\int|\nabla v|^2\right)^{n/n-2} + \left(\int|\nabla v|^2\right)^{3/2}\right) & \text{if } n < 6 \end{cases} \tag{3.124}$$

This yields G8.

Proof of Estimate G13: We have:

$$\left| \int_{\mathbb{R}^n} K(x) \left[(\Sigma \alpha_i \delta_i + v)^{\frac{2n}{n-2}} - (\Sigma \alpha_i \delta_i)^{\frac{2n}{n-2}} - \frac{2n}{n-2} (\Sigma \alpha_i \delta_i)^{\frac{n+2}{n-2}} v - \right.\right. \tag{3.125}$$

$$\left.\left. - \frac{2n(n+2)}{(n-2)^2} (\Sigma \alpha_i \delta_i)^{4/n-2} v^2 \right] \right| \leq C \left[\int_{\mathbb{R}^n} |v|^{\frac{2n}{n-2}} + \right.$$

$$\left. + \int_{\mathbb{R}^n} (\Sigma \alpha_i \delta_i)^{\frac{6-n}{n-2}} \operatorname{Inf} \left((\Sigma \alpha_i \delta_i)^3, |v|^3 \right) \right]$$

We thus estimate:

$$\int_{\mathbb{R}^n} (\Sigma \alpha_i \delta_i)^{\frac{6-n}{n-2}} \operatorname{Inf} \left((\Sigma \alpha_i \delta_i)^3, |v|^3 \right) \tag{3.126}$$

which we upperbound by:

$$\int_{\mathbb{R}^n} (\Sigma \alpha_i \delta_i)^{\frac{6-n}{n-2}} \operatorname{Inf} \left((\Sigma \alpha_i \delta_i)^3, |v|^3 \right) \leq \int_{\mathbb{R}^n} |v|^{\frac{2n}{n-2}} + \tag{3.127}$$

$$+ \int_{\{x | \Sigma \alpha_i \delta_i(x) \geq |v(x)|\}} (\Sigma \alpha_i \delta_i)^{\frac{6-n}{n-2}} |v|^3 dx$$

We now have:

$$(\Sigma \alpha_i \delta_i)^{\frac{6-n}{n-2}} |v|^3 \chi_{\{x | \Sigma \alpha_i \delta_i \geq |v(x)|\}} \leq |v|^{\frac{2n}{n-2}} \quad \text{for } n \geq 6 \tag{3.128}$$

while

$$3 < \frac{2n}{n-2} \quad \text{if } n < 6 \tag{3.129}$$

Thus, in the case $n < 6$, we can use a Hölder inequality to get:

$$\int_{\{x | \Sigma \alpha_i \delta_i(x) \geq |v(x)|\}} (\Sigma \alpha_i \delta_i)^{\frac{6-n}{n-2}} |v|^3 dx \leq C \left(\int |\nabla v|^2 \right)^{3/2} \tag{3.130}$$

Hence

$$\int_{\mathbb{R}^n} (\Sigma \alpha_i \delta_i)^{\frac{6-n}{n-2}} \operatorname{Inf} \left((\Sigma \alpha_i \delta_i)^3, |v|^3 \right) \leq \begin{cases} C(\int |\nabla v|^2)^{n/n-2} & \text{if } n \geq 6 \\ C(\int |\nabla v|^2)^{3/2} & \text{if } n < 6 \end{cases} \tag{3.131}$$

and we have:

$$\Big| \int_{\mathbb{R}^n} K(x) (\Sigma \alpha_i \delta_i + v)^{\frac{2n}{n-2}} - (\Sigma \alpha_i \delta_i)^{\frac{2n}{n-2}} - \frac{2n}{n-2} (\Sigma \alpha_i \delta_i)^{\frac{n+2}{n-2}} v - \tag{3.132}$$

$$- \frac{2n}{(n-2)^2} (n+2) (\Sigma \alpha_i \delta_i)^{4/n-2} v^2 \Big] \Big| \leq \begin{cases} 0(\int |\nabla v|^2)^{n/n-2} & \text{if } n \geq 6 \\ 0((\int |\nabla v|^2)^{3/2}) & \text{if } n < 6 \end{cases}$$

This yields G13.

<u>Proof of Estimates G5-G6-G7:</u> We want to estimate quantities of the type:

$$\int_{\mathbb{R}^n} \nabla \left(\Delta_{\Omega}^{-1} (K(x)) \left[(\Sigma \alpha_i \delta_i + v)^{\frac{n+2}{n-2}} - (\Sigma \alpha_i \delta_i)^{\frac{n+2}{n-2}} \right] \right) \nabla \varphi \tag{3.133}$$

where φ can be δ_j or $\frac{\partial \delta_j}{\partial x_j}$ or $\frac{\partial \delta_j}{\partial \lambda_j}$.

Hence satisfies, by (G0)

$$\int \nabla v \, . \, \nabla \varphi = 0 \tag{3.134}$$

On the other hand, as

$$\begin{cases} \Delta \varphi = -\delta_j^{\frac{n+2}{n-2}} = -\delta_j^{4/n-2} \varphi & \text{when } \varphi = \delta_j \\ \Delta \varphi = -\frac{n+2}{n-2} \delta_j^{4/n-2} \varphi & \text{when } \varphi = \frac{\partial \delta_j}{\partial x_j} \text{ or } \varphi = \frac{\partial \delta_j}{\partial \lambda_j} \end{cases} \tag{3.135}$$

(3.134) implies:

$$\int v \, \delta_j^{4/n-2} \varphi = 0 \tag{3.136}$$

We introduce:

$$\begin{cases} \Delta\psi = 0 \\ \psi = \varphi \mid \partial\Omega \end{cases} \tag{3.137}$$

We have

$$\int_{\mathbb{R}^n} \nabla(\Delta_\Omega^{-1} K(x)\left[(\Sigma\alpha_i\delta_i+v)^{\frac{n+2}{n-2}} - (\Sigma\alpha_i\delta_i)^{\frac{n+2}{n-2}}\right]\nabla\varphi = \tag{3.138}$$

$$= -\int_\Omega K(x)\left[(\Sigma\alpha_i\delta_i+v)^{\frac{n+2}{n-2}} - (\Sigma\alpha_i\delta_i)^{\frac{n+2}{n-2}}\right](\varphi - \psi)$$

On the other hand, for any ε, ε' such that $0 < \varepsilon < \frac{\varepsilon'}{4}$

$$\left|(v+\Sigma\alpha_i\delta_i)^{\frac{n+2}{n-2}} - (\Sigma\alpha_i\delta_i)^{\frac{n+2}{n-2}} - \frac{n+2}{n-2}(\alpha_j\delta_j)^{4/n-2}v\right| \leq C_\varepsilon|v|^{\frac{n+2}{n-2}} \tag{3.139}$$

if $|v| \geq \varepsilon\,\Sigma\alpha_i\delta_i$

$$\left|(v+\Sigma\alpha_i\delta_i)^{\frac{n+2}{n-2}} - (\Sigma\alpha_i\delta_i)^{\frac{n+2}{n-2}} - \frac{n+2}{n-2}(\alpha_j\delta_j)^{4/n-2}v\right| \leq \tag{3.140}$$

$$\leq C\left(\sum_{k\neq j}(\alpha_k\delta_k)^{4/n-2}|v| + (\alpha_j\delta_j)^{\frac{4}{n-2}-1}v^2 + \right.$$

$$\left. + (\alpha_j\delta_j)^{\frac{4}{n-2}-1}\operatorname{Inf}(\alpha_j\delta_j, \sum_{i\neq j}\alpha_i\delta_i)|v|\right) \quad \text{if } |v| \leq \varepsilon' \sum_{i\neq j}\alpha_i\delta_i$$

$$|(v+\Sigma\alpha_i\delta_i)^{\frac{n+2}{n-2}} - (\Sigma\alpha_i\delta_i)^{\frac{n+2}{n-2}} - \frac{n+2}{n-2}(\alpha_j\delta_j)^{\frac{4}{n-2}}v| \leq \tag{3.141}$$

$$\leq C\left((\alpha_j\delta_j)^{\frac{4}{n-2}-1}v^2 + (\alpha_j\delta_j)^{\frac{4}{n-2}-1}(\sum_{i\neq j}\alpha_i\delta_i)|v|\right)$$

if $\varepsilon' \sum_{i\neq j}\alpha_i\delta_i \leq |v| \leq \varepsilon\,\Sigma\alpha_i\delta_i$

(3.139) is immediate.

(3.140) and (3.141) require some work.

We have:

$$\left| (v+\Sigma\alpha_i\delta_i)^{\frac{n+2}{n-2}} - (\Sigma\alpha_i\delta_i)^{\frac{n+2}{n-2}} - \frac{n+2}{n-2}(\Sigma\alpha_i\delta_i)^{4/n-2} v \right| \tag{3.142}$$

$$\leq C(\Sigma\alpha_i\delta_i)^{\frac{4}{n-2}-1} v^2 \quad \text{as soon as } |v| \leq \varepsilon' \Sigma\alpha_i\delta_i$$

Now

$$\left| (\Sigma\alpha_i\delta_i)^{\frac{4}{n-2}} v - (\alpha_j\delta_j)^{\frac{4}{n-2}} v \right| \leq C\Big[\sum_{k\neq j} (\alpha_k\delta_k)^{\frac{4}{n-2}} + \tag{3.143}$$

$$+ (\alpha_j\delta_j)^{\frac{4}{n-2}-1} \operatorname{Inf}(\alpha_j\delta_j, \sum_{i\neq j} \alpha_i\delta_i) \Big] |v|$$

and

$$(\Sigma\alpha_i\delta_i)^{\frac{4}{n-2}-1} v^2 \leq C\Big[(\sum_{i\neq j} \alpha_i\delta_i)^{\frac{4}{n-2}-1} |v| \Big] |v| + C(\alpha_j\delta_j)^{\frac{4}{n-2}-1} v^2 \tag{3.144}$$

$$\leq C\,\varepsilon'(\sum_{i\neq j} \alpha_i\delta_i)^{\frac{4}{n-2}} |v| + C(\alpha_j\delta_j)^{\frac{4}{n-2}-1} v^2$$

if $|v| \leq \varepsilon' \sum_{i\neq j} \alpha_i\delta_i$

Thus, (3.140) follows.

For (3.141) we notice that if $\varepsilon' \sum_{i\neq j} \alpha_i\delta_i \leq |v| \leq \varepsilon \Sigma\alpha_i\delta_i$, we have:

$$\alpha_j\delta_j \geq \frac{\varepsilon'-\varepsilon}{\varepsilon} \sum_{i\neq j} \alpha_i\delta_i; \text{ hence } \sum_{i\neq j} \alpha_i\delta_i < \frac{\varepsilon}{\varepsilon'-\varepsilon} \alpha_j\delta_j < \frac{1}{2}\alpha_j\delta_j \tag{3.145}$$

Thus

$$\left| (\Sigma\alpha_i\delta_i)^{\frac{4}{n-2}} v - (\alpha_j\delta_j)^{\frac{4}{n-2}} v \right| \leq (\alpha_j\delta_j)^{\frac{4}{n-2}} |v| \left[\left(1+\frac{\sum_{i\neq j} \alpha_i\delta_i}{\alpha_j\delta_j} \right)^{\frac{4}{n-2}} -1 \right] \leq$$

$$\leq C(\alpha_j\delta_j)^{\frac{4}{n-2}-1} (\sum_{i\neq j} \alpha_i\delta_i) |v| \tag{3.146}$$

and

$$(\Sigma\alpha_i\delta_i)^{\frac{4}{n-2}-1} v^2 = (\alpha_j\delta_j)^{\frac{4}{n-2}-1}\left(1+\frac{\sum_{i\neq j}\alpha_i\delta_i}{\alpha_j\delta_j}\right)^{\frac{4}{n-2}-1} v^2 \le \qquad (3.147)$$

$$\le C(\alpha_j\delta_j)^{\frac{4}{n-2}-1} v^2$$

(3.142) together with (3.146) and (3.147) yield (3.141).

Turning back to (3.138), we have:

$$\int_{\mathbb{R}^n} \nabla\left(\Delta_\Omega^{-1} K(x)\left[(\Sigma\alpha_i\delta_i+v)^{\frac{n+2}{n-2}} - (\Sigma\alpha_i\delta_i)^{\frac{n+2}{n-2}}\right]\right)\nabla\varphi =$$

$$= -\frac{n+2}{n-2}\int_\Omega (\alpha_j\delta_j)^{\frac{4}{n-2}} v(\varphi-\psi)K(x) + C_\varepsilon 0\left(\int_{|v|\ge\varepsilon\Sigma\alpha_i\delta_i} |\varphi-\psi|\,|v|^{\frac{n+2}{n-2}}\right) +$$

$$+0\left(\int_{|v|\le\varepsilon'\sum_{i\neq j}\alpha_i\delta_i} |\varphi-\psi|\left(\sum_{k\neq j}(\alpha_k\delta_k)^{\frac{4}{n-2}}|v| +\right.\right.$$

$$\left.\left.+(\alpha_j\delta_j)^{\frac{4}{n-2}-1}\times \mathrm{Inf}(\alpha_j\delta_j, \sum_{i\neq j}\alpha_i\delta_i)\,|v|\right)\right)+$$

$$+0\left(\int_{|v|\le\varepsilon\Sigma\alpha_i\delta_i} (\alpha_j\delta_j)^{\frac{4}{n-2}-1} v^2|\varphi-\psi|\right) +$$

$$+0\left(\int_{\varepsilon'\sum_{i\neq j}\alpha_i\delta_i \le |v| \le \varepsilon\Sigma\alpha_i\delta_i} (\alpha_j\delta_j)^{\frac{4}{n-2}-1}\times(\sum_{i\neq j}\alpha_i\delta_i)\,|v|\times|\varphi-\psi|\right)$$

$$= -\frac{n+2}{n-2}\int_\Omega (\alpha_j\delta_j)^{\frac{4}{n-2}} v(\varphi-\psi)K(x) + 0_\varepsilon\left(\int_{|v|\ge\Sigma\alpha_i\delta_i} |\varphi-\psi|\,|v|^{\frac{n+2}{n-2}}\right) +$$

$$+0_{\varepsilon'}\left(\int_{|v|\le\varepsilon\Sigma\alpha_i\delta_i} (\alpha_j\delta_j)^{\frac{4}{n-2}-1} v^2|\varphi-\psi|\right) +$$

$$+0\left(\int_{|v|\le\varepsilon'\sum_{i\neq j}\alpha_i\delta_i} |\varphi-\psi|\left(\sum_{k\neq j}(\alpha_k\delta_k)^{\frac{4}{n-2}} +\right.\right.$$

$$\left.\left.+(\alpha_j\delta_j)^{\frac{4}{n-2}-1}\mathrm{Inf}(\alpha_j\delta_j, \sum_{i\neq j}\alpha_i\delta_i)\right)|v|\right)$$

Depending on G5, G6 or G7, φ is δ_j, $\frac{\partial \delta_j}{\partial x_j}$ or $\frac{\partial \delta_j}{\partial \lambda_j}$.

Thus

$$|\varphi - \psi| \le C\,\delta_j \text{ for G5;} \quad |\varphi - \psi| \le C\,\lambda_j \delta_j \text{ for G6;} \tag{3.149}$$
$$|\varphi - \psi| \le C\,\frac{\delta_j}{\lambda_j} \text{ for G7.}$$

We thus have:

$$\int_{|v| \le \varepsilon' \sum_{i \neq j} \alpha_i \delta_i} |\varphi - \psi| \sum_{k \neq j} (\alpha_k \delta_k)^{\frac{4}{n-2}} |v| \le C\,\varepsilon' \int |\varphi - \psi| \left(\sum_{i \neq j} \delta_i^{\frac{n+2}{n-2}} \right) \le$$
$$\le \begin{cases} C\,\varepsilon'\,\varepsilon_{ij} & \text{for G5} \\ C\,\varepsilon'\,\lambda_j\,\varepsilon_{ij} & \text{for G6} \\ C\,\varepsilon'\,\frac{\varepsilon_{ij}}{\lambda_j} & \text{for G7} \end{cases} \tag{3.150}$$

and

$$\int_{|v| \le \varepsilon' \sum_{i \neq j} \alpha_i \delta_i} (\alpha_j \delta_j)^{\frac{4}{n-2} - 1} \operatorname{Inf}(\alpha_j \delta_j, \sum_{i \neq j} \alpha_i \delta_i) |v| |\varphi - \psi| \le \tag{3.151}$$
$$\le C\,\varepsilon' \int (\alpha_j \delta_j)^{\frac{4}{n-2} - 1} \left(\sum_{i \neq j} \operatorname{Inf}(\alpha_j \delta_j, \alpha_i \delta_i) \right) \left(\sum_{i \neq j} \alpha_i \delta_i \right) |\varphi - \psi| \le$$
$$\le \begin{cases} C\,\varepsilon'\,\varepsilon_{ij} & \text{for G5} \\ C\,\varepsilon'\,\lambda_j\,\varepsilon_{ij} & \text{for G6} \\ C\,\varepsilon'\,\frac{\varepsilon_{ij}}{\lambda_j} & \text{for G7} \end{cases}$$

while

$$\int_{|v|\leq\varepsilon\Sigma\alpha_i\delta_i}(\alpha_j\delta_j)^{\frac{4}{n-2}-1}v^2|\varphi-\psi| \leq \begin{cases} C\int_{|v|\leq\varepsilon\Sigma\alpha_i\delta_i}\delta_j^{4/n-2}v^2 & \text{for G5} \\ C\lambda_j\int_{|v|\leq\varepsilon\Sigma\alpha_i\delta_i}\delta_j^{4/n-2}v^2 & \text{for G6} \\ \frac{C}{\lambda_j}\int_{|v|\leq\varepsilon\Sigma\alpha_i\delta_i}\delta_j^{4/n-2}v^2 & \text{for G7} \end{cases}$$

(3.152)

$$\int_{|v|\geq\varepsilon\Sigma\alpha_i\delta_i}|\varphi-\psi|\,|v|^{\frac{n+2}{n-2}} \leq C \begin{cases} \int_{|v|\ \varepsilon\Sigma\alpha_i\delta_i}\delta_j|v|^{\frac{n+2}{n-2}} & \text{for G5} \\ \lambda_j \times \text{same thing for G6} \\ \frac{1}{\lambda_j}\times\text{ same thing for G7} \end{cases}$$

Thus

$$0_\varepsilon\left(\int_{|v|\geq\varepsilon\Sigma\alpha_i\delta_i}|\varphi-\psi|\,|v|^{\frac{n+2}{n-2}}\right) + 0_{\varepsilon'}\left(\int_{|v|\leq\varepsilon\Sigma\alpha_i\delta_i}(\alpha_j\delta_j)^{\frac{4}{n-2}-1}v^2|\varphi-\psi|\right) +$$

$$+ 0\left(\int_{|v|\leq\varepsilon'\sum_{i\neq j}\alpha_i\delta_i}|\varphi-\psi|\left(\sum_{k\neq j}(\alpha_k\delta_k)^{\frac{4}{n-2}-1}\operatorname{Inf}(\alpha_j\delta_j,\sum_{i\neq j}\alpha_i\delta_i)\qquad|v|\right)\right)$$

$$= 0_\varepsilon\left(\int_{|v|\geq\varepsilon\Sigma\alpha_i\delta_i}\delta_j|v|^{\frac{n+2}{n-2}}\right) + 0_{\varepsilon'}\left(\int_{|v|\leq\varepsilon\Sigma\alpha_i\delta_i}\delta_j^{\frac{4}{n-2}}v^2\right) +$$

$$+ 0\left(\int_{|v|\leq\varepsilon'\sum_{i\neq j}\alpha_i\delta_i}\delta_j|v|\left(\sum_{k\neq j}(\alpha_k\delta_k)^{\frac{4}{n-2}} + (\alpha_j\delta_j)^{\frac{4}{n-2}-1}\operatorname{Inf}(\alpha_j.\delta_j,\sum_{i\neq j}\alpha_i\delta_i)\right)\right)$$

$$= 0_\varepsilon\left(\int_{|v|\geq\varepsilon\Sigma\alpha_i\delta_i}\delta_j|v|^{\frac{n+2}{n-2}}\right) + 0_{\varepsilon'}\left(\int_{|v|\leq\varepsilon\Sigma\alpha_i\delta_i}\delta_j^{\frac{4}{n-2}}v^2\right) +$$

$+\, o_{\varepsilon'}(\varepsilon_{ij})$ by (3.150), (3.153)

(3.151) for G5

$= \lambda_j \times$ same quantity for G6

$$= \frac{1}{\lambda_j} \times \text{same quantity for G7}$$

To complete the proof of Estimates G5, G6 and G7, it thus remains to estimate:

$$\int_\Omega K(x)\, v\, \delta_j^{4/n-2}(\varphi - \psi) \tag{3.154}$$

For this, we first consider:

$$\int_\Omega v\delta_j^{4/n-2}(\varphi - \psi) = \int_{\mathbb{R}^n} v\delta_j^{4/n-2}\varphi - \int_\Omega v\delta_j^{4/n-2}\psi + \int_{\Omega^c} v\delta_j^{4/n-2}\varphi \tag{3.155}$$

Thus, by (3.136)

$$\int_\Omega v\delta_j^{4/n-2}(\varphi - \psi) = - \int_\Omega v\delta_j^{4/n-2}\psi + \int_{\Omega^c} v\delta_j^{4/n-2}\varphi \tag{3.156}$$

We estimate:

$$\int_{\Omega^c} |v|\delta_j^{\frac{n+2}{n-2}} + \operatorname*{Sup}_{\partial\Omega} \delta_j \int_\Omega |v|\delta_j^{4/n-2} \tag{3.157}$$

We have:

$$\int_{\Omega^c} |v|\delta_j^{\frac{n+2}{n-2}} \le \left(\int_{\Omega^c} \delta_j^{2n/n-2}\right)^{\frac{n+2}{2n}} \left(\int_{\Omega^c} |v|^{2n/n-2}\right)^{\frac{n-2}{2n}} \le \tag{3.158}$$

$$\le \frac{C}{d_j^{\frac{n+2}{2}}\lambda_j^{\frac{n+2}{2}}} \left(\int_{\Omega^c} |v|^{2n/n-2}\right)^{\frac{n-2}{2n}}$$

while

$$\int_\Omega |v|\delta_j^{4/n-2} \le C\left(\int |\nabla v|^2\right)^{1/2} \left(\int_\Omega \delta_j^{8n/n^2-4}\right)^{\frac{n+2}{2n}} \tag{3.159}$$

$$\le C\left(\int |\nabla v|^2\right)^{1/2} \frac{1}{\lambda_j^2}$$

and

$$\sup_{\partial\Omega} \delta_j \leq \frac{1}{\lambda_j^{\frac{n-2}{2}} d_j^{n-2}} \tag{3.160}$$

Thus, taking into account (3.148), (3.149), (3.150), (3.151) and (3.152) :

$$\left| \int_\Omega v \delta_j^{\frac{4}{n-2}} (\varphi - \psi) \right| \leq \frac{C}{d_j^{\frac{n+2}{2}} \lambda_j^{\frac{n+2}{2}}} \left(\int_{\Omega^c} |v|^{\frac{2n}{n-2}} \right)^{\frac{n-2}{2n}} + \tag{3.161}$$

$$+ \frac{C}{\lambda_j^{\frac{n+2}{2}} d_j^{n-2}} \left(\int |\nabla v|^2 \right)^{1/2} \quad \text{for G5}$$

$$\leq \frac{C}{d_j^{\frac{n+2}{2}} \lambda_j^{n/2}} \left(\int_{\Omega^c} |v|^{\frac{2n}{n-2}} \right)^{\frac{n-2}{2n}} + \frac{C}{\lambda_j^{n/2} d_j^{n-2}} \left(\int |\nabla v|^2 \right)^{1/2} \quad \text{for G6}$$

$$\leq \frac{C}{d_j^{\frac{n+2}{2}} \lambda_j^{\frac{n+4}{2}}} \left(\int_{\Omega^c} |v|^{\frac{2n}{n-2}} \right)^{\frac{n-2}{2n}} + \frac{C}{\lambda_j^{\frac{n+4}{2}} d_j^{n-2}} \left(\int |\nabla v|^2 \right)^{1/2} \quad \text{for G7}$$

We complete the proof of Estimates G5-G6-G7 by estimating:

$$\int_\Omega (K(x) - K(x_j)) v \delta_j^{4/n-2} (\varphi - \psi) \tag{3.162}$$

which is upperbounded by:

$$C \int_{B_j} \left[|DK(x_j)| |x - x_j| + |x - x_j|^2 \right] |v| \delta_j^{4/n-2} |\varphi - \psi| + \tag{3.163}$$

$$+ \int_{B_j^c} |v| \delta_j^{4/n-2} |\varphi - \psi|$$

This yields as in (3.108)-(3.109)-(3.110)

$$\int_\Omega K(x)\, v\, \delta_j^{4/n-2}(\varphi - \psi) = K(x_j) \int_\Omega v\, \delta_j^{4/n-2}(\varphi - \psi) + \tag{3.164}$$

$$+ 0_K\left(\left(\int |\nabla v|^2\right)^{1/2} \left(\frac{|DK(x_j)|}{\lambda_j} + \frac{1}{\lambda_j^2}\right)\right) \quad \text{for G5}$$

$$+ 0_K\left(\left(\int |\nabla v|^2\right)^{1/2} \left(|DK(x_j)| + \frac{1}{\lambda_j}\right)\right) \quad \text{for G6}$$

$$+ 0_K\left(\left(\int |\nabla v|^2\right)^{1/2} \left(\frac{|DK(x_j)|}{\lambda_j^2} + \frac{1}{\lambda_j^3}\right)\right) \quad \text{for G7}$$

Here again we point out that the 0_K in (3.164) is related to the case where K is non constant.

Before summing up our estimates, we point out that we have:

$$\left(\int_{\Omega^c} |v|^{2n/n-2}\right)^{\frac{n-2}{2n}} \leq C \Sigma \frac{1}{(\lambda_i d_i)^{\frac{n-2}{2}}} \quad \text{if } v = -\Sigma \alpha_i \delta_i \text{ on } \Omega^c \tag{3.165}$$

where $d_i = d(x_i, \partial\Omega)$; i.e. if $\Sigma \alpha_i \delta_i + v$ has support in Ω.

We thus have by (3.153), (3.161), (3.164) and (3.165)

$$\int_{\mathbb{R}^n} \nabla\left(\Delta_\Omega^{-1} K(x) \left[(\Sigma \alpha_i \delta_i + v)^{\frac{n+2}{n-2}} - (\Sigma \alpha_i \delta_i)^{\frac{n+2}{n-2}}\right]\right) \nabla\varphi \, dx \leq \tag{3.166}$$

$$\leq A_j \left[0_\varepsilon\left(\int_{|v| \geq \varepsilon \Sigma \alpha_i \delta_i} \delta_j |v|^{\frac{n+2}{n-2}}\right) + 0_{\varepsilon'}\left(\int_{|v| \leq \varepsilon \Sigma \alpha_i \delta_i} \delta_j^{4/n-2} v^2\right) + \right.$$

$$\left. + o_{\varepsilon'}(\varepsilon_{ij})\right] + 0\left(\frac{1}{\lambda_j^{\frac{n+2}{2}}} \frac{1}{d_j^{\frac{n+2}{2}}}\right) \left(\int_{\Omega^c} |v|^{2n/n-2}\right)^{\frac{n-2}{2n}} \times$$

$$\times \frac{1}{\lambda_j^{\frac{n+2}{2}} d_j^{n-2}} \left(\int |\nabla v|^2 \right)^{1/2} \Bigg) \Bigg] + A_j\, o_K \left(\left(\int |\nabla v|^2 \right)^{1/2} \left(\frac{|DK(x_j)|}{\lambda_j} + \frac{1}{\lambda_j^2} \right) \right)$$

where (3.165) holds if $\Sigma \alpha_i \delta_i + v$ has support in Ω and:

$$\theta = \operatorname{Inf}\left(\frac{4}{n-2}, 1\right) \tag{3.167}$$

$$A_j = 1 \text{ for G5; } A_j = \lambda_j \text{ for G6; } A_j = \frac{1}{\lambda_j} \text{ for G7} \tag{3.168}$$

$$o_K\left(\left(\int |\nabla v|^2 \right)^{1/2} \left(\frac{|DK(x_j)|}{\lambda_j} + \frac{1}{\lambda_j^2} \right) \right) \text{ is zero if } K \text{ is constant} \tag{3.169}$$

and

$$o_{\varepsilon'}(\varepsilon_{ij}) = 0 \left(\int_{|v| \le \varepsilon' \sum_{i \ne j} \alpha_i \delta_i} \delta_j |v| \left(\sum_{k \ne j} (\alpha_k \delta_k)^{4/n-2} + \right.\right. \tag{3.169'}$$

$$\left.\left. + (\alpha_j \delta_j)^{\frac{4}{n-2}-1} \right) \operatorname{Inf}\left(\alpha_j \delta_j, \sum_{i \ne j} \alpha_i \delta_i\right) \right)$$

3.3 The flow; a first normal form

We consider now a function u in the following set:

$$\Sigma^+ = \{ u > 0;\ \int_\Omega |\nabla v|^2 = 1;\ u = 0_{|\partial\Omega} \text{ if } \Omega \ne \mathbb{R}^n \} \tag{3.170}$$

We assume such a u to be split in

$$u = \sum_{i=1}^{p} \alpha_i \delta_i + v\ ; \quad \alpha_i > 0 \tag{3.171}$$

where

> v satisfies (G0) with respect to the δ_i's. v is by (3.170) equal to $-\Sigma \alpha_i \delta_i$ on Ω^c. (3.172)

We assume in this section that, with $J(u) = 1/\int K(x)\, u^{2n/n-2}$:

$$v \text{ is small in the norm } \left(\int |\nabla v|^2\right)^{1/2} ; \tag{2.173}$$

$(J(u)) \times \alpha_i^{4/n-2} K(x_i) = 1 + 0(1)$; and the

ε_{ij}'s for $i \neq j$ are small.

Let

$$\partial J(u) \text{ be the gradient of } u \text{ with respect to the scalar-product } \left(\int \nabla . \nabla\right) \tag{3.174}$$

We assume we are dealing with a u, function of the time s and of x in Ω, satisfying (3.171) -(3.172) and (3.173) and also

$$\frac{\partial u}{\partial s} = -\partial J(u) = -\lambda(u)\, u - \lambda(u)^{\frac{n+2}{n-2}} \Delta_{\Omega}^{-1} \, K(x)\, u^{\frac{n+2}{n-2}} \tag{3.175}$$

where

$$\lambda(u) = \frac{1}{\left(\int K(x)\, u^{2n/n-2}\right)^{\frac{n-2}{4}}} = (J(u))^{n-2/4} \tag{3.176}$$

3.3.1 First estimates

We wish to understand the dynamical evolution of v, α_i, x_i and λ_i. For this, we begin with a rough estimate on v.

We have, starting from (3.175) :

$$\frac{\partial v}{\partial s} = -\lambda(u)\, v - \lambda(u)^{\frac{n+2}{n-2}} \Delta_{\Omega}^{-1} K(x)\, (\Sigma \alpha_i \delta_i + v)^{\frac{n+2}{n-2}} - \lambda(u)\, (\Sigma\, \alpha_i \delta_i) - \tag{3.177}$$

$$- \Sigma \frac{\partial \alpha_i}{\partial s} \delta_i - \Sigma\, \alpha_i \frac{\partial \delta_i}{\partial x_i} \cdot \frac{\partial x_i}{\partial s} - \Sigma\, \alpha_i \frac{\partial \delta_i}{\partial \lambda_i} \frac{\partial \lambda_i}{\partial s}$$

Using (G0), we derive from (3.177) after multiplication by $-\Delta v$ and integration:

$$\frac{1}{2}\frac{\partial}{\partial s}\int_{\mathbb{R}^n}|\nabla v|^2 dx = -\lambda(u)\int_{\mathbb{R}^n}|\nabla v|^2 dx - \frac{n+2}{n-2}\lambda(u)^{\frac{n+2}{n-2}} \times \tag{3.178}$$

$$\times \int_{\mathbb{R}^n}\nabla(\Delta_\Omega^{-1}K(x)(\Sigma\alpha_i\delta_i)^{4/n-2}v)\nabla v\, dx$$

$$-\lambda(u)^{\frac{n+2}{n-2}}\int_{\mathbb{R}^n}\nabla\left(\Delta_\Omega^{-1}(K(x)(\Sigma\alpha_i\delta_i)^{\frac{n+2}{n-2}})\right)\nabla v\, dx + \text{remainder term}$$

where the remainder term is

$$-\lambda(u)^{\frac{n+2}{n-2}}\int_{\mathbb{R}^n}\nabla\left(\Delta_\Omega^{-1}K(x)\left[(\Sigma\alpha_i\delta_i+v)^{\frac{n+2}{n-2}} - (\Sigma\alpha_i\delta_i)^{\frac{n+2}{n-2}} - \right.\right. \tag{3.179}$$

$$\left.\left. -\frac{n+2}{n-2}(\Sigma\alpha_i\delta_i)^{4/n-2}v\right)\nabla v\, dx$$

We estimated this term in (G8).

Thus (3.178) becomes

$$\frac{1}{2}\frac{\partial}{\partial s}\int_{\mathbb{R}^n}|\nabla v|^2 dx = -\lambda(u)\int_{\mathbb{R}^n}|\nabla v|^2 dx - \frac{n+2}{n-2}\lambda(u)^{\frac{n+2}{n-2}} \times \tag{3.180}$$

$$\times \int_{\mathbb{R}^n}\nabla(\Delta_\Omega^{-1}K(x)(\Sigma\alpha_i\delta_i)^{4/n-2}v)\nabla v\, dx$$

$$-\lambda(u)^{\frac{n+2}{n-2}}\int_{\mathbb{R}^n}\nabla\left(\Delta_\Omega^{-1}\left(K(x)(\Sigma\alpha_i\delta_i)^{\frac{n+2}{n-2}}\right)\right)\nabla v\, dx + \begin{cases} 0((\int|\nabla v|^2)^{n/n-2}) & \text{if } n\geq 6 \\ 0((\int|\nabla v|^2)^{3/2}) & \text{if } n<6 \end{cases}$$

We now have, with θ_j defined in (2.3):

$$-\int_{\mathbb{R}^n}\nabla(\Delta_\Omega^{-1}K(x)(\Sigma\alpha_i\delta_i)^{4/n-2}v)\nabla v\, dx = \tag{3.181}$$

$$= -\int_\Omega\nabla(\Delta_\Omega^{-1}K(x)(\Sigma\alpha_i\delta_i)^{4/n-2}v.\nabla(v-\Sigma\alpha_j\theta_j))dx$$

$$= \int_{\mathbb{R}^n}K(x)(\Sigma\alpha_i\delta_i)^{4/n-2}v^2 dx - \int_{\Omega^c}K(x)(\Sigma\alpha_i\delta_i)^{\frac{2n}{n-2}} +$$

$$+ \int_{\mathbb{R}^n} K(x) (\Sigma \alpha_i \delta_i)^{4/n-2} v(\Sigma \alpha_j \theta_j)\, dx$$

(3.181) is obtained through integration by parts and uses also the fact that the support of θ_j is in Ω (in order to replace $\int_\Omega K(x)(\Sigma\alpha_i\delta_i)^{4/n-2} v(\Sigma\alpha_j\theta_j)\,dx$ by $\int_{\mathbb{R}^n} K(x)(\Sigma\alpha_i\delta_i)^{4/n-2} v(\Sigma\alpha_j\theta_j)\,dx$).

It is easy to check that:

$$\int_{\Omega^c} K(x) (\Sigma \alpha_i \delta_i)^{2n/n-2} = 0 \left(\Sigma \frac{1}{d_i^n \lambda_i^n} \right) \tag{3.182}$$

Thus, using Estimates G9 and G11, we have:

$$-\int_{\mathbb{R}^n} \nabla (\Delta_\Omega^{-1} K(x) (\Sigma \alpha_i \delta_i)^{\frac{4}{n-2}} v) \nabla v \, dx = \Sigma \alpha_i^{4/n-2} K(x_i) \int_{\mathbb{R}^n} \delta_i^{\frac{4}{n-2}} v^2 dx + R + R_K \tag{3.183}$$

where

$$R = \begin{cases} 0\left(\left(\Sigma \frac{1}{d_j^{n-2}} \right) \left(\Sigma \frac{1}{\lambda_j^{\frac{n+2}{2}}} \right) \left(\int |\nabla v|^2 \right)^{1/2} + \left(\int |\nabla v|^2 \right) \left(\sum_{i \neq j} \varepsilon_{ij}^{2/n-2} \left(\mathrm{Log}\, \varepsilon_{ij}^{-1} \right)^{2/} \right. \right. \\ \qquad \left. \left. + \Sigma \frac{1}{\lambda_i^n} \right) \right) \quad \text{for } n \neq 6;\ n \neq 3 \\ 0\left(\left(\Sigma \frac{1}{d_j^4} \right) \left(\Sigma\ \mathrm{Log}\, \lambda_j \right)^{2/3} / \lambda_j^4 \left(\int |\nabla v|^2 \right)^{1/2} \right. \\ \qquad \left. + \left(\int |\nabla v|^2 \right) \left(\sum_{i \neq j} \varepsilon_{ij}^{1/2} \left(\mathrm{Log}\, \varepsilon_{ij}^{-1} \right)^{1/3} + \Sigma \frac{1}{\lambda_i^6} \right) \right) \quad \text{for } n = 6 \\ 0\left(\left(\Sigma \frac{1}{d_j} \right) \left(\Sigma \frac{1}{\lambda_j^{5/2}} \right) \left(\int |\nabla v|^2 \right)^{1/2} + \left(\int |\nabla v|^2 \right) \left(\sum_{i \neq j} \varepsilon_{ij} \mathrm{Log}\, \varepsilon_{ij}^{-1} + \Sigma \frac{1}{\lambda_i^3} \right) \right) \\ \qquad \text{for } n = 3 \end{cases} \tag{3.184}$$

and

$$R_K = 0_K \left(\int |\nabla v|^2 \left(\frac{|DK(x_i)|}{\lambda_i} + \frac{(\mathrm{Log}\, \lambda_i)^{2/n}}{\lambda_i^2} \right) \right) \tag{3.185}$$

appears when K is non constant.

Turning to the other term in (3.180), we have:

$$-\int_{\mathbb{R}^n} \nabla \Delta_\Omega^{-1} K(x) (\Sigma \alpha_i \delta_i)^{\frac{n+2}{n-2}} \nabla v \, dx = \int_{\mathbb{R}^n} K(x) (\Sigma \alpha_i \delta_i)^{\frac{n+2}{n-2}} v \, dx - \tag{3.186}$$

$$-\int_{\Omega^c} K(x) (\Sigma \alpha_i \delta_i)^{2n/n-2} dx + \int_\Omega K(x) (\Sigma \alpha_i \delta_i)^{\frac{n+2}{n-2}} (\Sigma \alpha_j \theta_j) \, dx$$

Using (F8) and (F9), we have:

$$\int_\Omega K(x) (\Sigma \alpha_i \delta_i)^{\frac{n+2}{n-2}} (\Sigma \alpha_j \theta_j) = \sum_{i,j} \int_\Omega K(x) \Sigma \alpha_i^{\frac{n+2}{n-2}} \alpha_j \delta_i^{\frac{n+2}{n-2}} \theta_j + \tag{3.187}$$

$$+ 0 \left(\sum_{i \neq j} \frac{1}{\lambda_j^{\frac{n-2}{2}} d_j^{n-2}} \times \frac{1}{\lambda_i^{\frac{(n-2)^2}{2(n+2)}}} \left(\varepsilon_{ij}^{4/n+2} + \frac{1}{(\lambda_i \lambda_j)^{2 \cdot \frac{n-2}{n+2}}} \right) \right)$$

$$= \sum_i \alpha_i^{\frac{n+2}{n-2}} \alpha_j \frac{K(x_i) \theta_j (x_i)}{\lambda_i^{\frac{n-2}{2}}} c_4 + 0 \left(\Sigma \frac{1}{\lambda_i^{\frac{n+2}{2}}} \frac{1}{\lambda_j^{\frac{n-2}{2}}} \frac{\operatorname{Log} \lambda_i}{d_j^n} + \sum_{i \neq j} \frac{1}{\lambda_j^{\frac{n-2}{2}} d_j^{n-2}} \times \right.$$

$$\left. \times \frac{1}{\lambda_i^{\frac{(n-2)^2}{2(n+2)}}} \left(\varepsilon_{ij}^{4/n+2} + \frac{1}{(\lambda_i \lambda_j)^{2 \cdot \frac{n-2}{n+2}}} \right) \right)$$

with

$$|\theta_j (x_i)| \leq \operatorname*{Sup}_{\partial\Omega} \theta_j \leq \frac{1}{\lambda_j^{\frac{n-2}{2}} d_j^{n-2}} \tag{3.188}$$

Thus

$$\left| \int_\Omega K(x) (\Sigma \alpha_i \delta_i)^{\frac{n+2}{n-2}} (\Sigma \alpha_j \theta_j) \right| \leq 0 \left(\Sigma \frac{1}{(\lambda_i \lambda_j)^{\frac{n-2}{2}}} \frac{1}{d_j^{n-2}} + \right. \tag{3.189}$$

$$+ \sum_{i,j} \frac{1}{\lambda_i^{\frac{n+2}{2}}} \frac{1}{\lambda_j^{\frac{n-2}{2}}} \frac{\operatorname{Log} \lambda_i}{d_j^n} + \sum_{i \neq j} \frac{1}{\lambda_j^{\frac{n-2}{2}} d_j^{n-2}} \times$$

$$\times \frac{1}{\frac{(n-2)^2}{\lambda_i^{2(n+2)}}} \left(\varepsilon_{ij}^{4/n+2} + \frac{1}{(\lambda_i \lambda_j)^{2 \cdot \frac{n-2}{n+2}}} \right) \Bigg)$$

Using (G10), we thus have:

$$-\int_{\mathbb{R}^n} \nabla (\Delta_\Omega^{-1} K(x) (\Sigma \alpha_i \delta_i)^{\frac{n+2}{n-2}} \nabla v \, dx = 0 \left(\Sigma \frac{1}{(\lambda_i \lambda_j)^{\frac{n-2}{2}}} \frac{1}{d_j^{n-2}} + \right. \tag{3.190}$$

$$+ \sum_{i,j} \frac{1}{\lambda_i^{\frac{n+2}{2}}} \frac{1}{\lambda_j^{\frac{n-2}{2}}} \frac{\operatorname{Log} \lambda_i}{d_j^n} + \sum_{i \neq j} \frac{1}{\lambda_j^{\frac{n-2}{2}} d_j^{n-2}} \cdot \frac{1}{\frac{(n-2)^2}{\lambda_i^{2(n+2)}}} \left(\varepsilon_{ij}^{\frac{4}{n+2}} + \frac{1}{(\lambda_i \lambda_j)^{2 \cdot \frac{n-2}{n+2}}} \right) +$$

$$\left. + \Sigma \frac{1}{\lambda_i^n d_i^n} + \begin{cases} (\int |\nabla v|^2)^{1/2} (\sum_{i \neq j} \varepsilon_{ij}^{\frac{n+2}{2(n-2)}} (\operatorname{Log} \varepsilon_{ij}^{-1})^{\frac{n+2}{2n}}) & \text{if } n \geq 6 \\ (\int |\nabla v|^2)^{1/2} (\sum_{i \neq j} \varepsilon_{ij} (\operatorname{Log} \varepsilon_{ij}^{-1})^{\frac{n-2}{n}}) & \text{if } n < 6 \end{cases} \right) +$$

$$+ 0_K \left(\left(\int |\nabla v|^2 \right)^{1/2} \times \left(\Sigma \frac{|DK(x_i)|}{\lambda_i} + \frac{1}{\lambda_i^2} \right) \right)$$

where 0_K appears when K is non constant.

Therefore, (3.180) becomes after these estimates:

$$\frac{1}{2} \frac{\partial}{\partial s} \int_{\mathbb{R}^n} |\nabla v|^2 dx = \frac{n+2}{n-2} \lambda(u)^{\frac{n+2}{n-2}} \Sigma \alpha_i^{4/n-2} K(x_i) \times \tag{3.191}$$

$$\times \int_{\mathbb{R}^n} \delta_i^{4/n-2} v^2 dx - \lambda(u) \int_{\mathbb{R}^n} |\nabla v|^2 dx + 0 + 0_K$$

where

$$0_K = 0_K\left(\int|\nabla v|^2\right)^{1/2}\left(\Sigma\frac{|DK(x_i)|}{\lambda_i}+\frac{1}{\lambda_i^2}\right) \quad \text{which appears} \qquad (3.192)$$

when K is non constant

and

$$0 = 0 \text{ of } (3.189) + 0 \text{ of } (3.184) + \begin{cases} 0\left(\left(\int|\nabla v|^2\right)^{n/n-2}\right) & \text{if } n \geq 6 \\ 0\left(\left(\int|\nabla v|^2\right)^{3/2}\right) & \text{if } n < 6 \end{cases} \qquad (3.193)$$

Using now the assumption (3.173), we derive:

$$\frac{1}{2}\frac{\partial}{\partial s}\int_{\mathbb{R}^n}|\nabla v|^2 dx = \lambda(u)\left(\frac{n+2}{n-2}\sum_{i=1}^{p}\int_{\mathbb{R}^n}\delta_i^{4/n-2}v^2 dx - \int_{\mathbb{R}^n}|\nabla v|^2 dx\right) +$$

$$+ o\left(\int|\nabla v|^2\right)dx + 0_K\left(\left(\int|\nabla v|^2\right)^{1/2}\left(\Sigma\frac{|DK(x_i)|}{\lambda_i}+\frac{1}{\lambda_i^2}\right)\right) +$$

$$+ 0\left(\left(\int|\nabla v|^2\right)^{1/2} \cdot R_n\right) + 0(R_n') \qquad (3.194)$$

with

$$R_n = \sum_{i\neq j}\varepsilon_{ij}^{\frac{n+2}{2(n-2)}}\left(\text{Log}\,\varepsilon_{ij}^{-1}\right)^{\frac{n+2}{2n}} + \left(\Sigma\frac{1}{d_j^{n-2}}\right)\left(\Sigma\frac{1}{\lambda_j^{\frac{n+2}{2}}}\right) \quad \text{for } n > 6 \qquad (3.195)$$

$$R_6 = \sum_{i\neq j}\varepsilon_{ij}\left(\text{Log}\,\varepsilon_{ij}^{-1}\right)^{2/3} + \left(\Sigma\frac{1}{d_j^4}\right)\left(\Sigma\frac{(\text{Log}\,\lambda_j)^{2/3}}{\lambda_j^4}\right)$$

$$n=4;\ n=5 \quad R_n = \sum_{i\neq j}\varepsilon_{ij}\left(\text{Log}\,\varepsilon_{ij}^{-1}\right)^{\frac{n-2}{n}} + \left(\Sigma\frac{1}{d_j^{n-2}}\right)\left(\Sigma\frac{1}{\lambda_j^{\frac{n+2}{2}}}\right)$$

$$R_3 = \sum_{i\neq j}\varepsilon_{ij}\left(\text{Log}\,\varepsilon_{ij}^{-1}\right)^{1/3} + \left(\Sigma\frac{1}{d_j}\right)\left(\Sigma\frac{1}{\lambda_j^{5/2}}\right)$$

and

$$R'_n = \Sigma \frac{1}{\lambda_i^n d_i^n} + \Sigma \frac{1}{(\lambda_i \lambda_j)^{\frac{n-2}{2}} d_j^{n-2}} + \Sigma \frac{1}{\lambda_i^{\frac{n+2}{2}}} \frac{1}{\lambda_j^{\frac{n-2}{2}}} \frac{\operatorname{Log} \lambda_i}{d_j^n} + \tag{3.196}$$

$$+ \sum_{i \neq j} \frac{1}{\lambda_j^{\frac{n-2}{2}} d_j^{n-2}} \frac{1}{\lambda_i^{\frac{(n-2)^2}{2(n+2)}}} \left(\varepsilon_{ij}^{4/n+2} + \frac{1}{(\lambda_i \lambda_j)^{2 \cdot \frac{n-2}{n+2}}} \right)$$

We now have:

Lemma 3.3: either $\int |\nabla v|^2 dx^{1/2} (s) \leq 0 \left(R_n + \sqrt{R'_n} \right) +$

$+ 0_K \left(\Sigma \frac{|DK(x_i)|}{\lambda_i} + \frac{1}{\lambda_i^2} \right)$ or $\frac{\partial}{\partial s} \int |\nabla v|^2 dx \leq -\bar{\alpha}_0 \int |\nabla v|^2 dx;$ $\bar{\alpha}_0 > 0$ fixed constant.

Proof: It is an easy consequence of Proposition 3.1 and of (3.194). The $\bar{\alpha}_0$ can be taken to be $\alpha_0/2$ where α_0 is the constant appearing in Proposition 3.1.

From Lemma 3.3, we see that the quantity which decides the nature of the gradient line defined by (3.175) is $\Sigma \frac{1}{\lambda_i}(s)$ and its evolution.

We thus have to come now to a finer analysis of such a gradient line, which amounts to finding a normal form to the flow near the set of potential critical points at infinity defined by $\sum_{i=1}^{p} \delta_i(x_i, \lambda_i)$, with small ε_{ij} for $i \neq j$ and small $\frac{1}{\lambda_i d(x_i, \partial \Omega)}$. This is done in the next Chapter (Chapter 4).

3.4 $\int \delta_i^{4/n-2} v^2 dx$

In this section, we prove a dynamical estimate on $\int \delta_i^{4/n-2} v^2 dx$.

For this, we introduce

$$(h_k, \mu_k) ; \quad k \in \mathbb{N}; \quad \mu_k > 0; \quad \int_{\mathbb{R}^n} |\nabla h_k|^2 dx = 1 \tag{3.197}$$

the eigenvectors and positive eigenvalues of:

$$-\left[\Delta + \frac{n+2}{n-2} \delta_i^{4/n-2} \right] \tag{3.198}$$

acting on functions w belonging to:

$$\{ w | \nabla w \in L^2(\mathbb{R}^n) \quad w \in L^{2n/n-2}(\mathbb{R}^n) \} = H \tag{3.199}$$

with respect to the scalar product on H:

$$(w, w_1)_H = \int_{\mathbb{R}^n} \nabla w . \nabla w_1 dx \tag{3.200}$$

We thus have

$$(1-\mu_k) \Delta h_k + \frac{n+2}{n-2} \delta_i^{4/n-2} h_k = 0 \tag{3.201}$$

Let

$$L \tag{3.202}$$

be the Yamabe operator on S^n

and

$$q : S^n - \{\text{south pole}\} \rightarrow \mathbb{R}^n \tag{3.203}$$

be the stereographic projection sending the north pole onto zero.

q is a conformal transformation.

Let

τ_i be the conformal transformation obtained by dilation of order λ_i along the unit vector field ξ on S^n singular at both north and south poles whose integral curves are the meridians of S^n. (3.204)

$$(3.205)$$

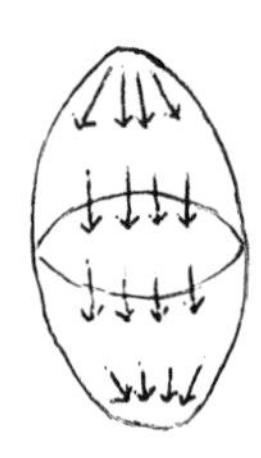

Let

$$\tilde{\delta} \text{ be the constant solution to } L\tilde{\delta} + \tilde{\delta}^{\frac{n+2}{n-2}} = 0 \text{ on } S^n \tag{3.206}$$

We have:

$$\delta_i(q(y) + x_i) = \text{Jac}_y \tau_i \quad \tilde{\delta}(\tau_i(y)) \tag{3.207}$$

Using (3.203)-(3.207) and the invariance of the Yamabe equation under conformal transformation, we derive:

Lemma 3.4: The operator $-\left[\Delta + \frac{n+2}{n-2}\delta_i^{4/n-2}\right]$ acting on H has a negative and null eigenspace spanned by δ_i, $\frac{\partial \delta_i}{\partial x_i}$, $\frac{\partial \delta_i}{\partial \lambda_i}$ (with respect to the scalar product $\int_{\mathbb{R}^n} \nabla w. \nabla w_1$). The positive eigenvalues μ_k increase to 1 when k goes to $+\infty$. The eigenvectors h_k satisfy the following inequalities:

$$|h_k(x)| \leq C_k \delta_i(x) ; \quad \int_{\mathbb{R}^n} |h_k|^{2n/n-2} \leq C_k^{2n/n-2} ; \quad \int_{\mathbb{R}^n} |\nabla h_k|^2 = 1 \tag{3.208}$$

$$\left|\frac{\partial h_k}{\partial \lambda_i}\right|(x) \leq C_k \frac{1}{\lambda_i} \delta_i(x) ; \quad \int_{\mathbb{R}^n} \left|\nabla \frac{\partial h_k}{\partial \lambda_i}\right|^2 dx \leq \frac{C_k^2}{\lambda_i^2} \tag{3.209}$$

$$\left|\frac{\partial h_k}{\partial x_i}\right|(x) \leq C_k \lambda_i \delta_i(x) ; \quad \int_{\mathbb{R}^n} \left|\nabla \frac{\partial h_k}{\partial x_i}\right|^2 dx \leq C_k^2 \lambda_i^2 \tag{3.210}$$

Proof of Lemma 3.4: (3.208), (3.209) and (3.210) are easily derived, by conformal transformation, from estimates on the eigenvectors of the operator $-L - \frac{n+2}{n-2}\tilde{\delta}^{4/n-2}$.

We now introduce the components of v on the h_k's.

v satisfies (G0); hence v splits:

$$v = \sum_i^{\infty} \beta_k h_k \tag{3.211}$$

and we have:

$$\int_{\mathbb{R}^n} |\nabla v|^2 dx = \Sigma \beta_k^2 ; \quad \int_{\mathbb{R}^n} \nabla v . h_k = \beta_k \tag{3.212}$$

$$\frac{n+2}{n-2} \int_{\mathbb{R}^n} \delta_i^{4/n-2} v^2 = \frac{n+2}{n-2} \int_{\mathbb{R}^n} (\sum_k \delta_i^{4/n-2} h_k \beta_k)(\sum_k h_k \beta_k) \tag{3.213}$$

$$= \int_{\mathbb{R}^n} (\Sigma \beta_k (\mu_k - 1) \Delta h_k)(\sum_k h_k \beta_k)$$

$$= \sum_k (1 - \mu_k) \beta_k^2 .$$

We want now to derive a dynamical estimate on β_k.

We have:

$$\beta_k = \int_{\mathbb{R}^n} \nabla v . \nabla h_k \tag{3.214}$$

Hence

$$\frac{\partial \beta_k}{\partial s} = \int_{\mathbb{R}^n} \nabla (\frac{\partial v}{\partial s}) . \nabla h_k + \int_{\mathbb{R}^n} \nabla v . \nabla \frac{\partial h_k}{\partial s} \tag{3.215}$$

Hence

$$\frac{\partial \beta_k}{\partial s} = \int_{\mathbb{R}^n} \nabla \frac{\partial v}{\partial s} . \nabla h_k + \int_{\mathbb{R}^n} \nabla v . \nabla \left(\frac{\partial h_k}{\partial x_i} \dot{x}_i + \frac{\partial h_k}{\partial \lambda_i} \dot{\lambda}_i \right) \tag{3.216}$$

Thus, by (3.208), (3.209) and (3.210):

$$\frac{\partial \beta_k}{\partial s} = \int_{\mathbb{R}^n} \nabla (\frac{\partial v}{\partial s}) . \nabla h_k + 0 \left(|\dot{x}_i| \lambda_i \left(\int_{\mathbb{R}} |\nabla v|^2 \right)^{1/2} + \right. \tag{3.217}$$

$$+ \frac{|\dot{\lambda}_i|}{\lambda_i} \left(\int_{\mathbb{R}^n} |\nabla v|^2 \right)^{1/2} \Bigg)$$

we need now to estimate $\int_{\mathbb{R}^n} \nabla(\frac{\partial v}{\partial s}) \cdot \nabla h_k$.

For this, we take the scalar product of (3.177) with h_k.

We get:

$$\int_{\mathbb{R}^n} \nabla \frac{\partial v}{\partial s} \cdot \nabla h_k = -\lambda(u) \int_{\mathbb{R}^n} \nabla v \cdot \nabla h_k - \lambda(u)^{\frac{n+2}{n-2}} \int_{\mathbb{R}^n} \nabla (\Delta_\Omega^{-1} K(x) (\Sigma \alpha_j \delta_j + v)^{\frac{n+2}{n-2}}) \nabla h_k$$

$$-\lambda(u) \int_{\mathbb{R}^n} \nabla(\Sigma \alpha_j \delta_j) \nabla h_k - \Sigma \dot{\alpha}_j \int \nabla \delta_j \nabla h_k \tag{3.218}$$

$$- \Sigma \alpha_j \int \nabla \left(\frac{\partial \delta_j}{\partial x_j} \dot{x}_j \right) \nabla h_k - \Sigma \dot{\lambda}_j \alpha_j \int \nabla \frac{\partial \delta_j}{\partial \lambda_j} \nabla h_k$$

We now have the following estimates.

<u>Estimate H0</u>: $\int_{\mathbb{R}^n} \nabla h_k \nabla \delta_i = \int_{\mathbb{R}^n} \nabla h_k \nabla \frac{\partial \delta_i}{\partial x_i} = \int_{\mathbb{R}^n} \nabla h_k \nabla \frac{\partial \delta_i}{\partial \lambda_i} = 0$

<u>Estimate H1</u>: $\int_{\mathbb{R}^n} \nabla \delta_j \nabla h_k = C_k 0(\varepsilon_{ij})$; $\int_{\mathbb{R}^n} \nabla \frac{\partial \delta_j}{\partial x_j} \nabla h_k = C_k 0(\lambda_j \varepsilon_{ij})$

$$\int_{\mathbb{R}^n} \nabla \frac{\partial \delta_j}{\partial \lambda_j} \nabla h_k = C_k 0 \left(\frac{1}{\lambda_j} \varepsilon_{ij} \right)$$

<u>Estimate H2</u>:

$$\int_{\mathbb{R}^n} \nabla \left(\Delta_\Omega^{-1} K(x) (\Sigma \alpha_j \delta_j + v)^{\frac{n+2}{n-2}} \right) \nabla h_k = -\beta_k \alpha_i^{4/n-2} (1-\mu_k) K(x_i) +$$

$$+ C_k \Bigg\{ 0_K \left(\left[\frac{1}{\lambda_i} + \frac{(\int |\nabla v|^2)^{1/2}}{\lambda_i} \right] \left[|DK(x_i)| + \frac{1}{\lambda_i} \right] \right) +$$

$$+ 0_\varepsilon\left(\int_{|v|\geq \varepsilon \Sigma \alpha_j \delta_j} \delta_i |v|^{\frac{n+2}{n-2}}\right) +$$

$$+ 0_{\varepsilon'}\left(\int_{|v|\leq \varepsilon \Sigma \alpha_j \delta_j} \delta_i^{4/n-2} v^2\right) + o_{\varepsilon'}(\varepsilon_{ij}) + 0\left(\sum_j \frac{1}{(\lambda_i d_i)^{\frac{n-2}{2}}} \frac{1}{(\lambda_j d_j)^{\frac{n+2}{2}}} + \right.$$

$$\left. + \frac{1}{(\lambda_i d_i)^{\frac{n+2}{2}}} \sum_j \frac{1}{(\lambda_j d_j)^{\frac{n-2}{2}}} + \frac{(\int |\nabla v|^2)^{1/2}}{\lambda_i^{\frac{n+2}{2}} d_i^{n-2}} + \sum_{i\neq j} \varepsilon_{ij} + \sum_j \frac{1}{(\lambda_i \lambda_j)^{\frac{n-2}{2}}} \frac{1}{d_i^{n-2}}\right) \Bigg\}$$

with $\theta = \text{Inf}(4/n-2,\ 1)$; $d_j = d(x_j;\ \partial\Omega)$

0_K appears when K is non constant.

$$o_{\varepsilon'}(\varepsilon_{ij}) = 0\left(\int_{|v|\leq \varepsilon' \sum_{j\neq i} \alpha_j \delta_j} \delta_i |v| \left(\sum_{k\neq j} (\alpha_k \delta_k)^{4/n-2} + \right.\right.$$

$$\left.\left. + (\alpha_i \delta_i)^{\frac{4}{n-2}-1} \text{Inf}(\alpha_i \delta_i,\ \sum_{j\neq i} \alpha_j \delta_j)\right)\right)$$

Proof of Estimate H0: $\delta_i, \frac{\partial \delta_i}{\partial x_i}, \frac{\partial \delta_i}{\partial \lambda_i}$ span the negative or null eigenspace of $-[\Delta + \frac{n+2}{n-2}\delta_i^{4/n-2}]$ while h_k is in the positive eigenspace of the same operator. Hence H0.

Proof of Estimate H1: We have, by (3.208)

$$\left|\int_{\mathbb{R}^n} \nabla\delta_j \nabla h_k\right| = \left|-\int_{\mathbb{R}^n} \delta_j^{\frac{n+2}{n-2}} h_k\right| \leq C_k \int_{\mathbb{R}^n} \delta_j^{\frac{n+2}{n-2}} \delta_i = 0(\varepsilon_{ij}) \qquad (3.219)$$

$$\left|\int_{\mathbb{R}^n} \nabla \frac{\partial \delta_j}{\partial x_j} \nabla h_k\right| = \left|\int_{\mathbb{R}^n} \frac{n+2}{n-2} \delta_j^{\frac{4}{n-2}} \frac{\partial \delta_j}{\partial x_j} h_k\right| \leq \qquad (3.220)$$

$$\leq C\, C_k \lambda_j \int_{\mathbb{R}^n} \delta_j^{\frac{n+2}{n-2}} \delta_i = 0(\lambda_j \varepsilon_{ij})$$

$$\left|\int_{\mathbb{R}^n} \nabla \frac{\partial \delta_j}{\partial \lambda_j} \nabla h_k\right| = \left|\int_{\mathbb{R}^n} \frac{n+2}{n-2} \delta_j^{4/n-2} \frac{\partial \delta_j}{\partial \lambda_j} h_k\right| \leq C \frac{C_k}{\lambda_j} \int_{\mathbb{R}^n} \delta_j^{\frac{n+2}{n-2}} \delta_i$$

$$= 0(\frac{\varepsilon_{ij}}{\lambda_j}) \tag{3.221}$$

Hence (H1).

Proof of Estimate H2: Let

$$w_k \text{ be the solution of } \begin{cases} \Delta w_k = 0 \\ w_k = h_k|_{\partial\Omega} \end{cases} \tag{3.222}$$

we have

$$\int \nabla \left(\Delta_{\Omega}^{-1} K(x) (\Sigma \alpha_j \delta_j + v)^{\frac{n+2}{n-2}} \right) \nabla h_k = - \int_{\Omega} K(x) (\Sigma \alpha_j \delta_j + v)^{\frac{n+2}{n-2}} (h_k - w_k) \tag{3.223}$$

We notice that (3.223) looks like the quantity we estimated in G5. With respect to (G5), we see that we need to replace δ_j by h_k. This does not greatly change the estimates as, by (3.208), h_k is upperbounded by $C_k \delta_i$ (Note that j has to be replaced by i in our present notations).

Looking back in the proof of (G5), there is a modification with respect to setting h_k as φ in (3.133).

Indeed then (3.134) fails and has to be replaced by:

$$\int_{\mathbb{R}^n} \nabla v. \nabla h_k = \beta_k \tag{3.224}$$

This modifies (3.155)-(3.156) as follows:

$$\int_{\Omega} v \delta_i^{4/n-2} (\varphi - \psi) = \int_{\Omega} v \delta_i^{4/n-2} (h_k - w_k) = \frac{n-2}{n+2} (1-\mu_k) \beta_k - \tag{3.225}$$

$$- \int_{\Omega} v \delta_i^{4/n-2} w_k + \int_{\Omega^c} v \delta_i^{4/n-2} h_k$$

and yields a corresponding modification in (3.164).

Up to this modification the estimate holds.

We thus have:

$$\int_{\mathbb{R}^n} \nabla \Delta_{\Omega}^{-1}\left(K(x)(\Sigma\alpha_j\delta_j+v)^{\frac{n+2}{n-2}}\right)\nabla h_k = \int_{\mathbb{R}^n} \nabla \Delta_{\Omega}^{-1}\left(K(x)(\Sigma\alpha_j\delta_j)^{\frac{n+2}{n-2}}\right)\nabla h_k -$$

$$- \alpha_i^{4/n-2} K(x_i)(1-\mu_k)\beta_k + C_k\left[0_\varepsilon\left(\int_{|v|\geq\varepsilon\Sigma\alpha_j\delta_j}\delta_i |v|^{\frac{n+2}{n-2}}\right)+\right.$$

$$+ 0_{\varepsilon'}\left(\int_{|v|\leq\varepsilon\Sigma\alpha_j\delta_j}\delta_i^{4/n-2}v^2\right) + o_{\varepsilon'}(\varepsilon_{ij}) + 0\left(\frac{1}{\lambda_i^{\frac{n+2}{2}} d_i} \Sigma \frac{1}{(\lambda_j d_j)^{\frac{n-2}{2}}} +\right.$$

$$\left.+ \frac{(\int|\nabla v|^2)^{1/2}}{\lambda_i^{\frac{n+2}{2}} d_i^{n-2}}\right) + 0_K\left(\left(\int|\nabla v|^2\right)^{1/2}\left(\frac{|DK(x_i)|}{\lambda_i} + \frac{1}{\lambda_i^2}\right)\right) \tag{3.226}$$

where

$$\theta = \mathrm{Inf}(4/n-2, 1)\,; \quad d_i = d(x_i, \partial\Omega)\,; \quad 0_K((\int|\nabla v|^2)^{1/2} 1/\lambda_i) \tag{3.227}$$

appears when K is non constant.

Now, using the fact that $|h_k| \leq \delta_i$ (thus $|w_k| \leq \theta_i$):

$$\int_{\mathbb{R}^n} \nabla \Delta_{\Omega}^{-1} K(x)(\Sigma\alpha_j\delta_j)^{\frac{n+2}{n-2}} \nabla h_k = \int_{\Omega} K(x)(\Sigma\alpha_j\delta_j)^{\frac{n+2}{n-2}}(h_k - \tilde{w}_k) \tag{3.228}$$

$$= \int_{\mathbb{R}^n} K(x)(\Sigma\alpha_j\delta_j)^{\frac{n+2}{n-2}} h_k + C_k 0\left(\int_{\Omega}(\Sigma\alpha_j\delta_j)^{\frac{n+2}{n-2}}\theta_i + \int_{\Omega^c}(\Sigma\alpha_j\delta_j)^{\frac{n+2}{n-2}}\delta_i\right)$$

$$= \left(\int_{\mathbb{R}^n} K(x)\delta_i^{\frac{n+2}{n-2}} h_k\right)\alpha_i^{\frac{n+2}{n-2}} + C_k 0\left(\sum_{j\neq i}\int \delta_j^{\frac{n+2}{n-2}}\delta_i + \int_{\Omega}\left(\Sigma\alpha_j\delta_j\right)^{\frac{n+2}{n-2}}\theta_i +\right.$$

$$\left.+ \int_{\Omega^c}\left(\Sigma\alpha_j\delta_j\right)^{\frac{n+2}{n-2}}\delta_i\right)$$

$$= \int_{\mathbb{R}^n} K(x)\,\delta_i^{\frac{n+2}{n-2}} h_k \alpha_i^{\frac{n+2}{n-2}} + C_k\, 0\left(\sum_{j\neq i} \varepsilon_{ij} + \sum_j \frac{1}{(\lambda_i\lambda_j)^{\frac{n-2}{2}}}\,\frac{1}{d_i^{n-2}} + \right.$$

$$\left. + \sum_j \frac{1}{\lambda_i^{\frac{n-2}{2}} d_i^{\frac{n-2}{2}}}\;\frac{1}{\lambda_j^{\frac{n+2}{2}} d_j^{\frac{n+2}{2}}}\right)$$

Now

$$\int_{\mathbb{R}^n} \delta_i^{\frac{n+2}{n-2}} h_k = -\int_{\mathbb{R}^n} \nabla\delta_i \nabla h_k = 0 \tag{3.229}$$

Thus

$$\int_{\mathbb{R}^n} K(x)\,\delta_i^{\frac{n+2}{n-2}} h_k = \int_{B_i} (K(x) - K(x_i))\,\delta_i^{\frac{n+2}{n-2}} h_k + 0\left(\int_{B_i^c} \delta_i^{2n/n-2}\right) \tag{3.230}$$

$$= C_k\, 0_K\left(\frac{|DK(x_i)|}{\lambda_i} + \frac{1}{\lambda_i^2}\right)$$

where 0_K appears whenever K is non constant.

Thus

$$\int_{\mathbb{R}^n} \nabla\,\Delta_\Omega^{-1} K(x)\,(\Sigma\alpha_j\delta_j)^{\frac{n+2}{n-2}} \nabla h_k = C_k\, 0_K\left(\frac{|DK(x_i)|}{\lambda_i} + \frac{1}{\lambda_i^2}\right) + \tag{3.231}$$

$$+ C_k\, 0\left(\sum_{j\neq i} \varepsilon_{ij} + \sum_j \frac{1}{(\lambda_i\lambda_j)^{\frac{n-2}{2}}}\,\frac{1}{d_i^{n-2}} + \sum_j \frac{1}{\lambda_i^{\frac{n-2}{2}} d_i^{\frac{n-2}{2}}} \times \frac{1}{\lambda_j^{\frac{n+2}{2}} d_j^{\frac{n+2}{2}}}\right)$$

This yields

$$\int_{\mathbb{R}^n} \nabla\,\Delta_\Omega^{-1} K(x)\,(\Sigma\alpha_j\delta_j + v)^{\frac{n+2}{n-2}} \nabla h_k = -K(x_i)\,\alpha_i^{4/n-2}(1-\mu_k)\beta_k + \tag{3.232}$$

$$+ C_k\, 0_K\left(\frac{|DK(x_i)|}{\lambda_i} + \frac{1}{\lambda_i^2} + \left(\int |\nabla v|^2\right)^{1/2}\left(\frac{|DK(x_i)|}{\lambda_i} + \frac{1}{\lambda_i^2}\right)\right) +$$

$$+ C_k\Big[0_\varepsilon \Big(\int_{|v| \geq \Sigma \alpha_j \delta_j} \delta_i |v|^{\frac{n+2}{n-2}} \Big) + 0_{\varepsilon'} \Big(\int_{|v| \leq \varepsilon \Sigma \alpha_j \delta_j} \delta_i^{4/n-2} v^2 \Big) +$$

$$+ o_{\varepsilon'}(\varepsilon_{ij}) + 0\Bigg(\sum_j \frac{1}{(\lambda_i d_i)^{\frac{n-2}{2}}} \cdot \frac{1}{(\lambda_j d_j)^{\frac{n+2}{2}}} +$$

$$+ \frac{1}{\lambda_i^{\frac{n+2}{2}} d_i^{\frac{n+2}{2}}} \sum \frac{1}{(\lambda_j d_j)^{\frac{n-2}{2}}} + \frac{(\int |\nabla v|^2)^{1/2}}{\lambda_i^{\frac{n+2}{2}} d_i^{n-2}} \Bigg) + 0\Bigg(\sum_{j \neq i} \varepsilon_{ij} + \sum_j \frac{1}{(\lambda_i \lambda_j)^{\frac{n-2}{2}}} \frac{1}{d_i^{n-2}} \Bigg) \Big]$$

with θ and d_i defined in (3.227).

(3.232) is (H2).

We point out here that 0_K and 0 can be taken to be independent of k; i.e. these quantities, which a priori depend on k, are uniformly bounded by quantities of the same type independent of k. We will study further the dependence of 0 and 0_K on k.

Turning back to (3.217), we thus have:

$$\frac{\partial \beta_k}{\partial s} = -\lambda(u)\beta_k + \lambda(u)^{\frac{n+2}{n-2}} \alpha_i^{4/n-2} (1-\mu_k) K(x_i) \beta_k + \text{remainder term} \qquad (3.233)$$

The remainder term will be stated later.

Using now (3.173) and (3.174), we derive:

$$\frac{\partial \beta_k}{\partial s} = -\lambda(u) \mu_k \beta_k (1+o(1)) + \text{remainder term} \qquad (3.234)$$

This yields:

<u>Lemma 3.4</u>: either $|\beta_k(s)| \leq C_k[0(R_n'') + 0_K(R_n''')]$

or $-\overline{\alpha}_2 \leq \frac{1}{\beta_k} \frac{\partial \beta_k}{\partial s} \leq -\overline{\alpha}_1$; $\overline{\alpha}_1$ and $\overline{\alpha}_2$ fixed positive constants where

$$R_n'' = |\dot{x}_i| \lambda_i \left(\int_{\mathbb{R}^n} |\nabla v|^2 \right)^{1/2} + \frac{|\dot{\lambda}_i|}{\lambda_i} \left(\int_{\mathbb{R}^n} |\nabla v|^2 \right)^{1/2} +$$

$$+ \sum_{j\neq i} \varepsilon_{ij}\left[1 + |\dot{\alpha}_j| + \lambda_j|\dot{x}_j| + \frac{|\dot{\lambda}_j|}{\lambda_j}\right] + 0_\varepsilon\left(\int_{|v|\geq\varepsilon\Sigma\alpha_j\delta_j} \delta_i |v|^{\frac{n+2}{n-2}}\right) +$$

$$+ 0_{\varepsilon'}\left(\int_{|v|\leq\varepsilon\Sigma\alpha_j\delta_j} \delta_i^{4/n-2} v^2\right) +$$

$$+ 0\left(\sum_j \frac{1}{(\lambda_i d_i)^{\frac{n-2}{2}}} \frac{1}{(\lambda_j d_j)^{\frac{n-2}{2}}}\left(\frac{1}{(\lambda_i d_i)^2} + \frac{1}{(\lambda_j d_j)^2}\right) + \frac{(\int_{\mathbb{R}^n} |\nabla v|^2)^{1/2}}{\lambda_i^{\frac{n+2}{2}} d_i^{n-2}}\right.$$

$$\left. + \sum_j \frac{1}{(\lambda_i\lambda_j)^{\frac{n-2}{2}}} \frac{1}{d_i^{n-2}}\right)$$

and $R'''_n = \frac{|DK(x_i)|}{\lambda_i} + \frac{(\int|\nabla v|^2)^{1/2}}{\lambda_i}|DK(x_i)| + \frac{1}{\lambda_i^2}$

and 0, 0_K do not depend on k.

Now

$$\int_{\mathbb{R}^n} \delta_i^{4/n-2} v^2 = \Sigma(1-\mu_k)\beta_k^2 \ ; \ \mu_k \nearrow 1 \text{ when } k \to +\infty \tag{3.235}$$

Thus:

<u>Lemma 3.6</u>: (Dynamical estimate on $\int_{\mathbb{R}^n} \delta_i^{4/n-2} v^2$)

For any $\varepsilon > 0$, we have either:

$$\int_{\mathbb{R}^n} \delta_i^{4/n-2} v^2(s) \leq \varepsilon \int|\nabla v|^2(s) + [0(R''^2_n) + 0_K(R'''^2_n)](s) \times \tag{3.236}$$

$$\times \sum_{1-\mu_k\geq\varepsilon} (1-\mu_k)C_k^2$$

or

$$\sum_{1-\mu_k\geq\varepsilon} (1-\mu_k)C_k^2 \times [0(R''^2_n) + 0_K(R'''^2_n)](s) \leq \int_{\mathbb{R}^n} \delta_i^{4/n-2} v^2(s) \leq$$

$$\leq \varepsilon \int |\nabla v|^2(s) + e^{-\bar{\alpha}_1(s-\bar{s})} \times [0(R_n''^2)+0_K(R_2'''^2)](\bar{s}) \times \sum_{1-\mu_k \geq \varepsilon} (1-\mu_k) c_k^2$$

where $\bar{s}$ is the first time before time s such that:

$$|\beta_k(s)| \geq c_k[0(R_n') + 0_K(R_n''')] \text{ for an integer } k \text{ satisfying} \tag{3.237}$$

$$1 - \mu_k \geq \varepsilon \tag{3.238}$$

Proof: straightforward deduction from Lemma 3.4 and the fact that $\mu_k \nearrow 1$.

3.5 Differentiation along a gradient line

Let u be in Σ^+ and satisfy

$$u = \sum_{i=1}^{p} \alpha_i \delta_i + v; \quad \alpha_i > 0 \tag{3.171}$$

$$v \text{ satisfies (G0) with respect to the } \delta_i\text{'s} \tag{3.172}$$

$$\frac{\partial u}{\partial s} = -\partial J(u) = -\lambda(u)u - \lambda(u)^{\frac{n+2}{n-2}} \Delta_\Omega^{-1} K(x) u^{\frac{n+2}{n-2}} \tag{3.175}$$

We thus have functions $v(s)$, $\alpha_i(s)$, $\delta_i(x(s), \lambda_i(s))$ (x) which we assume to be well defined and differentiable with respect to s. We will justify this later on in the book.

We differentiate (G0) :

$$
(H0)\begin{cases}
-\lambda(u)\int_{\mathbb{R}^n}\nabla\left(u+\lambda(u)^{4/n-2}\Delta_\Omega^{-1}\left(K(x)\,u^{\frac{n+2}{n-2}}\right)\right)\nabla\delta_i=\int_{\mathbb{R}^n}\nabla\left(\frac{\partial}{\partial s}(\Sigma\,\alpha_j\delta_j)\right)\nabla\delta_i-\\
\qquad -\int_{\mathbb{R}^n}\nabla v.\,\nabla\frac{\partial\delta_i}{\partial s}\\
-\lambda(u)\int_{\mathbb{R}^n}\nabla\left(u+\lambda(u)^{4/n-2}\Delta_\Omega^{-1}\left(K(x)\,u^{\frac{n+2}{n-2}}\right)\right)\nabla\frac{\partial\delta_i}{\partial x_i}=\int_{\mathbb{R}^n}\nabla\left(\frac{\partial}{\partial s}(\Sigma\alpha_j\delta_j)\right)\nabla\frac{\partial\delta_i}{\partial x_i}-\\
\qquad -\int_{\mathbb{R}^n}\nabla v.\,\nabla\frac{\partial}{\partial s}\left(\frac{\partial\delta_i}{\partial x_i}\right)\\
-\lambda(u)\int_{\mathbb{R}^n}\nabla\left(u+\lambda(u)^{4/n-2}\Delta_\Omega^{-1}\left(K(x)\,u^{\frac{n+2}{n-2}}\right)\right)\nabla\frac{\partial\delta_i}{\partial\lambda_i}=\\
\qquad =\int_{\mathbb{R}^n}\nabla\left(\frac{\partial}{\partial s}(\Sigma\alpha_j\delta_j)\right)\nabla\frac{\partial\delta_i}{\partial\lambda_i}-\int_{\mathbb{R}^n}\nabla v.\,\nabla\frac{\partial}{\partial s}\left(\frac{\partial\delta_i}{\partial\lambda_i}\right)
\end{cases}
$$

We now use Lemma 3.6 and Estimates F7 to F19, G1 to G13, to derive a first "implicit" normal form to (H0). In the next section, we derive the normal form of (H0).

The idea is that, in (H0), there is an intrinsic part which is due to $\sum_{i=1}^{p}\alpha_i\delta_i$, which we isolate, and another part, involving v which we consider as a remainder term. By (G5)-(G6)-(G7), we have:

$$
\int_{\mathbb{R}^n}\nabla\,\Delta_\Omega^{-1}\left(K(x)\,u^{\frac{n+2}{n-2}}\right)\nabla\delta_i=\int_{\mathbb{R}^n}\nabla\,\Delta_\Omega^{-1}K(x)\,(\Sigma\alpha_j\delta_j)^{\frac{n+2}{n-2}}\nabla\delta_i+ \tag{3.239}
$$

$$
+0_\varepsilon\left(\int_{|v|\geq\varepsilon\Sigma\alpha_j\delta_j}\delta_i\,|v|^{\frac{n+2}{n-2}}\right)+0_{\varepsilon'}\left(\int_{|v|\leq\varepsilon\Sigma\alpha_j\delta_j}\delta_i^{4/n-2}v^2\right)+o_{\varepsilon'}(\varepsilon_{ij})+
$$

$$
+0\left(\frac{1}{\lambda_i^{\frac{n+2}{2}}}\frac{1}{d_i^{\frac{n+2}{2}}}\left(\int_{\Omega^c}|v|^{2n/n-2}\right)^{\frac{n-2}{2n}}+\frac{1}{\lambda_i^{\frac{n+2}{2}}}\frac{1}{d_i^{n-2}}\left(\int|\nabla v|^2\right)^{1/2}\right)
$$

$$+ 0_K\left(\left(\int|\nabla v|^2\right)^{1/2}\left(\frac{|DK(x_i)|}{\lambda_i}+\frac{1}{\lambda_i^2}\right)\right)$$

$$\int_{\mathbb{R}^n}\nabla\ \Delta_\Omega^{-1}\left(K(x)\,u^{\frac{n+2}{n-2}}\right)\nabla\frac{\partial\delta_i}{\partial x_i}=\int_{\mathbb{R}^n}\nabla\left(\Delta_\Omega^{-1}K(x)\,(\Sigma\,\alpha_j\delta_j)^{\frac{n+2}{n-2}}\right)\nabla\frac{\partial\delta_i}{\partial x_i}$$

$$(3.240)$$

$$+\quad\lambda_i 0_\varepsilon\left(\int_{|v|\geq\varepsilon\Sigma\alpha_j\delta_j}\delta_i|v|^{\frac{n+2}{n-2}}\right)+\lambda_i 0_{\varepsilon'}\left(\int_{|v|\leq\varepsilon\Sigma\,\alpha_j\delta_j}\delta_i^{4/n-2}v^2\right)+$$

$$+\lambda_i o_{\varepsilon'}(\varepsilon_{ij})+0\left(\frac{1}{\lambda_i^{n/2}}\frac{1}{d_i^{\frac{n+2}{2}}}\left(\int_{\Omega^c}|v|^{2n/n-2}\right)^{\frac{n-2}{2n}}+\frac{(|\nabla v|^2)^{1/2}}{\lambda_i^n d_i^{n-2}}\right)$$

$$+ 0_K\left(\left(\int|\nabla v|^2\right)^{1/2}\left(|DK(x_i)|+\frac{1}{\lambda_i}\right)\right)$$

$$\int_{\mathbb{R}^n}\nabla\ \Delta_\Omega^{-1}\left(K(x)\,u^{\frac{n+2}{n-2}}\right)\nabla\frac{\partial\delta_i}{\partial\lambda_i}=\int_{\mathbb{R}^n}\nabla\ \Delta_\Omega^{-1}\left(K(x)(\Sigma\,\alpha_j\delta_j)^{\frac{n+2}{n-2}}\right)\nabla\frac{\partial\delta_i}{\partial\lambda_i}$$

$$(3.241)$$

$$+\ 0\left(\frac{1}{\lambda_i}0_\varepsilon\left(\int_{|v|\geq\varepsilon\Sigma\,\alpha_j\delta_j}\delta_i|v|^{\frac{n+2}{n-2}}\right)+\frac{1}{\lambda_i}\int_{|v|\leq\varepsilon\Sigma\,\alpha_j\delta_j}\delta_i^{4/n-2}v^2+\right.$$

$$\left.+\frac{1}{\lambda_i}o_{\varepsilon'}(\varepsilon_{ij})+\frac{1}{\lambda_i^{\frac{n+4}{2}}}\frac{1}{d_i^{\frac{n+2}{2}}}\left(\int_{\Omega^c}|v|^{2n/n-2}\right)^{\frac{n-2}{2n}}\frac{(\int|\nabla v|^2)^{1/2}}{\lambda_i^{\frac{n+4}{2}}d_i^{n-2}}\right)+$$

$$+ 0_K\left(\left(\int|\nabla v|^2\right)^{1/2}\left(\frac{|DK(x_i)|}{\lambda_i^2}+\frac{1}{\lambda_i^3}\right)\right)$$

Let

$$R = \text{remainder term in } (3.239) \qquad (3.242)$$

Using now Lemma 3.6, we derive:

$$\int_{\mathbb{R}^n} \nabla \Delta_{\Omega}^{-1}\left(K(x)\, u^{\frac{n+2}{n-2}}\right)\nabla\delta_i = -\Sigma\, \alpha_j^{\frac{n+2}{n-2}} \int_{\Omega} K(x)\, \delta_j^{\frac{n+2}{n-2}}(\delta_i - \theta_i) - \tag{3.243}$$

$$- \frac{n+2}{n-2}\, \alpha_i^{4/n-2} \int_{\Omega} K(x)\, \delta_i^{\frac{n+2}{n-2}}\Big(\sum_{j\neq i} \alpha_j \delta_j\Big) + \frac{n+2}{n-2}\, \alpha_i^{4/n-2} \int_{\Omega} K(x)\, \delta_i^{4/n-2}\, \theta_i\Big(\sum_{j\neq i} \alpha_j \delta_j\Big)$$

$$+ R + \begin{cases} + 0\left(\sum\limits_{j\neq k} \varepsilon_{jk}^{n/n-2} \operatorname{Log} \varepsilon_{jk}^{-1}\right) & \text{for } n \geq 4 \\ + 0\left(\sum\limits_{\substack{j\neq k \\ j\neq i}} \varepsilon_{jk}^{3} \operatorname{Log} \varepsilon_{jk}^{-1} + \sum\limits_{j\neq i} \varepsilon_{ij}^{2} (\operatorname{Log} \varepsilon_{ij}^{-1})^{2/3}\right) & \text{for } n = 3 \end{cases}$$

We denote

R' the second remainder term in (3.243) (3.244)

We also have:

$$\int_{\mathbb{R}^n} \nabla \left(\Delta_{\Omega}^{-1} K(x)\, u^{\frac{n+2}{n-2}}\right) \nabla \frac{\partial \delta_i}{\partial x_i} = \tag{3.245}$$

$$-\Sigma\, \alpha_j^{\frac{n+2}{n-2}} \int_{\Omega} K(x)\, \delta_j^{\frac{n+2}{n-2}} \left(\frac{\partial \delta_i}{\partial x_i} - \frac{\partial \theta_i}{\partial x_i}\right) -$$

$$- \frac{n+2}{n-2}\, \alpha_j^{4/n-2} \int_{\Omega} K(x)\, \delta_i^{4/n-2} \Big(\sum_{j\neq i} \alpha_j \delta_j\Big) \left(\frac{\partial \delta_i}{\partial x_i} - \frac{\partial \theta_i}{\partial x_i}\right) + \lambda_i (R + R')$$

$$\int_{\mathbb{R}^n} \nabla \left(\Delta_{\Omega}^{-1} K(x)\, u^{\frac{n+2}{n-2}}\right) \nabla \frac{\partial \delta_i}{\partial \lambda_i} = \tag{3.246}$$

$$-\Sigma\, \alpha_j^{\frac{n+2}{n-2}} \int_{\Omega} K(x)\, \delta_j^{\frac{n+2}{n-2}} \left(\frac{\partial \delta_i}{\partial \lambda_i} - \frac{\partial \theta_i}{\partial \lambda_i}\right) - \frac{n+2}{n-2}\, \alpha_j^{4/n-2} \int_{\Omega} K(x)\, \delta_i^{4/n-2} \Big(\sum_{j\neq i} \alpha_j \delta_j\Big) \left(\frac{\partial \delta_i}{\partial \lambda_i} - \right.$$

$$\left. - \frac{\partial \theta_i}{\partial \lambda_i}\right) + \frac{R+R'}{\lambda_i}$$

We now use the F estimates to replace (3.243), (3.245) and (3.246) by:

$$\int_{\mathbb{R}^n} \nabla \Delta_{\Omega}^{-1}\left(K(x)u^{\frac{n+2}{n-2}}\right)\nabla\delta_i = -\alpha_i^{\frac{n+2}{n-2}} c_0^{2n/n-2}\left\{K(x_i)\int \frac{r^{n-1}dr\,d\sigma}{(1+r^2)^n} + \right. \tag{3.247}$$

$$\left. + \frac{\Delta K(x_i)}{\lambda_i^2}\int \frac{|x|^2 dx}{(1+|x|^2)^n} + o_K\left(\frac{1}{\lambda_i^2}\right) - c_1\frac{K(x_i)}{\lambda_i^{n-2}}H(x_i,x_i)\right\} -$$

$$-\sum_{j\neq i}\alpha_j^{\frac{n+2}{n-2}} c_0^{2n/n-2} c_1\left\{K(x_j)\quad \varepsilon_{ij} - \frac{H(x_i,x_j)}{(\lambda_i\lambda_j)^{\frac{n-2}{2}}}\right\} -$$

$$-\frac{n+2}{n-2}\sum_{j\neq i} c_0^{2n/n-2} c_1\alpha_i^{4/n-2}\alpha_j K(x_i)\varepsilon_{ij} + 0\left(\sum_{i\neq j}\varepsilon_{ij}^{n/n-2}\operatorname{Log}\varepsilon_{ij}^{-1} + \right.$$

$$+\sum_{i\neq j}\frac{1}{\lambda_i^{\frac{n-2}{2}} d_i^{n-2}}\cdot\frac{1}{(n-2)^2\lambda_j^{2(n+2)}}\left(\varepsilon_{ij}^{4/n+2} + \frac{1}{(\lambda_i\lambda_j)^{2.\frac{n-2}{n+2}}}\right) +$$

$$\left. + \sum_j \frac{1}{(\lambda_j d_j)^{\frac{n+2}{2}}}\cdot\frac{1}{(\lambda_i d_i)^{\frac{n-2}{2}}}\right) + 0\left(\left[\frac{\operatorname{Log}\lambda_j}{\lambda_j^{\frac{n+2}{2}}\lambda_i^{\frac{n-2}{2}}} + \frac{1}{\lambda_j^{\frac{n-2}{2}}}\frac{1}{\lambda_i^{\frac{n+2}{2}}}\right]\frac{1}{d_i^n}\right) +$$

$$+ o_K(\varepsilon_{ij}) + 0_K\left(\frac{1}{\lambda_i^{\frac{n+2}{2}}\lambda_j^{\frac{n-2}{2}}}\right) + R + R' + (\text{if } n=3)\ 0\left(\sum_{i\neq j}\varepsilon_{ij}^2(\operatorname{Log}\varepsilon_{ij}^{-1})^{2/3}\right)$$

We took care of the integration over Ω instead of the integration over $\mathbb{R}^n$ by adding an

$$0\left(\sum_j \frac{1}{(\lambda_j d_j)^{\frac{n+2}{2}}}\ \frac{1}{(\lambda_i d_i)^{n-2}}\right).$$

We also replaced

$$\frac{\theta_i(x_j)}{\lambda_i^{\frac{n-2}{2}}}$$

by

$$\frac{H(x_i, x_j)}{\lambda_j^{\frac{n-2}{2}} \lambda_i^{\frac{n-2}{2}}}$$

which yields an

$$0\left(\Sigma \frac{1}{\lambda_j^{\frac{n-2}{2}}} \frac{1}{\lambda_i^{\frac{n+2}{2}} d_i^n}\right)$$ (see later on in the book in (5.23)).

Here the term

$$o_K(\varepsilon_{ij}) + 0_K\left(\frac{1}{\lambda_j^{\frac{n+2}{2}}} \frac{1}{\lambda_i^{\frac{n-2}{2}}}\right)$$

appears when K is not constant. This term comes from the

$$o(\varepsilon_{ij}) + 0\left(\frac{1}{\lambda_j^{\frac{n+2}{2}}} \frac{1}{\lambda_i^{\frac{n-2}{2}}}\right)$$

in F7 and thus satisfies:

$$o_K(\varepsilon_{ij}) + 0_K\left(\frac{1}{\lambda_j^{\frac{n+2}{2}}} \frac{1}{\lambda_i^{\frac{n-2}{2}}}\right) \leq 0\left(\sum_{i \neq j}\left[\varepsilon_{ij}\left(\operatorname{Log} \varepsilon_{ij}^{-1}\right)^{\frac{n-2}{2}}\left(\frac{1}{\lambda_j}\left(\frac{\operatorname{Log} \lambda_j}{\lambda_j}\right)^{2/n}\right)+\right.\right.$$

$$\left.\left.+\sum_{i \neq j} \frac{\varepsilon_{ij}}{\lambda_j}\right]\left(|DK(x_i)| + \frac{1}{\lambda_j}\right) + \sum_j \frac{1}{\lambda_j^{\frac{n+2}{2}} \lambda_i^{\frac{n-2}{2}}}\right) \qquad (3.248)$$

From (3.246), we derive:

$$\int_{\mathbb{R}^n} \nabla\left(\Delta_\Omega^{-1} K(x)\, u^{\frac{n+2}{n-2}}\right) \nabla \frac{\partial \delta_i}{\partial \lambda_i} = -\alpha_i^{\frac{n+2}{n-2}} c_0^{2n/n-2} \left\{ \frac{-\Delta K(x_i)}{\lambda_i^3} \frac{n-2}{n} \int \frac{|x|^2 dx}{(1+|x|^2)^n} \right.$$

$$\left. + o_K\left(\frac{1}{\lambda_i^3}\right) + \frac{n-2}{2} \cdot c_1 K(x_i) \frac{H(x_i, x_i)}{\lambda_i^{n-1}} \right\} - \tag{3.249}$$

$$-\sum_{j \neq i} \alpha_j^{\frac{n+2}{n-2}} c_0^{2n/n-2} c_1 \left\{ K(x_j) \left[\frac{\partial}{\partial \lambda_i} (\varepsilon_{ij}) + \frac{n-2}{2} \frac{H(x_i, x_j)}{\lambda_i (\lambda_i \lambda_j)^{\frac{n-2}{2}}} \right] \right\} -$$

$$- \sum_{j \neq i} K(x_i)\, c_0^{2n/n-2} c_1\, \alpha_i^{4/n-2} \alpha_j \frac{\partial}{\partial \lambda_i} \varepsilon_{ij} + \frac{1}{\lambda_i} S_i$$

where S_i is the remainder in (3.247).

Here, we have:

$$\frac{\partial}{\partial \lambda_i} \varepsilon_{ij} = \frac{n-2}{2} \left\{ \frac{1}{\lambda_j} - \frac{\lambda_j}{\lambda_i^2} + \lambda_j |x_i - x_j|^2 \right\} \varepsilon_{ij}^{n/n-2} \tag{3.250}$$

Finally, we derive from (3.245):

$$\int_{\mathbb{R}^n} \nabla\left(\Delta_\Omega^{-1} K(x)\, u^{\frac{n+2}{n-2}}\right) \nabla \frac{\partial \delta_i}{\partial x_i} = -\alpha_i^{\frac{n+2}{n-2}} c_0^{2n/n-2} \frac{n-2}{2n} \left\{ DK(x_i) \int \frac{r^{n-1} dr\, d\sigma}{(1+r^2)^n} + \right.$$

$$\left. + \frac{\Delta(DK)}{\lambda_i^2}(x_i) \int \frac{|x|^2 dx}{(1+|x|^2)^n} + o_K\left(\frac{1}{\lambda_i^2}\right) + \frac{2n}{n-2} c_1 \frac{K(x_i)}{\lambda_i^{n-2}} \frac{\partial H}{\partial y}(x_i, x_i) \right\} \tag{3.251}$$

$$- \sum_{j \neq i} \alpha_j^{\frac{n+2}{n-2}} c_0^{2n/n-2} c_1 \left\{ K(x_j) \left(\frac{\partial}{\partial x_i} \varepsilon_{ij} - \frac{\partial H}{\partial x} \frac{(x_i, x_j)}{(\lambda_i \lambda_j)^{\frac{n-2}{2}}} \right) \right\}$$

$$- \sum_{j \neq i} c_0^{2n/n-2} c_1\, \alpha_i^{4/n-2} \alpha_j \left\{ DK(x_i)\, \varepsilon_{ij} + K(x_i) \frac{\partial}{\partial x_i} \varepsilon_{ij} \right\} + \lambda_i S_i$$

where S_i is the remainder term in (3.247).

Here, we have:

$$\frac{\partial}{\partial x_i} \varepsilon_{ij} = \frac{n+2}{n} \lambda_i \lambda_j (x_j - x_i) \varepsilon_{ij}^{n/n-2} \tag{3.252}$$

Now, we turn to the right-hand side of the equations (H0).

We have:

$$\int_{\mathbb{R}^n} \nabla \frac{\partial}{\partial s} (\Sigma \alpha_j \delta_j) \nabla \delta_i = \Sigma \dot{\alpha}_j \int_{\mathbb{R}^n} \nabla \delta_j \nabla \delta_i + \Sigma \alpha_j \dot{\lambda}_j \int_{\mathbb{R}^n} \nabla \frac{\partial \delta_j}{\partial \lambda_j} \nabla \delta_i + \tag{3.253}$$

$$+ \Sigma \alpha_j \int_{\mathbb{R}^n} \nabla \delta_i \nabla \frac{\partial \delta_j}{\partial x_j} \cdot \dot{x}_j =$$

$$= + \Sigma \dot{\alpha}_j \int_{\mathbb{R}^n} \delta_i \delta_j^{\frac{n+2}{n-2}} + \Sigma \alpha_j \dot{\lambda}_j \int_{\mathbb{R}^n} \frac{n+2}{n-2} \delta_j^{4/n-2} \frac{\partial \delta_j}{\partial \lambda_j} \delta_i +$$

$$+ \Sigma \alpha_j \int_{\mathbb{R}^n} \frac{n+2}{n-2} \delta_i \delta_j^{4/n-2} \frac{\partial \delta_j}{\partial x_j} \cdot \dot{x}_j$$

Thus, using (E1), (F11) and (F16):

$$\int_{\mathbb{R}^n} \nabla \left(\frac{\partial}{\partial s} (\Sigma \alpha_j \delta_j) \right) \nabla \delta_i = \sum_{j \neq i} \dot{\alpha}_j \left(\varepsilon_{ij} + 0(\varepsilon_{ij}^{n/n-2}) \right) + \tag{3.254}$$

$$+ \sum_{j \neq i} \alpha_j \frac{n+2}{n} \lambda_i \lambda_j c_7 \varepsilon_{ij}^{n/n-2} (x_i - x_j) \cdot \dot{x}_j + \sum_{j \neq i} 0 \left(\varepsilon_{ij}^{\frac{n+1}{n-2}} \lambda_i \lambda_j |x_i - x_j| |\dot{x}_j| \right) +$$

$$- \sum_{j \neq i} \alpha_j \dot{\lambda}_j \frac{n-2}{2} c_1 \left\{ \left(\frac{1}{\lambda_i} - \frac{\lambda_j}{\lambda_i} + \lambda_j |x_i - x_j|^2 \right) \varepsilon_{ij}^{n/2} + \frac{1}{\lambda_j} 0 \left(\varepsilon_{ij}^{n/n-2} \operatorname{Log} \varepsilon_{ij}^{-1} \right) \right\}$$

$$+ \dot{\alpha}_i \int_{\mathbb{R}^n} |\nabla \delta_i|^2 dx$$

$\int_{\mathbb{R}^n} |\nabla \delta_i|^2 dx$ is a constant c independent of i.

On the other hand, by (G0), we have:

$$\int_{\mathbb{R}^n} \nabla v \,.\, \nabla \frac{\partial \delta_i}{\partial s} = 0 \tag{3.255}$$

Thus

$$\int_{\mathbb{R}^n} \nabla \left(\frac{\partial}{\partial s} (\Sigma \alpha_j \delta_j) \right) \nabla \delta_i - \int_{\mathbb{R}^n} \nabla v \,.\, \nabla \frac{\partial \delta_i}{\partial s} \tag{3.256}$$

$$= c\, \dot{\alpha}_i + \sum_{j \neq i} \dot{\alpha}_j c_0^{2n/n-2} c_1 \left(\varepsilon_{ij} + 0(\varepsilon_{ij}^{n/n-2}) \right) - \sum_{j \neq i} \dot{\lambda}_j \alpha_j \frac{n-2}{n} c_1 c_0^{2n/n-2} \times$$

$$\times \left\{ \left(\frac{1}{\lambda_i} - \frac{\lambda_i}{\lambda_j^2} + \lambda_i |x_i - x_j|^2 \right) \varepsilon_{ij}^{\frac{n}{n-2}} + \frac{1}{\lambda_j} 0 \left(\varepsilon_{ij}^{n/n-2} \operatorname{Log} \varepsilon_{ij}^{-1} \right) \right\} +$$

$$+ \sum_{j \neq i} \dot{x}_j \left\{ \alpha_j (n-2) \lambda_i \lambda_j c_0^{2n/n-2} c_1 \varepsilon_{ij}^{n/n-2} (x_i - x_j) + 0\left(\varepsilon_{ij}^{n+1/n-2} \lambda_i \lambda_j |x_i - x_j| \right) \right\}$$

We consider now

$$\int_{\mathbb{R}^n} \nabla \left(\frac{\partial}{\partial s} (\Sigma \alpha_j \delta_j) \right) \nabla \frac{\partial \delta_i}{\partial x_i} - \int_{\mathbb{R}^n} \nabla v \,.\, \nabla \frac{\partial}{\partial s} \left(\frac{\partial \delta_i}{\partial x_i} \right) \tag{3.257}$$

We then use Estimates F1-F23 and we derive:

$$\int_{\mathbb{R}^n} \nabla \left(\frac{\partial}{\partial s} (\Sigma \alpha_j \delta_j) \right) \nabla \frac{\partial \delta_i}{\partial x_i} = \sum_{j \neq i} \dot{\alpha}_j \left[\frac{\partial \varepsilon_{ij}}{\partial x_i} + 0 \left(\varepsilon_{ij}^{\frac{n+1}{n-2}} \lambda_i \lambda_j |x_i - x_j| \right) \right] \tag{3.258}$$

$$+ \sum_{j \neq i} \alpha_j \dot{\lambda}_j 0 \left(\lambda_i \left(\varepsilon_{ij}^{n/n-2} |x_i - x_j| \right) \right) \sum_{i \neq j} \alpha_j 0 \left(|\dot{x}_j| \left[\lambda_i \lambda_j \varepsilon_{ij}^{n/n-2} \right] \right) +$$

$$+ \frac{a_0}{n} \lambda_i^2 \alpha_i \dot{x}_i$$

On the other hand, using Estimates G1-G5, we derive

$$\int_{\mathbb{R}^n} \nabla v \,.\, \nabla \frac{\partial}{\partial s}\left(\frac{\partial \delta_i}{\partial x_i}\right) = 0\left(|\dot{x}_i|\lambda_i^2\left(\int|\nabla v|^2\right)^{1/2}\right) + \dot{\lambda}_i 0\left(\left(\int|\nabla v|^2\right)^{1/2}\right) \tag{3.259}$$

Thus

$$\int_{\mathbb{R}^n} \nabla\left(\frac{\partial}{\partial s}(\Sigma\, \alpha_j \delta_j)\right) \nabla \frac{\partial \delta_i}{\partial x_i} - \int_{\mathbb{R}^n} \nabla v \,.\, \nabla \frac{\partial}{\partial s}\left(\frac{\partial \delta_i}{\partial x_i}\right) = \tag{3.260}$$

$$\lambda_i^2 \frac{a_0 \alpha_i}{n} \dot{x}_i + 0\left(|\dot{x}_i|\lambda_i^2\left(\int|\nabla v|^2\right)^{1/2}\right) + \dot{\lambda}_i\, 0\left(\left(\int|\nabla v|^2\right)^{1/2}\right)$$

$$+ \sum_{j\neq i}\left[\dot{\alpha}_j\left[\frac{\partial \varepsilon_{ij}}{\partial x_i} + 0\left(\varepsilon_{ij}^{\frac{n+1}{n-2}} \lambda_i \lambda_j |x_i - x_j|\right)\right]\right] +$$

$$+ \dot{\lambda}_j\, \alpha_j 0\left(\lambda_j\left(\varepsilon_{ij}^{n/n-2} \times |x_i - x_j|\right)\right)$$

$$+ 0\left(|\dot{x}_j|\left[\lambda_i \lambda_j \times \varepsilon_{ij}^{n/n-2}\right]\right)$$

Next, we study:

$$\int_{\mathbb{R}^n} \nabla\left(\frac{\partial}{\partial s}(\Sigma \alpha_j \delta_j)\right) \nabla \frac{\partial \delta_i}{\partial \lambda_i} - \int_{\mathbb{R}^n} \nabla v \,.\, \nabla \frac{\partial}{\partial s}\left(\frac{\partial \delta_i}{\partial \lambda_i}\right) \tag{3.261}$$

By (G1)-(G6), we have:

$$\int_{\mathbb{R}^n} \nabla v \,.\, \nabla \frac{\partial}{\partial s}\left(\frac{\partial \delta_i}{\partial \lambda_i}\right) = \frac{\dot{\lambda}_i}{\lambda_i^2}\; 0\left(\left(\int|\nabla v|^2\right)^{1/2}\right) + 0\left(\left(|\dot{x}_i|\right)\left(\int|\nabla v|^2\right)^{1/2}\right) \tag{3.262}$$

By (F1)-(F23), we have:

$$\int_{\mathbb{R}^n} \nabla\left(\frac{\partial}{\partial s}(\Sigma \alpha_j \delta_j)\right) \nabla \frac{\partial \delta_i}{\partial \lambda_i} = \sum_{i\neq j} \dot{\alpha}_j \left(c_1 c_0^{2n/n-2} \frac{\partial \varepsilon_{ij}}{\partial \lambda_i} + \right. \tag{3.263}$$

$$\left. + \frac{1}{\lambda_i} 0\left(\varepsilon_{ij}^{n/n-2} \operatorname{Log} \varepsilon_{ij}^{-1}\right)\right) + \sum_{i\neq j} \alpha_j \dot{\lambda}_j\, 0\left(\frac{\varepsilon_{ij}}{\lambda_i \lambda_j}\right) +$$

$$+ \sum_{i\neq j} \alpha_j 0\left(|\dot{x}_j| \lambda_j \left[\varepsilon_{ij}^{n/n-2} \times |x_i - x_j| \right]\right) + \frac{\dot{\lambda}_i \alpha_i}{\lambda_i^2} \gamma_1$$

Thus

$$\int_{\mathbb{R}^n} \nabla \frac{\partial}{\partial s} (\Sigma \alpha_j \delta_j) \quad \nabla \frac{\partial \delta_i}{\partial \lambda_i} = \gamma_1 \frac{\alpha_i \dot{\lambda}_i}{\lambda_i^2} + \sum_{i\neq j} \left[\dot{\alpha}_j \left(c_1 c_0^{2n/n-2} \frac{\partial \varepsilon_{ij}}{\partial \lambda_i} + \right.\right.$$

$$\left.\left. + \frac{1}{\lambda_i} 0 \left(\varepsilon_{ij}^{n/n-2} \mathrm{Log}\, \varepsilon_{ij}^{-1} \right)\right) \right] + \alpha_j \dot{\lambda}_j 0 \left(\frac{\varepsilon_{ij}}{\lambda_i \lambda_j} \right) +$$

$$+ \alpha_j 0 \left(|\dot{x}_j| \lambda_j [\varepsilon_{ij}^{n/n-2} \times |x_i - x_j|] \right)$$

and

$$\int_{\mathbb{R}^n} \nabla \left(\frac{\partial}{\partial s} (\Sigma\, \alpha_j \delta_j) \right) \quad \nabla \frac{\partial \delta_i}{\partial \lambda_i} - \int \nabla v \,.\, \nabla \frac{\partial}{\partial s} \left(\frac{\partial \delta_i}{\partial \lambda_i} \right) = \tag{3.265}$$

$$\frac{\dot{\lambda}_i}{\lambda_i^2} \left(\alpha_i \gamma_1 + 0 \left(\left(\int |\nabla v|^2 \right)^{1/2} \right)\right) + \sum_{i\neq j} \left[\dot{\alpha}_j \left(c_1 c_0^{2n/n-2} \frac{\partial \varepsilon_{ij}}{\partial \lambda_i} + \right.\right.$$

$$\left.\left. + \frac{1}{\lambda_i} 0 \left(\varepsilon_{ij}^{n/n-2} \mathrm{Log}\, \varepsilon_{ij}^{-1} \right)\right) + \alpha_j \dot{\lambda}_j 0 \left(\frac{\varepsilon_{ij}}{\lambda_i \lambda_j} \right) + \alpha_j 0 \left(|\dot{x}_j| \lambda_j \left[\varepsilon_{ij}^{\frac{n}{n-2}} \times |x_i - x_j| \right]\right)$$

$$+ 0 \left(|\dot{x}_i| \left(\int |\nabla v|^2 \right)^{1/2} \right)$$

4 The normal form of the dynamical system near infinity

4.1 The implicit form

We now derive the normal form of the dynamical system near infinity. For this, we first remark that

$$\int \nabla u . \nabla \delta_i = \int \nabla(\Sigma \alpha_j \delta_j) \nabla \delta_i = \int \Sigma \alpha_j \delta_j^{\frac{n+2}{n-2}} \delta_i \tag{4.1}$$

$$= \sum_{i \neq j} \alpha_j \varepsilon_{ij} c_0^{2n/(n-2)} c_1 + \alpha_i c_0^{2n/(n-2)} \int \frac{r^{n-2} dr\, d\sigma}{(1+r^2)} + 0(\varepsilon_{ij}^{n/(n-2)})$$

$$\int \nabla u . \frac{\nabla \partial \delta_i}{\partial x_i} = \int \nabla(\Sigma \alpha_j \delta_j) \frac{\nabla \partial \delta_i}{\partial x_i} = \int \Sigma \alpha_j \delta_j^{\frac{n+2}{n-2}} \frac{\partial \delta_i}{\partial x_i} \tag{4.2}$$

$$= \Sigma \alpha_j (n-2) c_0^{2n/(n-2)} c_1 (x_j - x_i) \lambda_i \lambda_j \varepsilon_{ij}^{n/(n-2)} + 0(\varepsilon_{ij}^{\frac{n+1}{n+2}} \lambda_i \lambda_j |x_i - x_j|)$$

$$\int \nabla u . \frac{\nabla \partial \delta_i}{\partial \lambda_i} = \int \nabla(\Sigma \alpha_j \delta_j) \frac{\nabla \partial \delta_i}{\partial \lambda_i} = \int \Sigma \alpha_j \delta_j^{\frac{n+2}{n-2}} \frac{\partial \delta_i}{\partial \lambda_i} \tag{4.3}$$

$$= - \sum_{i \neq j} \frac{n-2}{2} c_1 c_0^{2n/(n-2)} \alpha_j (1/\lambda_j - \lambda_j/\lambda_i^2 + \lambda_j |x_i - x_j|^2) \varepsilon_{ij}^{n/(n-2)}$$

$$+ \frac{1}{\lambda_j} 0(\varepsilon_{ij}^{n/n-2}) \operatorname{Log} \varepsilon_{ij}^{-1})$$

$$\lambda(\Sigma \alpha_i \delta_i + v)^{4/(n-2)} = J(\Sigma \alpha_i \delta_i + v) = \tag{4.4}$$

(by a first and easy expansion, see Chapter 5 for more details)

$$= \frac{(\Sigma \alpha_i^2)^{n/(n-2)}}{\Sigma \alpha_i^{2n/(n-2)} K(x_i)} \left(\int \delta^{2n/(n-2)}\right)^{2/(n-2)} \Big[1 + O\Big(\Sigma \frac{1}{\lambda_i^{n-2}} + \sum_{i \neq j} \varepsilon_{ij}$$

$$+ \int |\nabla v|^2 + \Sigma \frac{1}{\lambda_i^{n+2}} \frac{1}{d_i^{2(n+2)}} + \Sigma \frac{1}{\lambda_i^4 \lambda_j^{n-2}} \frac{1}{d_j^{2(n-2)}} + 0_k\Big(\Sigma \frac{1}{\lambda_i^2}\Big)\Big]$$

where 0_k appears when K is nonconstant.

From (3.256), (3.247), (4.2) and (4.4) we deduce:

$$i \begin{bmatrix} c & & \overset{j}{} & \\ & c_0^{2n/(n-2)} c_1 [\varepsilon_{ij} + 0(\varepsilon_{ij}^{n/(n-2)})] & & \\ & & & \\ & & & c \end{bmatrix} \begin{bmatrix} \dot{\alpha}_1 \\ \\ \\ \dot{\alpha}_p \end{bmatrix} \tag{4.5}$$

$$+ \; i \begin{bmatrix} 0 & & \overset{j}{} & \\ & c_0^{2n/(n-2)} c_1 \alpha_j \Big[\frac{\partial \varepsilon_{ij}}{\partial \lambda_j} + \frac{1}{\lambda_j} 0(\varepsilon_{ij}^{n/(n-2)} \operatorname{Log} \varepsilon_{ij}^{-1})\Big] & & \\ & & & \\ & & & 0 \end{bmatrix} \begin{bmatrix} \dot{\lambda}_1 \\ \cdot \\ \cdot \\ \cdot \\ \dot{\lambda}_p \end{bmatrix}$$

$$+ \; i \begin{bmatrix} 0 & & \overset{j}{} & \\ & c_0^{2n/(n-2)} c_1 \alpha_j \Big[\frac{\partial \varepsilon_{ij}}{\partial x_j} + 0(\varepsilon_{ij}^{\frac{n+1}{n-2}} \lambda_i \lambda_j |x_i - x_j|)\Big] & & \\ & & & \\ & & & 0 \end{bmatrix} \begin{bmatrix} \dot{x}_1 \\ \\ \\ \dot{x}_p \end{bmatrix}$$

$$= -\lambda(u)\left[\sum_{i\neq j}\alpha_j\varepsilon_{ij}c_1c_0^{2n/(n-2)}+\alpha_i c_0^{2n/(n-2)}\int\frac{r^{n-1}\,dr\,d\sigma}{(1+r^2)^n}+\right.$$

$$\left.+\frac{(\Sigma\alpha_r^2)^{n/(n-2)}}{\Sigma\alpha_r^{2n/(n-2)}K(x_r)}\left(\int\partial^{2n/(n-2)}\right)^{2/(n-2)}V_i\right]$$

where

$$V_i = -\alpha_i^{\frac{n+2}{n-2}}c_0^{2n/(n-2)}\left\{K(x_i)\int\frac{r^{n-1}\,dr\,d\sigma}{(1+r^2)^n}+\frac{\Delta K(x_i)}{\lambda_i^2}\int\frac{|x|^2dx}{(1+|x|^2)^n}\right. \tag{4.6}$$

$$\left.+o_K\left(\frac{1}{\lambda_i^2}\right)-c_1\frac{K(x_i)}{\lambda_i^{n-2}}H(x_i,x_i)\right\}-\sum_{j\neq i}\alpha_j^{\frac{n+2}{n-2}}c_0^{2n/(n-2)}c_1\left\{K(x_i)\left(\varepsilon_{ij}-\right.\right.$$

$$\left.\left.-\frac{H(x_i,x_j)}{(\lambda_i\lambda_j)^{(n-2)/2}}\right)\right\}-\frac{n+2}{n-2}\sum_{j\neq i}c_0^{2n/(n-2)}c_1\alpha_i^{4/(n-2)}\alpha_jK(x_i)\varepsilon_{ij}+$$

$$+S_i+0\left(\Sigma\frac{1}{\lambda_k^{n-2}}+\sum_{k\neq j}\varepsilon_{kj}+\int|\nabla v|^2+\Sigma\frac{1}{\lambda_k^{n+2}}\frac{1}{d_k^{2(n+2)}}+\right.$$

$$\left.+\Sigma\frac{1}{\lambda_k^4\lambda_j^{n-2}}\frac{1}{d_j^{2(n-2)}}+O_K\left(\Sigma\frac{1}{\lambda_k^2}\right)\right)\times\left[O_K(1)+0\left(\sum_{i\neq j}\varepsilon_{ij}+\right.\right.$$

$$\left.\left.+0_\Omega\left[\frac{1}{\lambda_i^{n-2}}+\frac{1}{\lambda_i^{(n-2)/2}}\Sigma\frac{1}{\lambda_j^{(n-2)/2}}\right]+S_i\right]$$

and S_i is the remainder term in (3.247). 0_Ω appears when $\Omega\in S^n$. Then

$$S_i = 0\left(\sum_{i\neq j}\varepsilon_{ij}^{n/(n-2)}\operatorname{Log}\varepsilon_{ij}^{-1}+\sum_{i\neq j}\frac{1}{\lambda_i^{(n-2)/2}d_i^{n-2}}\frac{1}{\lambda_j^{(n-2)^2/2(n+2)}}\right. \tag{4.7}$$

$$\times\left(\varepsilon_{ij}^{4/(n+2)}+\frac{1}{(\lambda_i\lambda_j)^{2(n-2)/(n+2)}}+\sum_j\frac{1}{(\lambda_jd_j)^{(n+2)/2}}\frac{1}{(\lambda_id_i)^{(n-2)/2}}\right.$$

$$+\sum \frac{1}{d_i^n}\left(\frac{\operatorname{Log}\lambda_j}{\lambda_j^{(n+2)/2}\lambda_i^{(n-2)/2}} + \frac{1}{\lambda_j^{(n-2)/2}}\ \frac{1}{\lambda_i^{(n+2)/2}}\right) +$$

$$o_K(\varepsilon_{ij}) + O_K\left(\frac{1}{\lambda_i^{(n+2)/2}\lambda_j^{(n-2)/2}}\right) + (\text{if } n=3)\ 0(\sum_{i\neq j}\varepsilon_{ij}^2(\operatorname{Log}\varepsilon_{ij}^{-1})^{2/3})$$

$$+ 0_\varepsilon\left(\int_{|v|\geq\varepsilon\Sigma\alpha_j\delta_j}\delta_i|v|^{\frac{n+2}{n-2}}\right) + 0_{\varepsilon'}(\int_{|v|\leq\varepsilon\Sigma\alpha_j\delta_j}\delta_i^{4/(n-2)}v^2) + o_{\varepsilon'}(\varepsilon_{ij})$$

$$+ 0\left(\frac{1}{\lambda_i^{(n+2)/2}}\ \frac{1}{d_i^{(n+2)/2}}(\int_{\Omega^c}|v|^{2n/(n-2)})^{\frac{n-2}{2n}}\right.$$

$$\left.+\frac{1}{\lambda_i^{(n+2)/2}}\ \frac{1}{d_i^{n-2}}(\int|\nabla v|^2)^{1/2}\right) + O_K\left((\int|\nabla v|^2)^{1/2}(\frac{|DK(x_i)|}{\lambda_i} + \frac{1}{\lambda_i^2}\right)$$

with

$$0_{\varepsilon'}(\varepsilon_{ij}) = \int_{|v|\leq\varepsilon'\sum_{j\neq i}\alpha_j\delta_j}\delta_i|v|(\sum_{k\neq j}(\alpha_k\delta_k)^{4/(n-2)}$$

$$+ (\alpha_i\delta_i)^{(4/(n-2))-1}\operatorname{Inf}(\alpha_i\delta_i, \sum_{j\neq i}\alpha_j\delta_j))$$

Next, from (4.2), (4.4), (3.260) and (3.251), we deduce

$$\begin{array}{c} \\ \\ \\ i \end{array}\overset{\qquad\qquad j}{\begin{bmatrix} & & \\ & \lambda_i^2\frac{a_0\alpha_i}{n} + 0(\lambda_i^2(\int|\nabla v|^2)^{1/2}) & \\ --- & 0(\lambda_i\lambda_j\varepsilon_{ij}^{n/(n-2)}) & --------- \end{bmatrix}}\begin{bmatrix}\dot{x}_1\\ \cdot\\ \cdot\\ \cdot\\ \dot{x}_p\end{bmatrix} \qquad (4.8)$$

$$+\begin{bmatrix} & j & \\ & 0((\int|\nabla v|^2)^{1/2}) & \\ & & \\ i & 0(\lambda_i(\varepsilon_{ij}^{n/n-2)} \times |x_i-x_j|) & \end{bmatrix}\begin{bmatrix}\dot\lambda_1\\ \\ \\ \dot\lambda_p\end{bmatrix}$$

$$+\begin{bmatrix}0 & j & \\ & & \\ & & \\ i & 0(\frac{\partial\varepsilon_{ij}}{\partial x_i}+0(\varepsilon_{ij}^{\frac{n+1}{n-2}}\lambda_i\lambda_j|x_i-x_j|)) & 0\end{bmatrix}\begin{bmatrix}\dot\alpha_1\\ \\ \\ \dot\alpha_p\end{bmatrix}$$

$$= -\lambda(u)\left[\sum_{i\neq j}\alpha_j\frac{\partial\varepsilon_{ij}}{\partial x_i}c_0^{2n/(n-2)}c_1+\frac{(\Sigma\alpha_r^2)^{n/(n-2)}}{\Sigma\alpha_r^{2n/(n-2)}K(x_r)}(\int\delta^{2n/(n-2)})^{\frac{2}{n-2}}W_i\right]$$

where

$$W_i = -\alpha_i^{\frac{n+2}{n-2}}c_0^{\frac{2n}{n-2}}\frac{n-2}{2n}\{DK(x_i)\int\frac{r^{n-1}dr\,d\sigma}{(1+r^2)^n}+\frac{\Delta(DK)}{\lambda_i^2}(x_i)\int\frac{|x|^2dx}{(1+|x|^2)^n}$$

$$+o_K(\frac{1}{\lambda_i^2})+\frac{2n}{n-2}c_1\frac{K(x_i)}{\lambda_i^{n-2}}\frac{\partial H}{\partial y}(x_i,x_i)\}-\sum_{j\neq i}c_0^{2n/(n-2)}c_1\alpha_j^{\frac{n+2}{n-2}}K(x_j) \tag{4.9}$$

$$\times(\frac{\partial}{\partial x_i}\varepsilon_{ij}-\frac{\frac{\partial H}{\partial x}(x_i,x_j)}{(\lambda_i\lambda_j)^{(n-2)/2}})-\sum_{j\neq i}c_0^{2n/(n-2)}c_1\alpha_i^{4/(n-2)}\alpha_j\{DK(x_i)\varepsilon_{ij}$$

$$+K(x_i)\frac{\partial}{\partial x_i}\varepsilon_{ij}\}+\lambda_iS_i+0(\Sigma\frac{1}{\lambda_k^{n-2}}+\sum_{k\neq j}\varepsilon_{ij}+\int|\nabla v|^2$$

$$+\Sigma\frac{1}{\lambda_k^{n+2}}\frac{1}{d_k^{2(n+2)}}+\Sigma\frac{1}{\lambda_k^4\lambda_j^{n-2}}\frac{1}{d_j^{2(n-2)}}+O_K(\Sigma\frac{1}{\lambda_k^2}))(O_K(1)+$$

$$+ 0\left(\sum_{i\neq j} \varepsilon_{ij} + \sum_{i\neq j} \left|\frac{\partial \varepsilon_{ij}}{\partial x_i}\right| + O_\Omega\left[\Sigma \frac{1}{\lambda_i^{(n-2)/2}} \frac{1}{\lambda_j^{(n-2)/2}} + \frac{1}{\lambda_i^{n-2}}\right] + \lambda_i S_i\right)\right).$$

O_Ω appears when $\Omega \Subset S^n$.

Finally, from (3.265), (3.249), (4.3), (4.4), we deduce:

$$\begin{bmatrix} [\alpha_i \gamma_1 + 0((\int |\nabla v|^2)^{1/2})]\frac{1}{\lambda_i} & \\ 0(\frac{\varepsilon_{ij}}{\lambda_i}) & \end{bmatrix} \begin{bmatrix} \frac{\dot{\lambda}_i}{\lambda_i} \\ \vdots \end{bmatrix}_i + \tag{4.10}$$

(rows indexed by i, columns by j)

$$+ \begin{bmatrix} 0 & c_0^{\frac{2n}{n-2}} c_1 \frac{\partial \varepsilon_{ij}}{\partial \lambda_j} + 0\left(\frac{\varepsilon_{ij}^{\frac{n}{n-2}}}{\lambda_j} \operatorname{Log} \varepsilon_{ij}^{-1}\right) \\ & 0 \end{bmatrix} \begin{bmatrix} \dot{\alpha}_1 \\ \vdots \\ \dot{\alpha}_j \end{bmatrix}$$

$$+ \begin{bmatrix} 0((\int |\nabla v|^2)^{1/2}) \\ 0(\lambda_j [\varepsilon_{ij}^{\frac{n}{n-2}} |x_i - x_j|]) \end{bmatrix} \begin{bmatrix} \dot{x}_1 \\ \vdots \\ \dot{x}_p \end{bmatrix} =$$

$$-\lambda(u) \left[+ \sum_{i\neq j} c_1 c_0^{\frac{2n}{n-2}} \alpha_j \frac{\partial \varepsilon_{ij}}{\partial \lambda_i} + \frac{(\Sigma \alpha_r^2)^{n/(n-2)}}{\Sigma \alpha_r^{2n/(n-2)} K(x_r)} (\int \delta^{2n/(n-2)})^{2/(n-2)} T_i \right]$$

where

$$T_i = -\alpha_i^{\frac{n+2}{n-2}} c_0^{2n/(n-2)} \left\{ -\frac{\Delta K(x_i)}{\lambda_i^3} \frac{n-2}{n} \int \frac{|x|^2 dx}{(1+|x|^2)^n} + o_K\left(\frac{1}{\lambda_i^3}\right) \right. \tag{4.11}$$

$$\left. + \frac{n-2}{2} c_1 K(x_i) \frac{H(x_i, x_i)}{\lambda_i^{n-1}} \right\} - \sum_{j \neq i} \alpha_j^{\frac{n+2}{n-2}} c_0^{\frac{2n}{n-2}} c_1 \left\{ K(x_j) \left[\frac{\partial}{\partial \lambda_i} (\varepsilon_{ij}) \right. \right.$$

$$\left. \left. + \frac{n-2}{2} \frac{H(x_i, x_j)}{\lambda_i (\lambda_i \lambda_j)^{(n-2)/2}} \right] \right\} - \sum_{j \neq i} K(x_i) c_0^{\frac{2n}{n-2}} c_1 \alpha_i^{\frac{4}{n-2}} \alpha_j \frac{\partial}{\partial \lambda_i} \varepsilon_{ij} + \frac{1}{\lambda_i} S_i$$

$$+ 0 \left(\Sigma \frac{1}{\lambda_j^{n-2}} + \sum_{k \neq j} \varepsilon_{kj} + \int |\nabla v|^2 + \Sigma \frac{1}{\lambda_k^{n+2}} \frac{1}{d_k^{2(n+2)}} \right.$$

$$\left. + \Sigma \frac{1}{\lambda_k^4 \lambda_j^{n-2}} \frac{1}{d_j^{2(n-2)}} + O_K\left(\Sigma \frac{1}{\lambda_k^2}\right) \right) \left(O_K\left(\frac{1}{\lambda_i^3}\right) + \right.$$

$$+ O_\Omega \left(\sum_{i \neq j} \frac{1}{\lambda_i (\lambda_i \lambda_j)^{(n-2)/2}} + \frac{1}{\lambda_i^{n-1}} \right) + 0 \left(\Sigma \left| \frac{\partial}{\partial \lambda_i} \varepsilon_{ij} \right| + \frac{S_i}{\lambda_i} \right) \Bigg) .$$

O_Ω appears when $\Omega \in S^n$.

4.2 The matricial form: inversion of the matrices

Thus, we have the following equation,

$$
\begin{array}{c}
\\ i \\ \\ \\ p+i \\ \\ 2p+i \\ \\
\end{array}
\begin{array}{c}
\begin{array}{ccc} j & p+j & 2p+j \end{array} \\
\left[\begin{array}{ccc}
D & E & F \\
 & & \\
 & & \\
G & H & I \\
 & & \\
J & K & L \\
 & &
\end{array}\right]
\end{array}
\left[\begin{array}{c}
\dot{\alpha}_1 \\ \vdots \\ \dot{\alpha}_p \\ \lambda_1 \dot{x}_1 \\ \vdots \\ \lambda_p \dot{x}_p \\ \dot{\lambda}_1/\lambda_1 \\ \lambda_p/\lambda_p
\end{array}\right] = \tag{4.12}
$$

$$
=-\lambda(u)
\begin{array}{c} i \\ \\ p+i \\ \\ 2p+1 \end{array}
\left[\begin{array}{l}
\sum_{j\neq i} \alpha_j \varepsilon_{ij} c_1 c_0^{\frac{2n}{n-2}} + \alpha_i c_0^{\frac{2n}{n-2}} \int \frac{r^{n-1} dr\, d\sigma}{(1+r^2)^n} + \frac{(\Sigma\alpha_r^2)^{n/(n-2)}}{\Sigma\alpha_r^{2n/(n-2)} K(x_r)} \left(\int \delta^{\frac{2n}{n-2}}\right)^{\frac{2}{n-2}} V_i \\
\\
\sum_{j\neq i} \alpha_j \frac{\partial \varepsilon_{ij}}{\partial x_i} c_0^{\frac{2n}{n-2}} c_1 \qquad + \frac{(\Sigma\alpha_r^2)^{n/(n-2)}}{\Sigma\alpha_r^{2n/(n-2)} K(x_r)} \left(\int \delta^{\frac{2n}{n-2}}\right)^{\frac{2}{n-2}} W_i \\
\\
\sum_{j\neq i} \alpha_j \frac{\partial \varepsilon_{ij}}{\partial x_i} c_0^{\frac{2n}{n-2}} c_1 \qquad + \frac{(\Sigma\alpha_r^2)^{n/(n-2)}}{\Sigma\alpha_r^{2n/(n-2)} K(x_r)} \left(\int \delta^{\frac{2n}{n-2}}\right)^{\frac{2}{n-2}} T_i
\end{array}\right]
$$

where

$$
D = i\left[\begin{array}{ccc}
C & & \\
 & c_0^{\frac{2n}{n-2}} c_1 \left[\varepsilon_{ij} + 0\left(\varepsilon_{ij}^{\frac{n}{n-2}}\right)\right] & \\
 & & C
\end{array}\right]
$$

and

$$E = \begin{bmatrix} 0 & & \\ & c_0^{\frac{2n}{n-2}} c_1 \frac{\alpha_j}{\lambda_j} \left[\frac{\partial \varepsilon_{ij}}{\partial x_j} + 0(\varepsilon_{ij}^{\frac{n+1}{n-2}} \lambda_i \lambda_j |x_i - x_j|) \right] & \\ & & 0 \end{bmatrix}$$

$$F = \begin{bmatrix} 0 & & \\ & c_0^{\frac{2n}{n+2}} c_1 \alpha_j \left[\lambda_j \frac{\partial \varepsilon_{ij}}{\partial \lambda_j} + 0(\varepsilon_{ij}^{\frac{n}{n-2}} \operatorname{Log} \varepsilon_{ij}^{-1}) \right] & \\ & & 0 \end{bmatrix}$$

$$G = \begin{bmatrix} 0 & & \\ & 0\left(\frac{\partial \varepsilon_{ij}}{\partial x_i} + 0(\varepsilon^{\frac{n+1}{n-2}} \lambda_i \lambda_j |x_i - x_j|)\right) & \\ & & 0 \end{bmatrix}$$

$$H = \begin{bmatrix} \lambda_1 \frac{a_0 \alpha_1}{n} + 0(\lambda_1 (\int |\nabla v|^2)^{1/2}) & & 0(\lambda_i \varepsilon_{ij}^{\frac{n}{n+2}}) \\ & \ddots & \\ & & \lambda_p \frac{a_0 \alpha_p}{n} + 0(\lambda_p (\int |\nabla v|^2)^{1/2}) \end{bmatrix}$$

$$I = \begin{bmatrix} \lambda_1 0((\int |\nabla v|^2)^{1/2}) & & 0(\lambda_i \lambda_j \varepsilon_{ij}^{\frac{n}{n-2}} |x_i - x_j|) \\ & \ddots & \\ & & \lambda_p 0((\int |\nabla v|^2)^{1/2}) \end{bmatrix}$$

$$J = \begin{bmatrix} 0 & & \\ & c_0^{\frac{2n}{n-2}} c_1 \frac{\partial \varepsilon_{ij}}{\partial \lambda_i} + \frac{1}{\lambda_i} 0(\varepsilon_{ij}^{\frac{n}{n-2}} \operatorname{Log} \varepsilon_{ij}^{-1}) & \\ & & 0 \end{bmatrix}$$

$$K = \begin{bmatrix} \frac{1}{\lambda_1} 0((\int|\nabla v|^2)^{1/2}) & & 0\left(\varepsilon_{ij}^{\frac{n}{n-2}} |x_i - x_j|\right) \\ & \ddots & \\ & & \frac{1}{\lambda_p} 0((\int|\nabla v|^2)^{1/2}) \end{bmatrix}$$

$$L = \begin{bmatrix} [\alpha_1\gamma_1 + 0((\int|\nabla v|^2)^{1/2})]\frac{1}{\lambda_1} & & 0(\frac{\varepsilon_{ij}}{\lambda_i}) \\ & \ddots & \\ & & [\alpha_p\gamma_1 + 0((\int|\nabla v|^2)^{1/2})]\frac{1}{\lambda_p} \end{bmatrix}$$

Let A be the following diagonal matrix,

$$A = \begin{bmatrix} C & & & & & & 0 \\ & C & & & & & \\ & & \lambda_1\left[\frac{a_0\alpha_1}{n} + (\int|\nabla v|^2)^{1/2}\right] & & & & \\ & & & \lambda_p\left[\frac{a_0\alpha_p}{n} + (\int|\nabla v|^2)^{1/2}\right] & & & \\ & & & & [\alpha_1\gamma_1 + 0((\int|\nabla v|^2)^{1/2})]\frac{1}{\lambda_1} & & \\ 0 & & & & & [\alpha_p\gamma_1 + 0((\int|\nabla v|^2)^{1/2})]\frac{1}{\lambda_p} \end{bmatrix} \quad (4.13)$$

Let

$$A' \text{ be the left-hand side matrix in } (4.12) \tag{4.13}$$

and

$$B = A' - A \tag{4.15}$$

We compute the matrix

$$C = A^{-1}B . \tag{4.16}$$

$$C = \begin{array}{c} \\ \\ i \\ \\ \\ p{+}i \\ \\ \\ 2p{+}i \\ \\ \end{array}\left[\begin{array}{cccccc}
 & j & & p{+}j & & 2p{+}j \\
0 & & 0 & & 0 & \\
 & D & & E & & F \\
 & 0 & & 0 & & 0 \\
0 & & 0 & & O((\int|\nabla v|^2)^{1/2}) & \\
 & G & & H & & I \\
 & 0 & & 0 & & O((\int|\nabla v|^2)^{1/2}) \\
0 & & O((\int|\nabla v|^2)^{1/2}) & & 0 & \\
 & J & & K & & L \\
 & 0 & & O((\int|\nabla v|^2)^{1/2}) & & 0
\end{array}\right] \tag{4.17}$$

where $D = c_0^{\frac{2n}{n-2}} c_1 \left[\varepsilon_{ij} + O(\varepsilon_{ij}^{\frac{n}{n-2}})\right]$; $E = c_0^{\frac{2n}{n-2}} c_1 \frac{\alpha_j}{c\lambda_j}\left[\frac{\partial\varepsilon_{ij}}{\partial x_j} + O\left(\varepsilon_{ij}^{\frac{n+2}{n-2}}\lambda_i\lambda_j|x_i - x_j|\right)\right]$; $F = c_0^{\frac{2n}{n-2}} c_1 \frac{\alpha_j}{c}\left[\lambda_j \frac{\partial\varepsilon_{ij}}{\partial\lambda_j} + O\left(\varepsilon_{ij}^{\frac{n}{n-2}} \operatorname{Log} \varepsilon_{ij}^{-1}\right)\right]$; $G = O\left(\frac{1}{\lambda_i}\frac{\partial\varepsilon_{ij}}{\partial x_i} + O\left(\varepsilon_{ij}^{\frac{n+1}{n-2}}\lambda_j|x_i - x_j|\right)\right)$; $H = O(\varepsilon_{ij}^{\frac{n}{n-2}})$; $I = O(\lambda_j\varepsilon_{ij}^{\frac{n}{n-2}}|x_i - x_j|)$; $J = c_0^{\frac{2n}{n-2}} c_1\lambda_i \frac{\partial\varepsilon_{ij}}{\partial\lambda_i} + O\left(\varepsilon_{ij}^{\frac{n}{n-2}} \operatorname{Log} \varepsilon_{ij}^{-1}\right)$; $K = O\left(\lambda_i\varepsilon_{ij}^{\frac{n}{n-2}}|x_i - x_j|\right)$; $L = O(\varepsilon_{ij})$.

Now let

$$C_2 = \left[\begin{array}{c|c|c} 0 & 0 & 0 \\ \hline 0 & 0 & \begin{matrix} 0(|v|) & & 0 \\ & \ddots & \\ 0 & & 0(|v|) \end{matrix} \\ \hline 0 & \begin{matrix} 0(|v|) & & 0 \\ & \ddots & \\ 0 & & 0(|v|) \end{matrix} & 0 \end{array}\right] \tag{4.18}$$

where

$$|v| = \left(\int |\nabla v|^2\right)^{1/2} \tag{4.19}$$

and

$$C_1 = C - C_2. \tag{4.20}$$

We first compute $(\mathrm{Id} + C_1)^{-1}$. We have

$$(\mathrm{Id} + C_1)^{-1} = \mathrm{Id} - C_1 + C_1^2 + \ldots + (-1)^j C_1^j + \ldots \tag{4.21}$$

The matrix C_1 is:

$$
C_1 = \begin{array}{c|ccc|ccc|ccc|}
 & & j & & & p+j & & & 2p+j & \\
 & 0 & & & 0 & & & 0 & & \\
i & & 0(\varepsilon_{ij}) & & & 0\left(\frac{1}{\lambda_j}\frac{\partial \varepsilon_{ij}}{\partial x_j}\right) & & & 0\left(\lambda_j \frac{\partial \varepsilon_{ij}}{\partial \lambda_j}\right) & \\
 & & & 0 & & & 0 & & & 0 \\
\hline
 & 0 & & & 0 & & & 0 & & \\
p+i & & 0\left(\frac{1}{\lambda_i}\frac{\partial \varepsilon_{ij}}{\partial x_i}\right) & & & 0\left(\varepsilon_{ij}^{\frac{n}{n-2}}\right) & & & 0\left(\frac{1}{\lambda_i}\frac{\partial \varepsilon_{ij}}{\partial x_i}\right) & \\
 & & & 0 & & & 0 & & & 0 \\
\hline
 & 0 & & & 0 & & & 0 & & \\
2p+i & & 0\left(\lambda_i \frac{\partial \varepsilon_{ij}}{\partial \lambda_j}\right) & & & 0\left(\frac{1}{\lambda_j}\frac{\partial \varepsilon_{ij}}{\partial x_j}\right) & & & 0(\varepsilon_{ij}) & \\
 & & & 0 & & & 0 & & & 0
\end{array} \tag{4.22}
$$

We have:

$$\frac{1}{\lambda_j}\frac{\partial \varepsilon_{ij}}{\partial x_j} = 0(\varepsilon_{ij}) \tag{4.23}$$

$$\lambda_j \frac{\partial \varepsilon_{ij}}{\partial \lambda_j} = 0(\varepsilon_{ij}) \tag{4.24}$$

Furthermore:

$$\varepsilon_{ik}\varepsilon_{kj} \leq C\varepsilon_{ij} \tag{4.25}$$

Thus, denoting:

$$a_{ij} = \text{the coefficient of } C_1 \text{ at column } j \text{ and line } i, \tag{4.26}$$

we have:

$$|a_{ij}| \leq c\varepsilon_{ij},\ i-j \not\equiv 0 \ (\mathrm{mod}\ p) \ (\varepsilon_{ij} = \varepsilon_{i,j+p} \text{ by convention}). \tag{4.27}$$

Some thought shows then that if:

$$(a_{ij})_p = \text{the coefficient of } C_1^p \text{ at column } j \text{ and line } i, \tag{4.28}$$

then:

$$(a_{ij})_p = 0(\varepsilon_{ij}), \quad i-j \not\equiv 0 \pmod p \tag{4.29}$$

In fact, we may upperbound $(a_{ij})_p$ as follows:

$$|(a_{ij})_p| \leq c^p[\varepsilon_{ii_1}\varepsilon_{i_1 i_2} \dots \varepsilon_{i_{p-1}j}], \text{ with } i_1 \neq i; \tag{4.30}$$

$i_2 \neq i_1, \dots, i_{p-1} \neq j.$

From the fact that the indices i_k are distinct and run into a finite subset of $\mathbb{N}$, namely $[1, \dots, p]$ and also from (4.25), we derive:

$$\Big|\sum_{p\geq 2} (a_{ij})_p\Big| = 0(\varepsilon_{ij}) \leq c[\varepsilon_{ij}^2 + \operatorname*{Sup}_{\substack{k\neq i \\ k\neq j}} (\varepsilon_{ik}\varepsilon_{kj})] \text{ if } i-j \not\equiv 0 \pmod p \tag{4.31}$$

while

$$\Big|\sum_{p\geq 2} (a_{ij})_p\Big| \leq c \operatorname*{Sup}_{k\neq i} (\varepsilon_{ik}^2) \qquad \text{if } i-j \equiv 0 \pmod p \tag{4.32}$$

We thus have:

$$(Id + C_1)^{-1} = Id - C_1 + C_1 . D \tag{4.33}$$

$$D = \begin{bmatrix} \begin{matrix} & 0(\varepsilon_{ij}) \\ 0(\operatorname*{Sup}_{k\neq i} \varepsilon_{ik}^2) & \end{matrix} & \begin{matrix} & 0(\varepsilon_{ij}) \\ 0(\operatorname*{Sup}_{k\neq i} \varepsilon_{ik}^2) & \end{matrix} & \begin{matrix} & 0(\varepsilon_{ij}) \\ 0(\operatorname*{Sup}_{k\neq i} \varepsilon_{ik}^2) & \end{matrix} \\ \hline \begin{matrix} & 0(\varepsilon_{ij}) \\ 0(\operatorname*{Sup}_{k\neq i} \varepsilon_{ik}^2) & \end{matrix} & \begin{matrix} & 0(\varepsilon_{ij}) \\ 0(\operatorname*{Sup}_{k\neq i} \varepsilon_{ik}^2) & \end{matrix} & \begin{matrix} & 0(\varepsilon_{ij}) \\ 0(\operatorname*{Sup}_{k\neq i} \varepsilon_{ik}^2) & \end{matrix} \\ \hline \begin{matrix} & 0(\varepsilon_{ij}) \\ 0(\operatorname*{Sup}_{k\neq i} \varepsilon_{ik}^2) & \end{matrix} & \begin{matrix} & 0(\varepsilon_{ij}) \\ 0(\operatorname*{Sup}_{k\neq i} \varepsilon_{ik}^2) & \end{matrix} & \begin{matrix} & 0(\varepsilon_{ij}) \\ 0(\operatorname*{Sup}_{k\neq i} \varepsilon_{ik}^2) & \end{matrix} \end{bmatrix}$$

Now:

$$(\mathrm{Id} + C)^{-1} = (\mathrm{Id} + C_1 + C_2)^{-1} \tag{4.34}$$

Observe that:

$C_1^2 =$

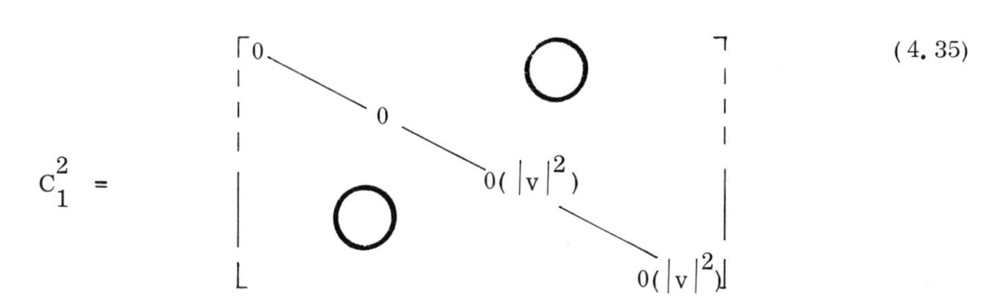

(4.35)

while

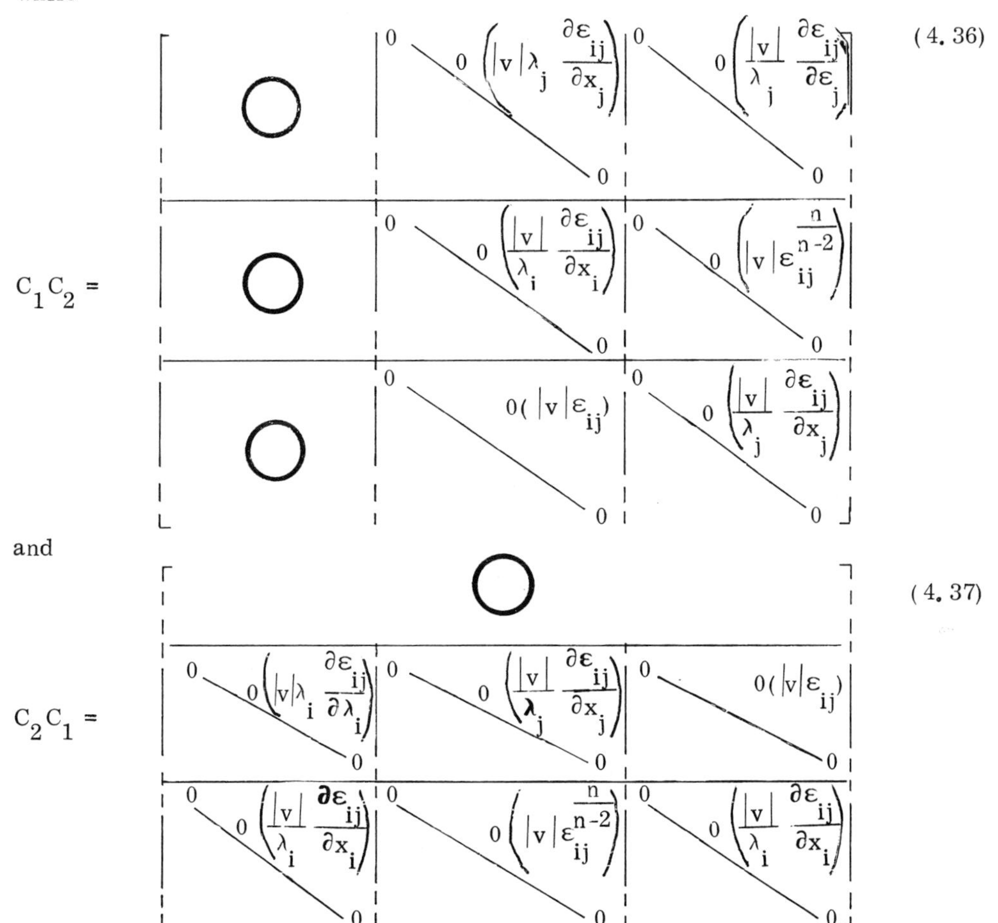

(4.36) (4.37)

In $(\mathrm{Id} + C_1 + C_2)^{-1}$, there is a part which is built up with sums of monomials involving C_1C_2 and C_2C_1. This part follows the same rules for the resulting coefficients, after the multiplication of the matrices and the sum taken on all coefficients at line i and column j, then the $(a_{ij})_p$ in $(\mathrm{Id} + C_1)^{-1}$. The only modification is here the multiplication of each coefficient of a single matrix (C_1C_2 or C_2C_1) by $|v|$.

Thus, we expect inequalities of the type (4.30), (4.31) and (4.32). Calling again:

$$(a_{ij})_q \text{ the coefficient at column j and line i of such a sum,} \tag{4.38}$$

we have

$$|(a_{ij})_q| \leq c^q |v| [\varepsilon_{ii_1} \ldots \varepsilon_{i_{p-1}j}], \quad i_1 \neq i, \ldots, i_{p-1} \neq j, \tag{4.39}$$

$$= O(\varepsilon_{ij} |v|), \qquad \text{if } i-j \not\equiv 0 \pmod p$$

$$|(a_{ij})_q| \leq c |v|^2 \operatorname*{Sup}_{k \neq i} \varepsilon_{ik}^2 \qquad \text{if } i-j \equiv 0 \pmod p. \tag{4.40}$$

We thus have:

$$(\mathrm{Id}+C_1+C_2)^{-1} = (\mathrm{Id}+C_1)^{-1} - C_2 - C_1C_2 - C_2C_1 - C_2^2 - (C_1+C_2)\, D_1 \tag{4.41}$$

$$D_1 = \left|\begin{array}{c|c|c}
\begin{matrix} \ddots & O(|v|\varepsilon_{ij}) \\ O(\operatorname*{Sup}_{i\neq k}|v|^2\varepsilon_{ik}^2) & \ddots \end{matrix} &
\begin{matrix} \ddots & O(|v|\varepsilon_{ij}) \\ O(\operatorname*{Sup}_{i\neq k}|v|^2\varepsilon_{ik}^2) & \ddots \end{matrix} &
\begin{matrix} \ddots & O(|v|\varepsilon_{ij}) \\ O(\operatorname*{Sup}_{i\neq k}|v|^2\varepsilon_{ik}^2) & \ddots \end{matrix} \\ \hline
\begin{matrix} \ddots & O(|v|\varepsilon_{ij}) \\ O(\operatorname*{Sup}_{i\neq k}|v|^2\varepsilon_{ik}^2) & \ddots \end{matrix} &
\begin{matrix} \ddots & O(|v|\varepsilon_{ij}) \\ O(\operatorname*{Sup}_{i\neq k}|v|^2\varepsilon_{ik}^2) & \ddots \end{matrix} &
\begin{matrix} \ddots & O(|v|\varepsilon_{ij}) \\ O(\operatorname*{Sup}_{i\neq k}|v|^2\varepsilon_{ik}^2) & \ddots \end{matrix} \\ \hline
\begin{matrix} \ddots & O(|v|\varepsilon_{ij}) \\ O(\operatorname*{Sup}_{i\neq k}|v|^2\varepsilon_{ik}^2) & \ddots \end{matrix} &
\begin{matrix} \ddots & O(|v|\varepsilon_{ij}) \\ O(\operatorname*{Sup}_{i\neq k}|v|^2\varepsilon_{ik}^2) & \ddots \end{matrix} &
\begin{matrix} \ddots & O(|v|\varepsilon_{ij}) \\ O(\operatorname*{Sup}_{i\neq k}|v|^2\varepsilon_{ik}^2) & \ddots \end{matrix}
\end{array}\right|$$

Using (4.33) and the fact that $|v|$ is small, this yields:

$$(Id+C_1+C_2)^{-1} = Id-C_1-C_2-C_2^2-C_1C_2-C_2C_1-C_1E + F \qquad (4.42)$$

where

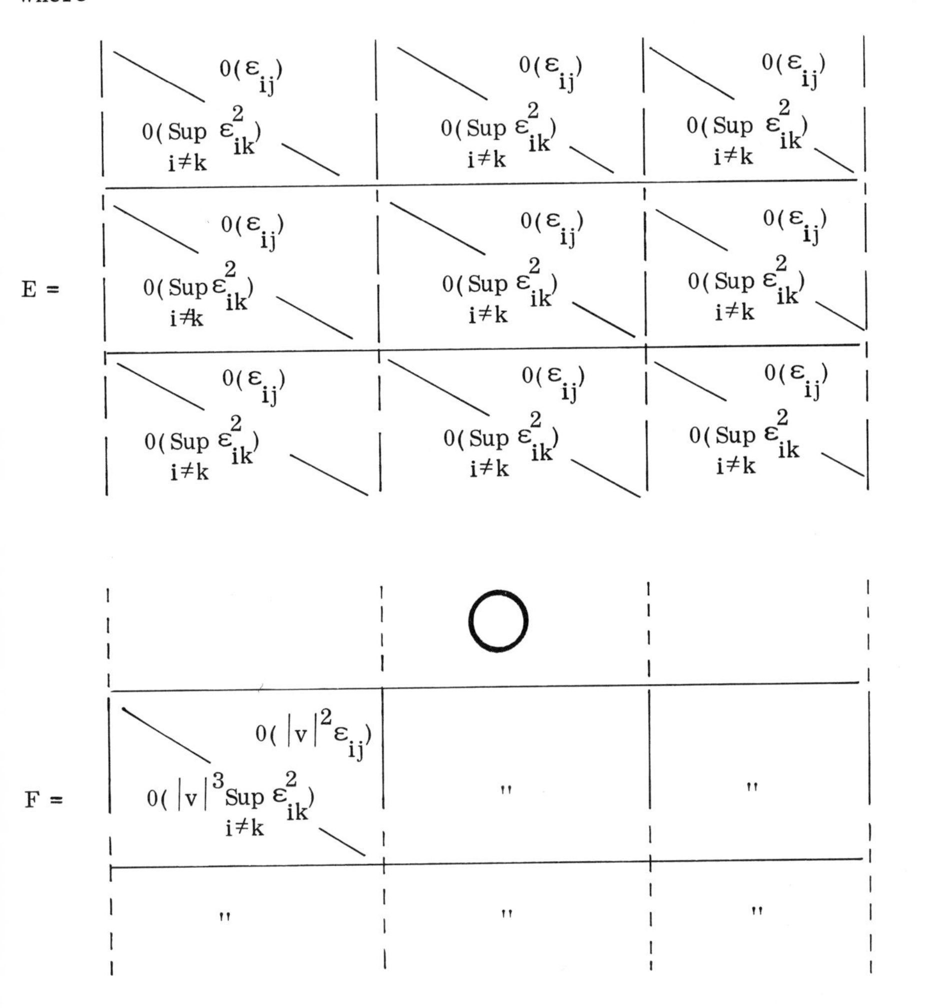

(where " means that we repeat the same matrix). We introduce the matrix

$0(\Sigma\varepsilon_{ik}\varepsilon_{kj} + \sup_{j\neq k}\varepsilon_{jk}^2\varepsilon_{ij})$ (above the diagonal); $0(\sup_{i\neq k}\varepsilon_{ik}^2)$ (below the diagonal)	same matrix as on the left	same matrix as on the left	(4.43)
Γ_{ij} (above the diagonal); $\Sigma 0(\varepsilon_{ik}\frac{1}{\lambda_i}\frac{\partial\varepsilon_{ik}}{\partial x_i} + 0\left(\varepsilon_{ik}^{1+\frac{n}{n-2}}\right)$ (below the diagonal)	same matrix as on the left	same matrix as on the left	
same matrix as above	same matrix as on the left	same matrix as on the left	

where $\Gamma_{ij} = \Sigma\, 0\left(\frac{1}{\lambda_i}\frac{\partial\varepsilon_{ik}}{\partial x_i}\varepsilon_{jk}\right) + 0\left(\varepsilon_{ik}^{\frac{n}{n-2}}\varepsilon_{jk}\right) + \sup_{j\neq k}\varepsilon_{jk}^2\left[0\left(\frac{1}{\lambda_i}\frac{\partial\varepsilon_{ij}}{\partial x_i}\right) + 0\left(\varepsilon_{ij}^{\frac{n}{n-2}}\right)\right]$. This matrix is the result of the product of C_1 by E of (4.42). Thus, to compute $(Id+C_1+C_2)^{-1}$, we have to add the matrices defined in (4.18), (4.22), (4.35), (4.36), (4.37), (4.43) and the last matrix F in (4.42). Adding (4.18) and (4.35) yields:

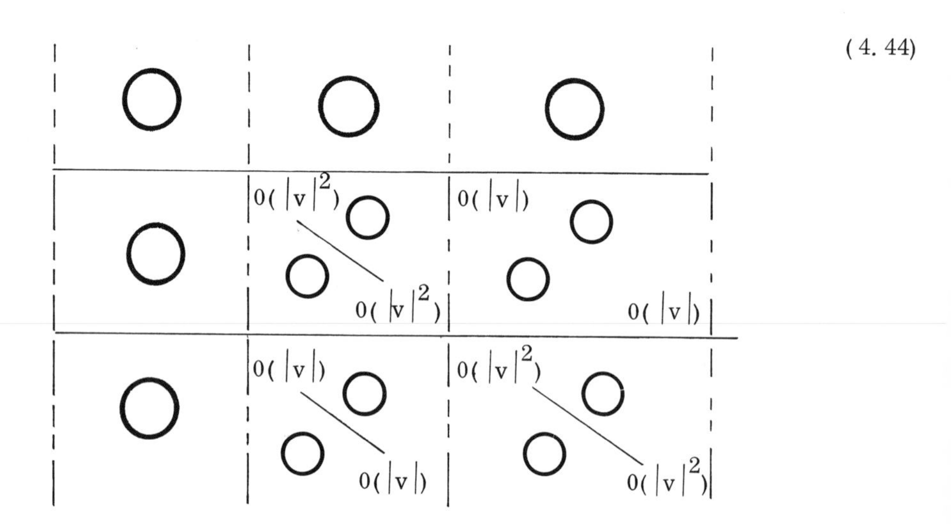

Adding (4.36) and (4.37) yields:

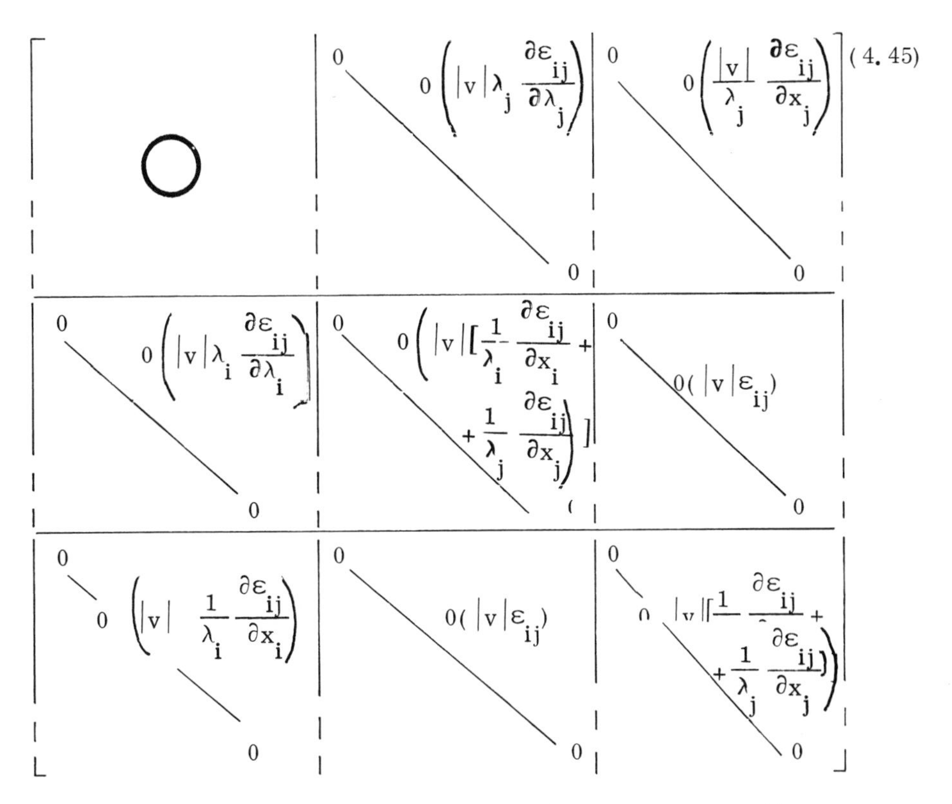

Adding (4.44), (4.45) and the last matrix in (4.42) (matrix F) yields:

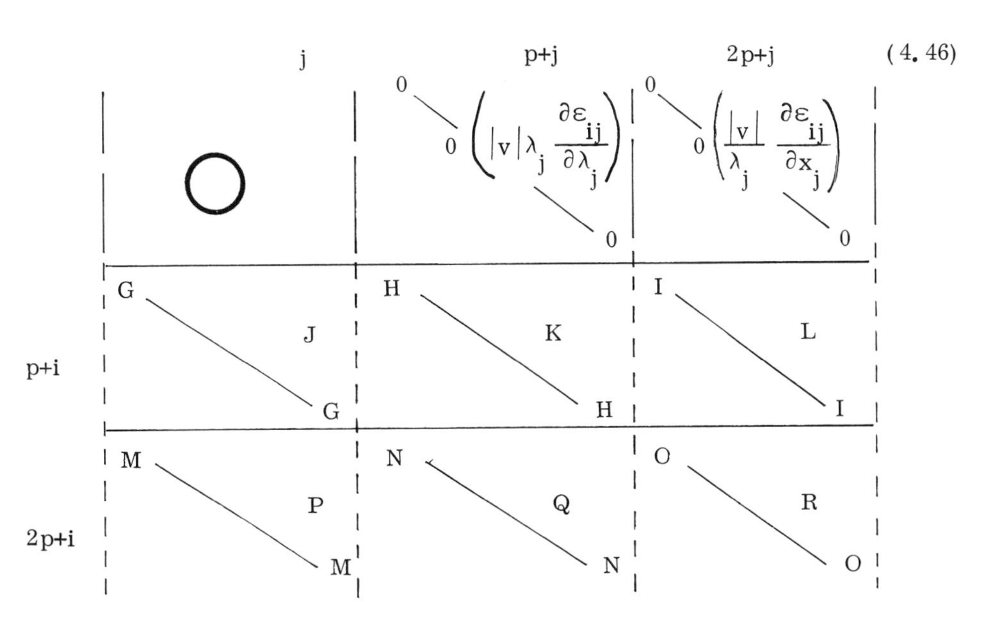

where $G = 0(|v|^2 \operatorname{Sup}_{i\neq k} \varepsilon_{ik}^2)$, $H = 0\left(|v|^2(1+0(\operatorname{Sup}_{i\neq k} \varepsilon_{ik}^2)\right)$, $I = 0\left(|v|(1+0|v|\operatorname{Sup}_{i\neq k} \varepsilon_{ik}^2)\right)$,

$J = 0(|v|^2\varepsilon_{ij} + |v|\lambda_i \frac{\partial \varepsilon_{ij}}{\partial \lambda_i})$, $K = 0(|v|^2\varepsilon_{ij} + |v|[\frac{1}{\lambda_j}\frac{\partial \varepsilon_{ij}}{\partial x_j} + \frac{1}{\lambda_i}\frac{\partial \varepsilon_{ij}}{\partial x_i}])$,

$L = 0(|v|\varepsilon_{ij})$, $M = 0(|v|^2 \operatorname{Sup}_{i\neq k} \varepsilon_{ik}^2)$, $N = 0\left(|v|(1+0|v|\operatorname{Sup}_{i\neq k} \varepsilon_{ik}^2)\right)$, $O = 0\left(|v|^2(1+0(\operatorname{Sup}_{i\neq k} \varepsilon_{ik}^2)\right)$

$P = 0\left(|v|^2\varepsilon_{ij} + |v|\frac{1}{\lambda_i}\frac{\partial \varepsilon_{ij}}{\partial x_i}\right)$, $Q = 0(|v|\varepsilon_{ij})$, $R = 0\left(|v|^2\varepsilon_{ij} + |v|\left[\frac{1}{\lambda_j}\frac{\partial \varepsilon_{ij}}{\partial x_j} + \frac{1}{\lambda_i}\frac{\partial \varepsilon_{ij}}{\partial x_i}\right]\right)$

Adding (4.22) and (4.43) yields:

(4.47)

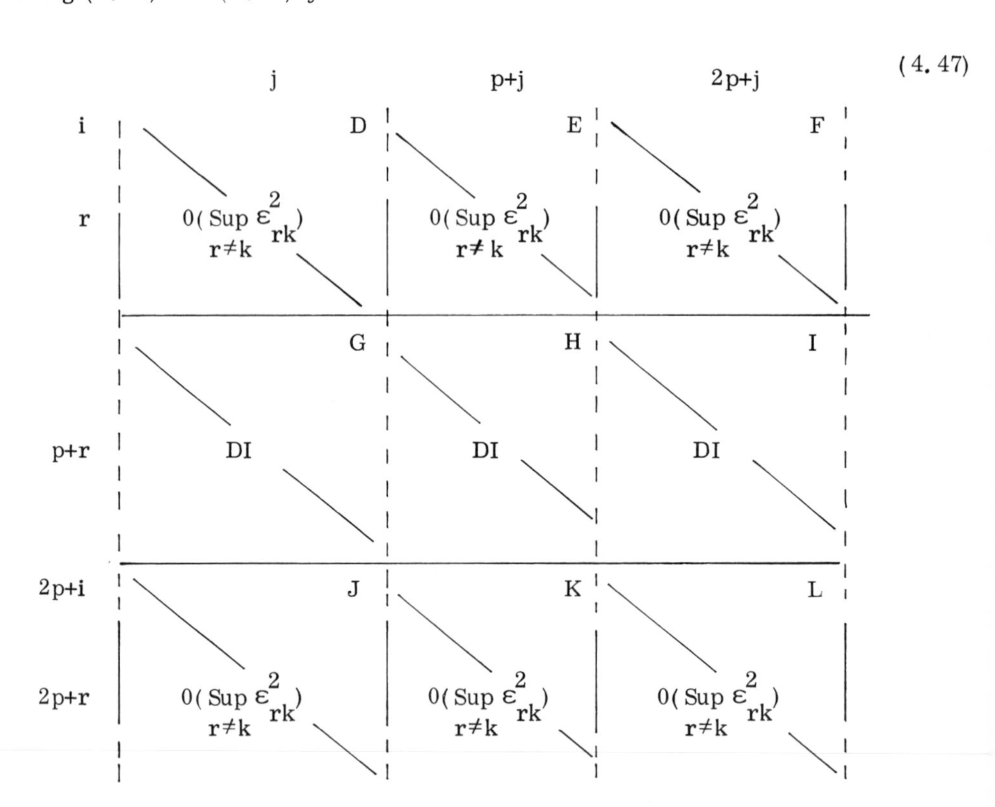

where $D = 0(\varepsilon_{ij})$; $E = 0\left(\Sigma \varepsilon_{ik}\varepsilon_{kj} + \operatorname{Sup}_{j\neq k} \varepsilon^2_{jk}\varepsilon_{ij} + \frac{1}{\lambda_j}\frac{\partial \varepsilon_{ij}}{\partial x_j}\right)$; $F = 0\left(\Sigma \varepsilon_{ik}\varepsilon_{kj} + \operatorname{Sup}_{j\neq k} \varepsilon^2_{jk}\varepsilon_{ij} + \lambda_j \frac{\partial \varepsilon_{ij}}{\partial \lambda_j}\right)$; $G = 0\left(\frac{1}{\lambda_i}\frac{\partial \varepsilon_{ij}}{\partial x_i}\right) + \operatorname{Sup}_{j\neq k} \varepsilon^2_{jk} 0 \quad \varepsilon_{ij}^{\frac{n}{n-2}} + \Sigma\, 0\left(\frac{1}{\lambda_i}\frac{\partial \varepsilon_{ik}}{\partial x_i}\varepsilon_{jk}\right)$; $H = 0\left(\varepsilon_{ij}^{\frac{n}{n-2}}\right) + \operatorname{Sup}_{j\neq k} \varepsilon^2_{jk} 0\left(\frac{1}{\lambda_i} \times \frac{\partial \varepsilon_{ij}}{\partial x_j}\right) + \Sigma\, 0\left(\frac{1}{\lambda_i}\frac{\partial \varepsilon_{ik}}{\partial x_i}\varepsilon_{jk} + 0\left(\varepsilon_{ik}^{\frac{n}{n-2}}\varepsilon_{jk}\right)\right)$; $I = 0\left(\frac{1}{\lambda_i}\frac{\partial \varepsilon_{ij}}{\partial x_i}\right) + \operatorname{Sup}_{j\neq k} \varepsilon^2_{jk} 0\left(\varepsilon_{ij}^{\frac{n}{n-2}}\right) + \Sigma\, 0\left(\frac{1}{\lambda_i}\frac{\partial \varepsilon_{ik}}{\partial x_i}\varepsilon_{jk}\right) + 0\left(\varepsilon_{ik}^{\frac{n}{n-2}}\varepsilon_{jk}\right)$; $DI = \Sigma 0\left(\varepsilon_{rk}\frac{1}{\lambda_r}\frac{\partial \varepsilon_{rk}}{\partial x_r}\right) + 0\left(\varepsilon_{rk}^{1+\frac{n}{n-2}}\right)$; $J = 0\left(\Sigma \varepsilon_{ik}\varepsilon_{kj} + \operatorname{Sup}_{j\neq k} \varepsilon^2_{jk}\varepsilon_{ij}\ \lambda_i \frac{\partial \varepsilon_{ij}}{\partial \lambda_i}\right)$; $K = 0\left(\Sigma\ \varepsilon_{ik}\varepsilon_{kj} + \operatorname{Sup}_{j\neq k} \varepsilon^2_{jk}\varepsilon_{ij} + \frac{1}{\lambda_j}\frac{\partial \varepsilon_{ij}}{\partial x_j}\right)$; $L = 0(\varepsilon_{ij})$.

Thus $(Id+C_1+C_2)^{-1}$ is the sum of the matrices defined in (4.46) and (4.47) and of the identity matrix. However, our aim was to compute A'^{-1}; and we have:

$$A' = A + B = A(Id + A^{-1}B) = A(Id + C) = A(Id + C_1 + C_2)\ . \tag{4.48}$$

Thus

$$A'^{-1} = (Id + C_1 + C_2)^{-1}A^{-1} \tag{4.49}$$

Now

$$A^{-1} = \begin{bmatrix} 1/C & & & & & & & \\ & \ddots & & & & & & \\ & & 1/C & & & & & \\ & & & \frac{1}{\lambda_1}\left[\frac{n}{a_0\alpha_1} + 0(|v|)\right] & & & & \\ & & & & \ddots & & & \\ & & & & & \frac{1}{\lambda_p}\left[\frac{n}{a_0\alpha_p} + 0(|v|)\right] & & \\ & & & & & & \lambda_1\left[\frac{1}{\alpha_1\gamma_1} + 0(|v|)\right] & \\ & & & & & & & \ddots \\ & & & & & & & & \lambda_p\left[\frac{1}{\alpha_p\gamma_1} + 0(|v|)\right] \end{bmatrix} \tag{4.50}$$

Thus A'^{-1} is the sum of A^{-1} and the two following matrices:

$$\begin{array}{c|ccc|ccc|ccc|} & & & j & & & p+j & & & 2p+j \\ \hline i & & & D & & & E & & & F \\ r & & 0(\operatorname{Sup}_{r\neq k} \varepsilon_{rk}^2) & & 0 & \frac{1}{\lambda_r}\operatorname{Sup}_{r\neq k} \varepsilon_{rk}^2 & & & 0(\lambda_r \operatorname{Sup}_{r\neq k} \varepsilon_{rk}^2) & \\ \hline p+i & & & G & & & H & & & I \\ p+r & & DI1 & & & DI2 & & & DI3 & \\ \hline 2p+i & & & J & & & K & & & L \\ 2p+r & & 0(\operatorname{Sup}_{r\neq k} \varepsilon_{rk}^2) & & 0 & \frac{1}{\lambda_r}\operatorname{Sup}_{r\neq k} \varepsilon_{rk}^2 & & & 0(\lambda_r \operatorname{Sup}_{r\neq k} \varepsilon_{rk}^2) & \\ \hline \end{array} \tag{4.51}$$

where $D = 0(\varepsilon_{ij})$; $E = \frac{1}{\lambda_j}\, 0\left(\Sigma\varepsilon_{ik}\varepsilon_{kj} + \operatorname{Sup}_{j\neq k} \varepsilon_{jk}^2 \varepsilon_{ij} + \frac{1}{\lambda_j}\frac{\partial\varepsilon_{ij}}{\partial x_j}\right)$;

$$F = 0\left(\lambda_j\Big(\Sigma\, \varepsilon_{ik}\varepsilon_{kj} + \operatorname*{Sup}_{j\neq k} \varepsilon_{jk}^2 \varepsilon_{ij} + \lambda_j \frac{\partial \varepsilon_{ij}}{\partial \lambda_j}\right) \; ; \; G = 0\left(\frac{1}{\lambda_i}\,\frac{\partial \varepsilon_{ij}}{\partial x_i}\right) + \operatorname*{Sup}_{j\neq k} \varepsilon_{jk}^2\, 0\left(\varepsilon_{ij}^{\frac{n}{n-2}}\right)$$

$$+ \Sigma\; 0\left(\frac{1}{\lambda_i}\,\frac{\partial \varepsilon_{ik}}{\partial x_i}\,\varepsilon_{jk}\right) \; ; \; H = \frac{1}{\lambda_j}\left[\,0\left(\varepsilon_{ij}^{\frac{n}{n-2}}\right) + \operatorname*{Sup}_{j\neq k} \varepsilon_{jk}^2\, 0\left(\frac{1}{\lambda_i}\,\frac{\partial \varepsilon_{ij}}{\partial x_i}\right) + \right.$$

$$\left. \Sigma\; 0\left(\frac{1}{\lambda_i}\,\frac{\partial \varepsilon_{ik}}{\partial x_i}\,\varepsilon_{jk}\right) + 0\left(\varepsilon_{ik}^{\frac{n}{n-2}}\,\varepsilon_{jk}\right)\right] \; ; \; I = \lambda_j\;\; 0\left[\left(\frac{1}{\lambda_i}\,\frac{\partial \varepsilon_{ij}}{\partial x_i}\right) + \operatorname*{Sup}_{j\neq k} \varepsilon_{jk}^2\, 0\left(\varepsilon_{ij}^{\frac{n}{n-2}}\right) + \right.$$

$$\left. \Sigma\; 0\left(\frac{1}{\lambda_i}\,\frac{\partial \varepsilon_{ik}}{\partial x_i}\,\varepsilon_{jk}\right) + 0\left(\varepsilon_{ik}^{\frac{n}{n-2}}\,\varepsilon_{jk}\right)\right] ; \; DI1 = \Sigma\; 0\left(\varepsilon_{rk}\,\frac{1}{\lambda_r}\,\frac{\partial \varepsilon_{rk}}{\partial x_r}\right) + 0\left(\varepsilon_{rk}^{1+\frac{n}{n-2}}\right) \; ;$$

$$DI2 = \frac{1}{\lambda_r}\;\Sigma\; 0\left(\varepsilon_{rk}\,\frac{1}{\lambda_r}\,\frac{\partial \varepsilon_{rk}}{\partial x_r}\right) + 0\left(\varepsilon_{rk}^{1+\frac{n}{n-2}}\right) \; ; \; DI3 = \Sigma\, \lambda_r\; 0\left(\varepsilon_{rk}\,\frac{1}{\lambda_r}\,\frac{\partial \varepsilon_{rk}}{\partial x_r}\right) +$$

$$\lambda_r\; 0\left(\varepsilon_{rk}^{1+\frac{n}{n-2}}\right) \; ; \; J = 0\left(\Sigma\, \varepsilon_{ik}\varepsilon_{kj} + \operatorname*{Sup}_{j\neq k} \varepsilon_{jk}^2 \varepsilon_{ij} + \lambda_i\,\frac{\partial \varepsilon_{ij}}{\partial x_i}\right) \; ;$$

$$K = 0\left(\frac{1}{\lambda_j}\left[\Sigma\varepsilon_{ik}\varepsilon_{kj} + \operatorname*{Sup}_{j\neq k} \varepsilon_{jk}^2 \varepsilon_{ij} + \lambda_j\,\frac{\partial \varepsilon_{ij}}{\partial x_j}\right]\right) \; ; \; L = 0(\lambda_j \varepsilon_{ij}) .$$

$$\begin{array}{c|c|c|c|}
 & j & p+j & 2p+j \\
\hline
i & \bigcirc & \begin{matrix} 0 & \searrow & E \\ & & 0 \end{matrix} & \begin{matrix} 0 & \searrow & F \\ & & 0 \end{matrix} \\
\hline
p+i & \searrow \quad G & \searrow \quad H & \searrow \quad I \\
p+r & 0(|v|^2 \operatorname*{Sup}_{r\neq k} \varepsilon_{rk}^2) \searrow & \frac{1}{\lambda_i} |v|^2 (1+0 \operatorname*{Sup}_{k\neq r} \varepsilon_{rk}^2)) \searrow & \lambda_r |v| (1+0(|v| \operatorname*{Sup}_{k\neq r} \varepsilon_{rk}^2) \searrow \\
\hline
2p+i & \searrow \quad J & \searrow \quad K & \searrow \quad L \\
2p+r & 0(|v|^2 \operatorname*{Sup}_{r\neq k} \varepsilon_{rk}^2) \searrow & \frac{|v|}{\lambda_r} (1+0(|v| \operatorname*{Sup}_{k\neq r} \varepsilon_{rk}^2)) \searrow & \lambda_r |v|^2 (1+0(\operatorname*{Sup}_{k\neq r} \varepsilon_{rk}^2)) \searrow \\
\end{array} \qquad (4.52)$$

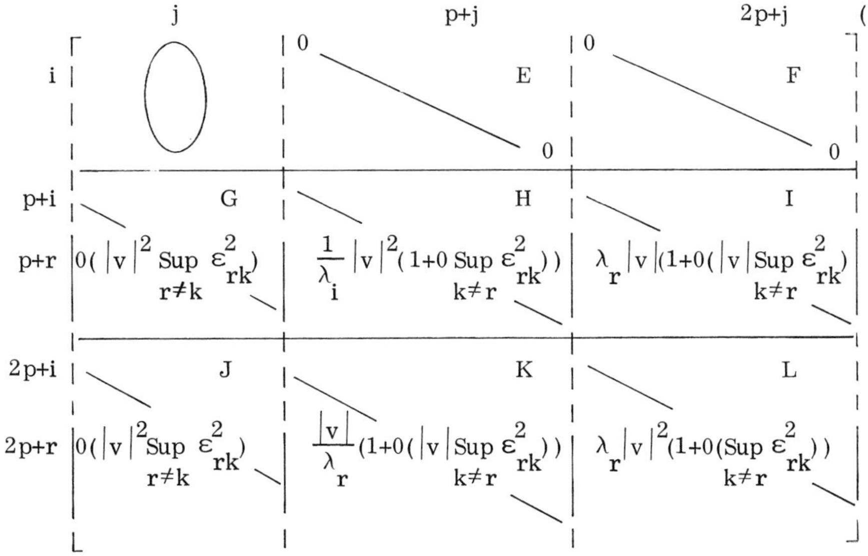

where $E = 0\left(|v| \frac{\partial \varepsilon_{ij}}{\partial \lambda_j}\right)$; $F = 0\left(|v| \frac{\partial \varepsilon_{ij}}{\partial x_j}\right)$; $G = 0\left(|v|^2 \varepsilon_{ij} + |v| \lambda_i \frac{\partial \varepsilon_{ij}}{\partial \lambda_i}\right)$;

$H=0\left(\frac{1}{\lambda_j}|v|^2 \varepsilon_{ij} + \frac{1}{\lambda_j}|v|\left(\frac{1}{\lambda_j} \frac{\partial \varepsilon_{ij}}{\partial x_j} + \frac{1}{\lambda_j} \frac{\partial \varepsilon_{ij}}{\partial x_i}\right)\right)$; $I = 0(|v| \lambda_j \varepsilon_{ij})$; $J = 0\left(|v|^2 \varepsilon_{ij} + \frac{|v|}{\lambda_i} \frac{\partial \varepsilon_{ij}}{\partial x_i}\right)$; $K = 0\left(|v| \frac{\varepsilon_{ij}}{\lambda_j}\right)$; $L = 0\left(\lambda_j \varepsilon_{ij} |v|^2 + \lambda_j \| v| \cdot \left(\frac{1}{\lambda_j} \frac{\partial \varepsilon_{ij}}{\partial x_j} + \frac{1}{\lambda_i} \frac{\partial \varepsilon_{ij}}{\partial x_i}\right)\right)$.

We now simplify our matrices in order to obtain a form of the inverse that we will be using most of the time, except when some further precisions are needed. We use for this (4.23), (4.24), (4.25) :

(4.53)

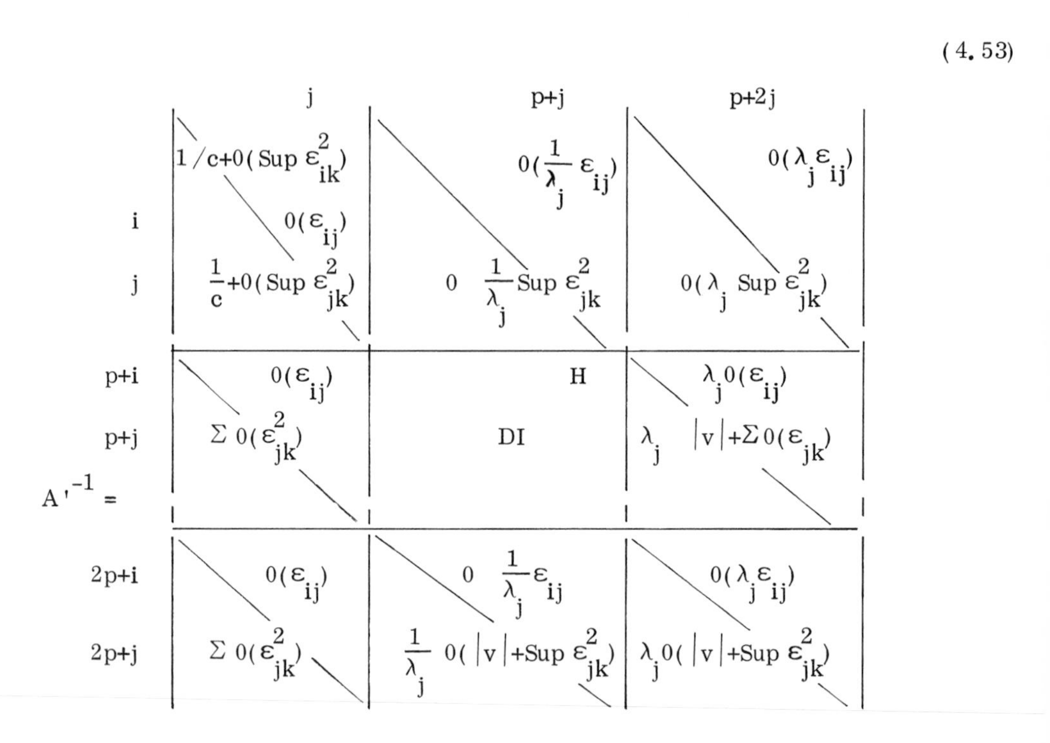

where

$$DI = \frac{1}{\lambda_j}\left[\frac{n}{a_0 \alpha_j} + 0(|v| + 0(\mathrm{Sup}\, \varepsilon_{jk}^2))\right] \tag{4.54}$$

$$H = 0\left(\frac{1}{\lambda_i}\varepsilon_{ij} + \frac{|v|\varepsilon_{ij}}{\lambda_j} + \frac{1}{\lambda_j}\varepsilon_{ij}^{\frac{n}{n-2}}\right) . \tag{4.55}$$

4.3 Further estimates on the v-part of u

Let v satisfy (G_0), i.e.:

$$\int \nabla v . \nabla \delta_i = \int \nabla v . \frac{\nabla \partial \delta_i}{\partial x_i} = \int \nabla v . \frac{\nabla \partial \delta_i}{\partial \lambda_i} = 0 \tag{4.56}$$

and $(\alpha_1, \ldots, \alpha_p)$ be such that:

$$J(\Sigma \alpha_i \delta_i)\, \alpha_i^{4/(n-2)} K(x_i) = 1 + 0(1). \tag{4.57}$$

Let

$$H(\delta_1, \ldots, \delta_p) = \left\{ w : \mathbb{R}^n \to \mathbb{R};\ w \in L^{\frac{2n}{n-2}}(\mathbb{R}^n).\ \int |\nabla w|^2 < +\infty;\ \int \nabla w . \nabla \delta_i = \int \nabla w . \frac{\nabla \partial \delta_i}{\partial x_i} = \int \nabla w . \frac{\nabla \partial \delta_i}{\partial \lambda_i} = 0 \right\} \tag{4.58}$$

Let

$q(x_1, \ldots, x_p)$ be the orthogonal projection onto $H(\delta_1, \ldots, \delta_p)$. (4.59)

We then have:

<u>Lemma 4.1</u>: Let $(x_1, \ldots, x_p)$ be given in a compact set in $\Omega^p \Subset \mathbb{R}^{np}$. Assume the $(\alpha_1, \ldots, \alpha_p)$ satisfy (4.57); assume the λ_i's are large enough, $i = 1, \ldots, p$; and ε_{ij} is small for any (i, j), $i \neq j$. Then the function

$$J(\sum_{i=1}^{p} \alpha_i \delta_i + v)$$

has a unique minimum $\overline{w}$ on a small neighbourhood of zero in $H(\delta_1, \ldots, \delta_p)$. The size of this neighbourhood only depends on the compact set where the

$(x_1, \ldots, x_p)$ are included, given the λ_i large enough and the ε_{ij} small enough ($\lambda_i > \lambda_0$; $\varepsilon_{ij} < \varepsilon_0$; depending also only on the compact set). Thus $\overline{w}$ satisfies the equation:

$$q(x_1, \ldots, x_p)(\partial J(\sum_{i=1}^{p} \alpha_i \delta_i + \overline{w})) = 0. \tag{4.60}$$

We have:

$$|\overline{w}|_H^2 \leq C(\sum_{i \neq j} \varepsilon_{ij}^{\frac{n+2}{n-2}} (\mathrm{Log}\, \varepsilon_{ij}^{-1})^{\frac{n+2}{n}} + 0_K\left(\sum \frac{|DK(x_i)|^2}{\lambda_i^2} + \frac{1}{\lambda_i^4}\right) \tag{4.61}$$

$$+ (\text{if } n < 6) \sum_{i \neq j} \varepsilon_{ij}^2 (\mathrm{Log}\, \varepsilon_{ij}^{-1})^{\frac{2(n-2)}{n}} .$$

Furthermore, for any v in this neighbourhood of zero in $H(\delta_1, \ldots, \delta_n)$, we have:

$$|v - \overline{w}|_H \leq C|q(\partial J(\sum_{i=1} \alpha_i \delta_i + v)]|; \quad C, \text{ a uniform constant.} \tag{4.62}$$

<u>Proof of Lemma 4.1</u>: Assuming the existence and uniqueness of $\overline{w}$ have been established, equation (4.60) of Lemma 4.1 follows immediately.
Next observe that by Lemma 3.2, we have on $H(\delta_1, \ldots, \delta_p)$ (see the end of Lemma 4.1 for full justification):

$$\partial^2 J(\sum_{i=1}^{p} \alpha_i \delta_i).v.v \geq \alpha|v|_H^2; \quad \alpha > 0 \text{ fixed}; \quad \forall v \in H(\delta_1, \ldots, \delta_p). \tag{4.63}$$

Now, for any v in $H(\delta_1, \ldots, \delta_p)$, we have:

$$q((\partial J(\Sigma \alpha_i \delta_i + v))).(v - \overline{w}) = (\partial J(\Sigma \alpha_i \delta_i + v) - \partial J(\Sigma \alpha_i \delta_i + \overline{w})).(v - \overline{w}) \tag{4.64}$$

$$= \partial^2 J(\Sigma \alpha_i \delta_i + t\overline{w} + (1-t)v) . (v - \overline{w}).(v - \overline{w})$$

for a suitable t in $(0, 1)$.
On the other hand, we have (see the end of the proof of Lemma 4.1):

$$|\partial^2 J(\Sigma\,\alpha_i\delta_i+\tau) - \partial^2 J(\Sigma\,\alpha_i\delta_i)| \le C(|\tau|_H+|\tau|_H^{4/(n-2)}). \qquad (4.65)$$

Thus, by (4.63):

$$q(\partial J(\Sigma\,\alpha_i\delta_i+v)).(v-\overline{w}) \ge \left(\alpha - C\left(|v|_H+|\overline{w}|_H+|v|_H^{\frac{4}{n-2}}+|\overline{w}|_H^{\frac{4}{n-2}}\right)\right)|v-\overline{w}|_H^2 \qquad (4.66)$$

This yields (4.62) as soon as:

$$|v|_H + |\overline{w}|_H + |v|_H^{\frac{4}{n-2}} + |\overline{w}|_H^{\frac{4}{n-2}} < \frac{\alpha}{2C} \qquad (4.67)$$

Taking into account (4.61), which has yet to be proved, (4.67) simply imposes a mild condition of smallness of v, independent of λ_i, x_i, etc.
We are left with proving the existence and uniqueness of $\overline{w}$ and the estimate (4.61). Expanding $J(\Sigma\,\alpha_i\delta_i+v)$ around zero in $H(\delta_1, \ldots, \delta_p)$, we have:

$$J(\Sigma\,\alpha_i\delta_i+v) = J(\Sigma\,\alpha_i\delta_i) + \partial J(\Sigma\,\alpha_i\delta_i).v + \frac{1}{2}\partial^2 J(\Sigma\,\alpha_i\delta_i+tv).v.v, \qquad (4.68)$$

$t \in (0,1)$.

This together with (4.65) yields:

$$J(\Sigma\,\alpha_i\delta_i+v) = J(\Sigma\,\alpha_i\delta_i) + \partial J(\Sigma\,\alpha_i\delta_i).v + \frac{1}{2}\partial^2 J(\Sigma\,\alpha_i\delta_i).v.v \qquad (4.69)$$
$$+ o(|v|_H^2),$$

and we also have, using again (4.65), after an expansion with integral rest:

$$\partial J(\Sigma\,\alpha_i\delta_i+v) = \partial J(\Sigma\,\alpha_i\delta_i) + \partial^2 J(\Sigma\,\alpha_i\delta_i).v + o(|v|_H). \qquad (4.70)$$

The o in (4.68) and (4.69) depend only on v satisfying:

$$|v|_H < \varepsilon_0;\ \varepsilon_0 \text{ fixed independent of } x_i \text{ and } \lambda_i. \qquad (4.71)$$

Indeed, they both rely on (4.65) and only on this inequality since we have:

$$o(|v|_H^2) = \frac{1}{2}(\partial^2 J(\Sigma\,\alpha_i\delta_i + tv) - \partial^2 J(\Sigma\,\alpha_i\delta_i)).v.v \tag{4.72}$$

$$\leq \frac{C}{2}\left[|v|_H^3 + |v|_H^{2+\frac{4}{n-2}}\right]$$

$$o(|v|_H) = \int_0^1(\partial^2 J(\Sigma\,\alpha_i\delta_i + tv) - \partial^2 J(\Sigma\,\alpha_i\delta_i)).v \leq \tag{4.73}$$

$$\leq C\left[|v|_H^2 + |v|_H^{1+\frac{4}{n-2}}\right]$$

Thus, by (4.68) and (4.69), $\phi(v) = J(\Sigma\,\alpha_i\delta_i + v)$ is C^1 close to the function:

$$\psi(v) = J(\Sigma\,\alpha_i\delta_i) + \partial J(\Sigma\,\alpha_i\delta_i).v + \frac{1}{2}\,\partial^2 J(\Sigma\,\alpha_i\delta_i).v.v\,. \tag{4.74}$$

By (4.63), $\frac{1}{2}\,\partial^2 J(\Sigma\,\alpha_i\delta_i).v.v$ is positive definite on $H(\delta_1, \ldots, \delta_p)$. Thus, ψ has a unique critical point $\overline{w}_1$, which is a minimum on $H(\delta_1, \ldots, \delta_p)$ and this minimum $\overline{w}_1$ satisfies:

$$\partial^2 J(\Sigma\,\alpha_i\delta_i).\overline{w}_1 = -q(\partial J(\Sigma\,\alpha_i\delta_i)). \tag{4.75}$$

Thus, using (4.63):

$$|\overline{w}_1|_H \leq \frac{1}{\alpha}\,|q(\partial J(\Sigma\,\alpha_i\delta_i))|_H\,. \tag{4.76}$$

By assumption the λ_i's are large, the ε_{ij}'s are small and the $J(\Sigma\,\alpha_i\delta_i)\,\alpha_i^{\frac{4}{n-2}}\,K(x_i)$ are close to 1; so that $\partial J(\Sigma\,\alpha_i\delta_i)$ is small and so is $|\overline{w}_1|_H$.

This implies at once the existence and uniqueness of $\overline{w}$ satisfying:

$$J(\Sigma\,\alpha_i\delta_i + \overline{w}) = \operatorname*{Min}_{w\in H(\delta_1, \ldots, \delta_p)} J(\Sigma\,\alpha_i\delta_i + w) \tag{4.77}$$

as ϕ is C^1 close to ψ on a neighbourhood (of uniform size, see (4.69), (4.70), (4.71)), of zero.

$\overline{w}$ satisfies:

$$q(\partial J(\Sigma \alpha_i \delta_i + \bar{w})) = 0 \tag{4.78}$$

and is in a neighbourhood of zero. Thus, by (4.70) :

$$0 = \partial J(\Sigma \alpha_i \delta_i + \bar{w}) . \bar{w} = \partial J(\Sigma \alpha_i \delta_i) . \bar{w} + \partial^2 J(\Sigma \alpha_i \delta_i) . \bar{w}. \bar{w} + \tag{4.79}$$

$$+ o(|\bar{w}|_H^2) ;$$

hence by (4.63) :

$$\alpha |\bar{w}|_H^2 - o(|\bar{w}|_H^2) \le -\partial J(\Sigma \alpha_i \delta_i) . \bar{w} \tag{4.80}$$

or else, provided the λ_i are large enough, the ε_{ij} are small enough and $J(\Sigma \alpha_i \delta_i) \alpha_i^{4/(n-2)} K(x_i)$ is close enough to 1:

$$\frac{\alpha}{2} |\bar{w}|_H^2 \le -(\partial J(\Sigma \alpha_i \delta_i)) . \bar{w} . \tag{4.81}$$

Thus

$$|\bar{w}|_H \le \frac{2}{\alpha} |(\partial J(\Sigma \alpha_i \delta_i)). \bar{w}|^{1/2} \tag{4.82}$$

We now have:

$$\partial J(\Sigma \alpha_i \delta_i) . \bar{w} \tag{4.83}$$

$$= -\lambda(\Sigma \alpha_i \delta_i) \int \nabla(\Sigma \alpha_i \delta_i + \lambda(\Sigma \alpha_i \delta_i)^{\frac{4}{n-2}} \Delta_\Omega^{-1} K(x) (\Sigma \alpha_i \delta_i)^{\frac{n+2}{n-2}}) \nabla \bar{w}$$

$$= \lambda(\Sigma \alpha_i \delta_i)^{\frac{n+2}{n-2}} \int K(x) (\Sigma \alpha_i \delta_i)^{\frac{n+2}{n-2}} \bar{w}$$

since $\bar{w}$, being in $H(\delta_1, \ldots, \delta_p)$ is orthogonal to the δ_i's. $\bar{w}$ satisfies (G.0). Thus (G.10) holds with $v = \bar{w}$ and we have:

$$\int_{\mathbb{R}^n} K(x) (\Sigma \alpha_i \delta_i)^{\frac{n+2}{n-2}} \bar{w} = 0 \left(|\bar{w}|_H \left[\sum_{i \neq j} \varepsilon_{ij}^{\frac{1}{2} \frac{n+2}{n-2}} (\text{Log } \varepsilon_{ij}^{-1})^{\frac{n+2}{2n}} \right. \right. \tag{4.84}$$

$$\left. \left. + 0_K \left(\Sigma \frac{|DK(x_i)|}{\lambda_i} + \frac{1}{\lambda_i^2} \right) \right] \right) \quad \text{if } n \ge 6,$$

$$= 0\left(|\overline{w}|_H\left[\sum_{i\neq j}\varepsilon_{ij}(\mathrm{Log}\,\varepsilon_{ij}^{-1})^{\frac{n-2}{n}}\right.\right.$$

$$\left.\left.+\,0_K\left(\Sigma\frac{|DK(x_i)|}{\lambda_i}+\frac{1}{\lambda_i^2}\right)\right]\right) \quad \text{if } n<6\,.$$

(4.84) together with (4.82) imply the estimate (4.61) of Lemma 4.1. We are thus left with proving (4.65) and fully justifying (4.63). For this, we need to compute $\partial^2 J(\sum_{i=1}^{p}\alpha_i\delta_i+\tau)$ for τ small in $H(\delta_1,\ldots,\delta_p)$.

To justify (4.63), we have:

$$-\partial J(\Sigma\,\alpha_i\delta_i+v) = -\lambda(\Sigma\,\alpha_i\delta_i+v)\left[(\Sigma\,\alpha_i\delta_i+v)\right. \tag{4.85}$$

$$\left.-\,\lambda(\Sigma\,\alpha_i\delta_i+v)^{\frac{4}{n-2}}\Delta_\Omega^{-1}K(x)\,(\Sigma\,\alpha_i\delta_i+v)^{\frac{n+2}{n-2}}\right].$$

Thus, for any w in $H(\delta_1,\ldots,\delta_p)$:

$$-\partial^2 J(\Sigma\,\alpha_i\delta_i+v).\,w = -\partial\lambda(\Sigma\,\alpha_i\delta_i+v).w\left[(\Sigma\,\alpha_i\delta_i+v)\right. \tag{4.86}$$

$$\left.-\,\frac{4}{n-2}\,\lambda(\Sigma\,\alpha_i\delta_i+v)^{\frac{4}{n-2}-1}\,\Delta_\Omega^{-1}K(\Sigma\,\alpha_i\delta_i+v)^{\frac{n+2}{n-2}}\right]$$

$$-\lambda(\Sigma\,\alpha_i\delta_i+v)\,w-\lambda(\Sigma\,\alpha_i\delta_i+v)^{\frac{n+2}{n-2}}\left(\frac{n+2}{n-2}\right)\Delta_\Omega^{-1}K(x)\,(\Sigma\,\alpha_i\delta_i+v)^{\frac{4}{n-2}}\,w$$

Taking $v = 0$ and applying (4.86) to w, we obtain:

$$-\partial^2 J(\Sigma\,\alpha_i\delta_i).\,w.\,w = \frac{4}{n-2}\,(\partial\lambda(\Sigma\,\alpha_i\delta_i).\,w)\,\lambda(\Sigma\,\alpha_i\delta_i)^{\frac{4}{n-2}-1} \tag{4.87}$$

$$\times\int_\Omega\nabla(\Delta_\Omega^{-1}K(\Sigma\,\alpha_i\delta_i))^{\frac{n+2}{n-2}}\,\nabla w$$

$$-\lambda(\Sigma\,\alpha_i\delta_i)\left[\int|\nabla w|^2-\frac{n+2}{n-2}\,\lambda(\Sigma\,\alpha_i\delta_i)^{\frac{4}{n-2}}\int K(x)\,(\Sigma\,\alpha_i\delta_i)^{\frac{4}{n-2}}w^2\right]$$

As $\int \nabla \delta_i \nabla w = 0$, we have:

$$\lambda(\Sigma \alpha_i \delta_i)^{\frac{n+2}{n-2}} \int_\Omega \nabla(\Delta_\Omega^{-1} K(\Sigma \alpha_i \delta_i)^{\frac{n+2}{n-2}}) \nabla w = \lambda(\Sigma \alpha_i \delta_i) \int_\Omega \nabla(\Sigma \alpha_i \delta_i) \nabla w$$

$$+ \lambda(\Sigma \alpha_i \delta_i)^{\frac{n+2}{n-2}} \int_\Omega \nabla(\Delta_\Omega^{-1} K(\Sigma \alpha_i \delta_i)^{\frac{n+2}{n-2}} \nabla w = \partial J(\Sigma \alpha_i \delta_i) . w . \tag{4.88}$$

On the other hand:

$$\lambda(u) = J(u)^{\frac{n-2}{4}} . \tag{4.89}$$

Thus

$$\partial \lambda(u) = \frac{n-2}{4} J(u)^{\frac{n-6}{4}} \partial J(u) . \tag{4.90}$$

Hence

$$-\partial^2 J(\Sigma \alpha_i \delta_i) . w. w = J(\Sigma \alpha_i \delta_i)^{-\frac{n+2}{4}} (\partial J(\Sigma \alpha_j \delta_i) . w)^2 \tag{4.91}$$

$$-\lambda(\Sigma \alpha_i \delta_i) \left[\int |\nabla w|^2 - \frac{n+2}{n-2} \lambda(\Sigma \alpha_i \delta_i)^{\frac{4}{n-2}} \int (K(x) (\Sigma \alpha_i \delta_i)^{\frac{4}{n-2}} w^2 \right] .$$

Using now G11, we derive:

$$-\partial^2 J(\Sigma \alpha_i \delta_i) . w. w = J(\Sigma \alpha_i \delta_i)^{-\frac{n+2}{4}} (\partial J(\Sigma \alpha_i \delta_i) . w)^2 \tag{4.92}$$

$$- \lambda(\Sigma \alpha_i \delta_i) [\int |\nabla w|^2 - \frac{n+2}{n-2} \lambda(\Sigma \alpha_i \delta_i)^{\frac{4}{n-2}} \Sigma \alpha_i^{\frac{4}{n-2}} K(x_i) \int \delta_i^{\frac{4}{n-2}} w^2]$$

$$+ o(1) \int |\nabla w|^2 .$$

Observe now that:

$$\partial J(\Sigma \alpha_i \delta_i) . w = o(|w|_H) \tag{4.93}$$

$$\lambda(\Sigma\,\alpha_i\delta_i)^{\frac{4}{n-2}}\alpha_i^{\frac{4}{n-2}}K(x_i) = J(\Sigma\,\alpha_i\delta_i)\,\alpha_i^{\frac{4}{n-2}}K(x_i) = 1 + o(1). \tag{4.94}$$

Thus

$$-\partial^2 J(\Sigma\,\alpha_i\delta_i).w.w = -\lambda(\Sigma\,\alpha_i\delta_i)\left(\int|\nabla w|^2 - \frac{n+2}{n-2}\sum_i\int\delta_i^{\frac{4}{n-2}}w^2\right) \tag{4.93'}$$
$$+ o(1)\int|\nabla w|^2.$$

Formula (4.63) follows then from Proposition 1.

We now establish (4.65). For this we turn back to (4.85)-(4.86). Observe first that:

$$\left|J\left(\sum_{i=1}^{p}\alpha_i\delta_i+v\right) - J\left(\sum_{i=1}^{p}\alpha_i\delta_i+\phi\right)\right| \le C|v-\phi|_H; \tag{4.94'}$$

$\forall\, v,\ \phi \in H$, v and ϕ small;

uniformly on the x_i, λ_i, α_i such that the ε_{ij}'s are small and (4.57) holds. In fact, even (4.57) is not necessary for (4.94') which is simply due to a first expansion of $J\left(\sum_{i=1}^{p}\alpha_i\delta_i+v\right)$ as follows:

$$J\left(\sum_{i=1}^{p}\alpha_i\delta_i+v\right) - J\left(\sum_{i=1}^{p}\alpha_i\delta_i+\phi\right) = \partial J\left(\sum_{i=1}^{p}\alpha_i\delta_i+tv+(1-t)\phi\right).(v-\phi) \tag{4.95}$$
$$t\in[0,1]$$

Now, $\partial J\left(\sum_{i=1}^{p}\alpha_i\delta_i+tv+(1-t)\phi\right)$ is bounded. Indeed:

$$\partial J\left(\sum_{i=1}^{p}\alpha_i\delta_i+tv+(1-t)\phi\right) = \lambda\left(\sum_{i=1}^{p}\alpha_i\delta_i+tv+(1-t)\phi\right)\left(\sum_{i=1}^{p}\alpha_i\delta_i+tv+(1-t)\phi\right)$$
$$-\lambda\left(\sum_{i=1}^{p}\alpha_i\delta_i+tv+(1-t)\phi\right)^{\frac{4}{n-2}}\Delta_\Omega^{-1}K(x)\left(\sum_{i=1}^{p}\alpha_i\delta_i+tv+(1-t)\phi\right)^{\frac{n+2}{n-2}}, \tag{4.96}$$

and we know that

$$\lambda\left(\sum_{i=1}^{p}\alpha_i\delta_i + tv + (1-t)\phi\right) = J\left(\sum_{i=1}^{p}\alpha_i\delta_i + tv + (1-t)\phi\right)^{\frac{n-2}{4}} \leq C(p). \quad (4.97)$$

$$\left|\sum_{i=1}^{p}\alpha_i\delta_i + tv + (1-t)\phi\right| \leq C(p), \quad (4.98)$$

assuming v and ϕ are bounded. Hence (4.94). (4.94) implies at once through (4.89) :

$$\left|\lambda\left(\sum_{i=1}^{p}\alpha_i\delta_i+v\right) - \lambda\left(\sum_{i=1}^{p}\alpha_i\delta_i+\phi\right)\right| \leq C|v-\phi|_H, \quad \forall v, \phi \in H; \quad (4.99)$$

v, ϕ small.

Finally, we also have:

$$\left|\nabla(\Delta_\Omega^{-1}K(x)\left(\sum_{i=1}^{p}\alpha_i\delta_i+v\right)^{\frac{n+2}{n-2}} - \Delta_\Omega^{-1}K(x)\left(\sum_{i=1}^{p}\alpha_i\delta_i+\phi\right)^{\frac{n+2}{n-2}}\right)\Bigg|_{L^2} \quad (4.100)$$

$$\leq C|v-\phi|_H.$$

$$\left|\nabla(\Delta_\Omega^{-1}K(x)\left(\sum_{i=1}^{p}\alpha_i\delta_i+v\right)^{\frac{4}{n-2}}w - \Delta_\Omega^{-1}K(x)\left(\sum_{i=1}^{p}\alpha_i\delta_i+\phi\right)^{\frac{4}{n-2}}w)\right|_{L^2}$$

$$\leq C\left(|v-\phi|_H+|v|_H^{\frac{4}{n-2}}+|\phi|_H^{\frac{4}{n-2}}\right)|w|_H; \quad \forall w \in H; \quad (4.101)$$

where

$$|\nabla\psi|_{L^2} = \left(\int_{\mathbb{R}^n}|\nabla\psi|^2\right)^{1/2}. \quad (4.102)$$

Indeed, we have:

$$\left|\left(\sum_{i=1}^{p}\alpha_i\delta_i+v\right)^{\frac{n+2}{n-2}} - \left(\sum_{i=1}^{p}\alpha_i\delta_i+\phi\right)^{\frac{n+2}{n-2}}\right| = \quad (4.103)$$

$$\left|\sum_{i=1}^{p}\alpha_i\delta_i+tv+(1-t)\phi\right|^{\frac{4}{n-2}}|v-\overline{w}|$$

$$\leq C|v-\phi|\left[\sum_{i=1}^{p}(\alpha_i\delta_i)^{\frac{4}{n-2}}+|v|^{\frac{4}{n-2}}+|\phi|^{\frac{4}{n-2}}\right],\quad t\in[0,1].$$

Thus:

$$\int\left|\nabla(\Delta_{\Omega}^{-1}K(x)\left[\sum_{i=1}^{p}\left(\alpha_i\delta_i+v\right)^{\frac{n+2}{n-2}}-\sum_{i=1}^{p}\left(\alpha_i\delta_i+\phi\right)^{\frac{n+2}{n-2}}\right|^2 \tag{4.104}$$

$$=-\int K(x)\left[\left(\sum_{i=1}^{p}\alpha_i\delta_i+v\right)^{\frac{n+2}{n-2}}-\left(\sum_{i=1}^{p}\alpha_i\delta_i+\phi\right)^{\frac{n+2}{n-2}}\right]\Delta_{\Omega}^{-1}\left(K(x)\left[\left(\sum_{i=1}^{p}\alpha_i\delta_i+v\right)^{\frac{n+2}{n-2}}\right.\right.$$

$$\left.\left.-\left(\sum_{i=1}^{p}\alpha_i\delta_i+\phi\right)^{\frac{n+2}{n-2}}\right]\right)$$

$$\leq -C\int|v-\phi|\left(\sum_{i=1}^{p}(\alpha_i\delta_i)^{\frac{4}{n-2}}+|v|^{\frac{4}{n-2}}+|\phi|^{\frac{4}{n-2}}\right)\Delta_{\Omega}^{-1}\left(|v-\phi|\left(\sum_{i=1}^{p}(\alpha_i\delta_i)^{\frac{4}{n-2}}\right.\right.$$

$$\left.+|v|^{\frac{4}{n-2}}+|\phi|^{\frac{4}{n-2}}\right).$$

Now let:

$$z=|v-\phi|\left(\sum_{i=1}^{p}(\alpha_i\delta_i)^{\frac{4}{n-2}}+|v|^{\frac{4}{n-2}}+|\phi|^{\frac{4}{n-2}}\right) \tag{4.105}$$

z belongs to $L^{2n/(n+2)}$. Hence $\nabla(\Delta_{\Omega}^{-1}z)$ belongs to L^2 and:

$$\int|\nabla\Delta_{\Omega}^{-1}z|^2\leq C\left(\int|z|^{\frac{2n}{n+2}}\right)^{\frac{n+2}{n}}. \tag{4.106}$$

(4.104) and (4.106) yield:

$$\int\left|\nabla\left(\Delta_{\Omega}^{-1}K(x)\left[\left(\sum_{i=1}^{p}\alpha_i\delta_i+v\right)^{\frac{n+2}{n-2}}-\left(\sum_{i=1}^{p}\alpha_i\delta_i+\phi\right)^{\frac{n+2}{n-2}}\right]\right)\right|^2 \tag{4.107}$$

$$\leq C\left[\int|v-\phi|^{\frac{2n}{n+2}}\left(\sum_{i=1}^{p}(\alpha_i\delta_i)^{\frac{8n}{n^2-4}}+|v|^{\frac{8n}{n^2-4}}+|\phi|^{\frac{8n}{n^2-4}}\right)\right]^{\frac{n+2}{n}}$$

$$\leq C_1 \left(\int |\nabla (v-\phi)|^2 \right) = C_1 |v-\phi|_H^2 ,$$

provided $|v|_H$ and $|\phi|_H$ are bounded. Hence (4.100). We turn now to (4.101). We have, after some thought:

$$\left| \left(\sum_{i=1}^{p} \alpha_i \delta_i + v \right)^{\frac{4}{n-2}} - \left(\sum_{i=1}^{p} \alpha_i \delta_i + \phi \right)^{\frac{4}{n-2}} \right| \leq C_\varepsilon \left(|v|^{\frac{4}{n-2}} + |\phi|^{\frac{4}{n-2}} \right) \tag{4.108}$$

$$\text{if } |v| + |\phi| \geq \varepsilon \sum_{i=1}^{p} \alpha_i \delta_i$$

$$\leq C |v-\phi| \left(\sum_{i=1}^{p} \alpha_i \delta_i \right)^{\frac{4}{n-2}-1} \quad \text{if } |v| + |\phi| \leq \varepsilon \sum_{i=1}^{p} \alpha_i \delta_i ;$$

$$\left| \left(\sum_{i=1}^{p} \alpha_i \delta_i + v \right)^{\frac{4}{n-2}} - \left(\sum_{i=1}^{p} \alpha_i \delta_i + \phi \right)^{\frac{4}{n-2}} \right| \leq C_\varepsilon \left(|v|^{\frac{4}{n-2}} + |\phi|^{\frac{4}{n-2}} + \right. \tag{4.109}$$

$$\left. + |v-\phi| \left(\sum_{i=1}^{p} \alpha_i \delta_i \right)^{\frac{4}{n-2}-1} \right) \quad \text{if } n \leq 6;$$

$$\left| \left(\sum_{i=1}^{p} \alpha_i \delta_i + v \right)^{\frac{4}{n-2}} - \left(\sum_{i=1}^{p} \alpha_i \delta_i + \phi \right)^{\frac{4}{n-2}} \right| \leq \tag{4.110}$$

$$\leq C_\varepsilon \left(|v|^{\frac{4}{n-2}} + |\phi|^{\frac{4}{n-2}} + |v-\phi|^{\frac{4}{n-2}} \right)$$

$$\leq C'_\varepsilon \quad |v|^{\frac{4}{n-2}} + |\phi|^{\frac{4}{n-2}} \quad \text{if } n > 6.$$

(4.109) and (4.110) imply then (4.101), in the same way as we have shown that (4.103) implies (4.101).

With (4.99), (4.100) and (4.101), (4.65) follows easily from (4.86) just by subtracting $\partial^2 J(\Sigma \alpha_i \delta_i + \phi) . w$ from $\partial^2 J(\Sigma \alpha_i \delta_i + v) . w$ and using also the boundedness of $\lambda \left(\sum_{i=1}^{p} \alpha_i \delta_i + v \right)$, $\lambda \left(\sum_{i=1}^{p} \alpha_i \delta_i + \phi \right)$, $\partial \lambda \left(\sum_{i=1}^{p} \alpha_i \delta_i + v \right)$,

$\partial\lambda\left(\sum_{i=1}^{p}\alpha_i\delta_i+\phi\right)$, using our hypothesis on α_i, λ_i, x_i and the boundedness of v and ϕ.

In fact, looking back on our arguments, we have shown a better inequality than (4.63), namely:

$$\left|\partial^2 J\left(\sum_{i=1}^{p}\alpha_i\delta_i+v\right)-\partial^2 J\left(\sum_{i=1}^{p}\alpha_i\delta_i+\phi\right)\right|\le C\left(|v-\phi|_H+|v|_H^{\frac{4}{n-2}}+|\phi|_H^{\frac{4}{n-2}}\right) \tag{4.111}$$

for any v and ϕ in $H(\delta_1,\ldots,\delta_p)$ such that $|v|_H<\varepsilon_0$; $|\phi|_H<\varepsilon_0$; ε_0 uniformly constant.

The proof of Lemma 4.1 is thereby complete.

We now take care of the remainder term S_i in (4.7).

<u>Lemma 4.2</u>: Let $0<\varepsilon<\varepsilon'/4$,

$$\int_{\substack{|v|\le\varepsilon\Sigma\alpha_i\delta_i\\|v|\le\varepsilon'\alpha_j\delta_j}}\delta_j^{\frac{4}{n-2}}v^2+\int_{|v|\ge\varepsilon(\Sigma\alpha_i\delta_i)}\delta_j|v|^{\frac{n+2}{n-2}}\le C_{\varepsilon,\varepsilon'}\left[|\partial J(\Sigma\alpha_i\delta_i+v)|^2\right.$$

$$\left.+\int_{\substack{|\overline{w}|\le 2\varepsilon\Sigma\alpha_i\delta_i\\|\overline{w}|\le 2\varepsilon'\alpha_j\delta_j}}\delta_j^{\frac{4}{n-2}}\overline{w}^2+\int_{|\overline{w}|\ge\frac{\varepsilon}{2}\Sigma\alpha_i\delta_i}\delta_j|\overline{w}|^{\frac{n+2}{n-2}}\right].$$

<u>Proof</u>: We have,

$$\int_{\substack{|v|\le\varepsilon\Sigma\alpha_i\delta_i\\|v|\le\varepsilon'\alpha_j\delta_j}}\delta_j^{\frac{4}{n-2}}v^2\le C|v-\overline{w}|_H^2+\int_{\substack{|v|\le\varepsilon\Sigma\alpha_i\delta_i\\|v|\le\varepsilon'\alpha_j\delta_j}}\delta_j^{\frac{4}{n-2}}\overline{w}^2. \tag{4.112}$$

Now

$$\int_{\substack{|v|\le\varepsilon\Sigma\alpha_i\delta_i \\ |v|\le\varepsilon'\alpha_j\delta_j}} \delta_j^{\frac{4}{n-2}} \bar{w}^2 \le \int_{\substack{|\bar{w}|\le 2\varepsilon\Sigma\alpha_i\delta_i \\ |\bar{w}|\le 2\varepsilon'\alpha_j\delta_j}} \delta_j^{4/(n-2)} \bar{w}^2 \tag{4.113}$$

$$+ \int_{\substack{|v|\le\varepsilon\Sigma\alpha_i\delta_i \\ |v|\le\varepsilon'\alpha_j\delta_j}} \delta_j^{\frac{4}{n-2}} \bar{w}^2 [\chi_{|\bar{w}|\ge 2\varepsilon\Sigma\alpha_i\delta_i} + \chi_{|\bar{w}|\ge 2\varepsilon'\alpha_j\delta_j}].$$

If $|v| \le \varepsilon\,\Sigma\,\alpha_i\delta_i$ while $|\bar{w}| \ge 2\varepsilon\,\Sigma\,\alpha_i\delta_i$, then $|v-\bar{w}| \ge \frac{1}{2}|\bar{w}|$. As well, if $|v| \le \varepsilon'\,\alpha_i\delta_i$, while $|\bar{w}| \ge 2\varepsilon'\,\alpha_j\delta_i$, then $|v-\bar{w}| \ge \frac{1}{2}|\bar{w}|$. Thus

$$\int_{\substack{|v|\le\varepsilon\Sigma\alpha_i\delta_i \\ |v|\le\varepsilon'\alpha_j\delta_j}} \delta_j^{\frac{4}{n-2}} \bar{w}^2 [\chi_{|\bar{w}|\ge 2\varepsilon\Sigma\alpha_i\delta_i} + \chi_{|\bar{w}|\ge 2\varepsilon'\alpha_j\delta_j}] \tag{4.114}$$

$$\le 4\int \delta_j^{\frac{4}{n-2}} |v-\bar{w}|^2 \le C|v-\bar{w}|_H^2$$

Summing up (4.112), (4.113) and (4.114), we have:

$$\int_{\substack{|v|\le\varepsilon\Sigma\alpha_i\delta_i \\ |v|\le\varepsilon'\alpha_j\delta_j}} \delta_j^{\frac{4}{n-2}} v^2 \le C\left[|v-\bar{w}|_H^2 + \int_{\substack{|\bar{w}|\le 2\varepsilon\Sigma\alpha_i\delta_i \\ |\bar{w}|\le 2\varepsilon'\alpha_j\delta_j}} \delta_j^{\frac{4}{n-2}} \bar{w}^2\right] \tag{4.115}$$

Next, we have:

$$\int_{|v|\ge\varepsilon(\Sigma\alpha_i\delta_i)} \delta_j |v|^{\frac{n+2}{n-2}} \le C\left[\int_{|v|\ge\varepsilon\Sigma\alpha_i\delta_i} \delta_j |\bar{w}|^{\frac{n+2}{n-2}} + \right. \tag{4.116}$$

$$\left. + \int_{|v|\ge\varepsilon\Sigma\alpha_i\delta_i} \delta_j |v-\bar{w}|^{\frac{n+2}{n-2}}\right]$$

Now:

$$\int_{|v|\geq \varepsilon \Sigma \alpha_i \delta_i} \delta_j |\bar{w}|^{\frac{n+2}{n-2}} \leq \int_{|\bar{w}|\geq \frac{\varepsilon}{2} \Sigma \alpha_i \delta_i} \delta_j |\bar{w}|^{\frac{n+2}{n-2}} + \tag{4.117}$$

$$+ \int_{\substack{|\bar{w}|\leq \frac{\varepsilon}{2} \Sigma \alpha_i \delta_i \\ |v|\geq \varepsilon \Sigma \alpha_i \delta_i}} \delta_j |\bar{w}|^{\frac{n+2}{n-2}}$$

$$\leq \int_{|\bar{w}|\geq \frac{\varepsilon}{2} \Sigma \alpha_i \delta_i} \delta_j |\bar{w}|^{\frac{n+2}{n-2}} + (2)^{\frac{n+2}{n-2}} \int_{|v|\geq \varepsilon \Sigma \alpha_i \delta_i} \delta_j |v-\bar{w}|^{\frac{n+2}{n-2}}]$$

Thus, from (4.116) and (4.117), we derive:

$$\int_{|v|\geq \varepsilon (\Sigma \alpha_i \delta_i)} \delta_j |v|^{\frac{n+2}{n-2}} \leq C \int_{|\bar{w}|\geq \frac{\varepsilon}{2} \Sigma \alpha_i \delta_i} \delta_j |\bar{w}|^{\frac{n+2}{n-2}} + \tag{4.118}$$

$$+ \int_{|v|\geq \varepsilon \Sigma \alpha_i \delta_i} \delta_j |v-\bar{w}|^{\frac{n+2}{n-2}}$$

Now:

$$\int_{|v|\geq \varepsilon \Sigma \alpha_i \delta_i} \delta_j |v-\bar{w}|^{\frac{n+2}{n-2}} \leq C_\varepsilon \Bigg[\int_{\substack{|v|\geq \varepsilon \Sigma \alpha_i \delta_i \\ \alpha_j \delta_j \leq 2/\varepsilon |v-\bar{w}|}} |v-\bar{w}|^{\frac{2n}{n-2}} + \tag{4.119}$$

$$+ \int_{\substack{|v|\geq \varepsilon \Sigma \alpha_i \delta_i \\ |v-\bar{w}|\leq \frac{\varepsilon}{2} \alpha_j \delta_j}} \delta_j |v-\bar{w}|^{\frac{n+2}{n-2}} \Bigg]$$

We also have

$$\int_{\substack{|v| \geq \varepsilon \Sigma \alpha_i \delta_i \\ |v-\overline{w}| \leq \frac{\varepsilon}{2} \alpha_j \delta_j}} \delta_j |v-\overline{w}|^{\frac{n+2}{n-2}} \leq \int_{|\overline{w}| \geq \frac{\varepsilon}{2}(\Sigma \alpha_i \delta_i)} \delta_j |\overline{w}|^{\frac{n+2}{n-2}} \tag{4.120}$$

Hence

$$\int_{|v| \geq \varepsilon \Sigma \alpha_i \delta_i} \delta_j |v-\overline{w}|^{\frac{n+2}{n-2}} \leq C_\varepsilon \Big(|v-\overline{w}|_H^{\frac{2n}{n-2}} + \int_{|\overline{w}| \geq \frac{\varepsilon}{2}(\Sigma \alpha_i \delta_i)} \delta_j |\overline{w}|^{\frac{n+2}{n-2}} \tag{4.121}$$

From (4.115), (4.118) and (4.121), we derive:

$$\int_{\substack{|v| \leq \varepsilon \Sigma \alpha_i \delta_i \\ |v| \leq \varepsilon' \alpha_j \delta_j}} \delta_j^{\frac{4}{n-2}} v^2 + \int_{|v| \geq \varepsilon(\Sigma \alpha_i \delta_i)} \delta_j |v|^{\frac{n+2}{n-2}} \tag{4.122}$$

$$\leq C_\varepsilon \Bigg[|v-\overline{w}|_H^2 + |v-\overline{w}|_H^{\frac{2n}{n-2}} + \int_{\substack{|\overline{w}| \leq 2\varepsilon \Sigma \alpha_i \delta_i \\ |\overline{w}| \leq 2\varepsilon' \alpha_j \delta_j}} \delta_j^{\frac{4}{n-2}} \overline{w}^2 +$$

$$+ \int_{|\overline{w}| \geq \frac{\varepsilon}{2}(\Sigma \alpha_i \delta_i)} \delta_j |\overline{w}|^{\frac{n+2}{n-2}} \Bigg]$$

Notice now that by (4.62) of Lemma 4.1, we have:

$$|v-\overline{w}|_H \leq C \left| \partial J(\Sigma \alpha_i \delta_i + v) \right| , \tag{4.123}$$

and also

$$|v-\overline{w}|_H = o(1) \tag{4.124}$$

as v and $\overline{w}$ are small. Thus:

$$|v-\overline{w}|_H^2 + |v-\overline{w}|_H^{\frac{2n}{n-2}} \leq C[\,\partial J(\Sigma\, \alpha_i \partial_i + v)\,|^2\,], \tag{4.125}$$

and Lemma 4.2 is thereby proved.

Part 2 The variational problem near infinity

5 Morse lemma near infinity

5.1 A first expansion of J near infinity

We derive here a normal form for J near infinity. Chapters 1 to 4 were devoted, as usual in variational theory, to the global study of the flow related to the variational problem. This allowed the study of the critical points at infinity in the variational problem. We now turn to the local study of the singularity of the flow at infinity, i.e. the equivalent of the Morse Lemma for a nondegenerate critical point of a gradient vector field.

For this, we will expand J near the set of potential critical points at infinity, which were analysed by J. Sacks and K. Uhlenbeck [8]. Let

$$H = \{w: \Omega \subset \mathbb{R}^n \to \mathbb{R} \text{ such that } \int_{\mathbb{R}^n} |\nabla w|^2 < +\infty ; \int_{\mathbb{R}^n} |w|^{\frac{2n}{n-2}} < +\infty \tag{5.1}$$

$$w = 0|_{\partial\Omega} \text{ if } \partial\Omega \neq \phi\}$$

H is a Hilbert space equipped with the scalar product:

$$(w, w_1)_H = \int_\Omega \nabla w \cdot \nabla w_1 ; \quad |w| = \sqrt{(w, w)_H} \tag{5.2}$$

Let

$$\Sigma = \{y \in H \mid |u| = 1\} \tag{5.3}$$

and

$$\Sigma^+ = \{u \in \Sigma \mid u \geq 0\}; \tag{5.4}$$

We introduced previously the functional

$$J(u) = \frac{(\int |\nabla u|^2)^{n/(n-2)}}{\int_\Omega K(x)\, u^{2n/(n-2)} dx} \quad ; \quad u \in H, \tag{5.5}$$

and the functions

$$\delta(a,\lambda) = c_0 \left(\frac{\lambda}{1+\lambda^2 |x-a|^2} \right)^{\frac{n-2}{2}} . \tag{5.6}$$

We consider J on Σ^+. Let

$$P \text{ be the orthogonal projection on } H_0^1(\Omega) \text{ if } \Omega \Subset \mathbb{R}^n; \tag{5.7}$$
$$P = \mathrm{Id} \text{ if } \Omega = \mathbb{R}^n.$$

P is defined by the equations:

$$P\phi = \phi - h \quad \text{with} \quad \Delta h = 0; \quad h = \phi|_{\partial\Omega}. \tag{5.8}$$

Let

$$S = \frac{\left(\int_{\mathbb{R}^n} |\nabla\delta(a,\lambda)|^2 \right)^{n/(n-2)}}{\int_{\mathbb{R}^n} \delta(a,\lambda))^{2n/(n-2)}} . \tag{5.9}$$

We introduce now the set of potential critical points at infinity. ε is a positive number and p is an integer.

$$V(p,\varepsilon) = \{u \in \Sigma \mid \exists (a_1,\dots,a_p) \in \Omega^p; \ \exists (\lambda_1,\dots,\lambda_p) \in \,]0,+\infty[^p \tag{5.10}$$

such that

$$\left.\begin{array}{l} \left| u - \frac{1}{\sqrt{p}} \sum_{i=1}^{p} \frac{1}{K(a_i)^{(n-2)/4}} P\delta(a_i,\lambda_i) \right| < \varepsilon; \quad \lambda_i d_i > \frac{1}{\varepsilon} \ \forall i; \\ \varepsilon_{ij}^{-1} > \frac{1}{\varepsilon} \ \forall i \neq j \\ V(0,\varepsilon) = \{ u \in \Sigma \mid |u| < \varepsilon \} \end{array} \right\} .$$

The following proposition is the analog, in the framework of these Yamabe type equations, of the analysis carried out by J. Sacks and K. Uhlenbeck [8] in the context of the harmonic map problem in dimension 2.

<u>Proposition 5.1</u>: Let (u_k) be a sequence in Σ_+ such that $\partial J(u_k) \to 0$ and $J(u_k)$ is bounded. There exists an extracted subsequence from (u_k), and a

function $\bar{u}$ in H, $\bar{u} \geq 0$, $\bar{u}/|\bar{u}|$ critical for J if $\bar{u} \neq 0$; there exist $p \in \mathbb{N}$ and a sequence $(\ _k)$, $\varepsilon_k \to 0$, such that $u_k - \bar{u}/(|u_k - \bar{u}|) \in V(p, \varepsilon_k)$.

The proof of Proposition 5.1 is available, up to minor modifications and improvements through papers by P. L. Lions [7], Y. T. Siu and S. T. Yau [10], M. Struwe [11] and Brezis-Coron [3]. Proposition 5.1 is related to J. Sacks and K. Uhlenbeck's work [8], showing the pointwise concentration of the singularities in the sequences violating condition (C) of Palais and Smale in these problems.

The technical proof of this proposition is left to the reader (see also A. Bahri - J. M. Coron [3]).

We now consider the following minimization problems for a function u in $V(p, \varepsilon)$ with ε small enough ($u \in \Sigma^+$ also).

$$\min \left| u - \sum_{i=1}^{p} \alpha_i \delta_i (x_i, \lambda_i) \right|; \quad \alpha_i > 0; \quad x_i \in \Omega; \quad \lambda_i > 0; \tag{5.11}$$

$$\min \left| u - \sum_{i=1}^{p} \alpha_i P\delta (x_i, \lambda_i) \right|; \quad \alpha_i > 0; \quad x_i \in \Omega; \quad \lambda_i > 0; \tag{5.12}$$

Let

$$B_\varepsilon = \{(\alpha_i, x_i, \lambda_i) ; \ i \in [1, p]; \ \alpha_i \in \mathbb{R}; \ x_i \in \Omega ; \ \lambda_i > 0 \tag{5.13}$$

such that

$$\lambda_i d_i \geq \frac{1}{\varepsilon} \ \forall\, i; \ \varepsilon_{ij}^{-1} \geq \forall\, i \neq j; \ \frac{1}{2\sqrt{p}} \frac{1}{K(a_i)^{(n-2)/4}} \leq \alpha_i \leq \frac{2}{K(a_i)^{(n-2)/4}} \}$$

We then have

<u>Proposition 5.2</u>: For any p, there exists $\varepsilon_0 > 0$ such that for any u in $V(p, \varepsilon_0)$, problems (5.11), (5.12) have a unique solution up to permutation, in $B_{4\varepsilon_0}$. The map therefore defined from $V(p, \varepsilon_0)$ into $B_{4\varepsilon_0}/\sigma_p$ is continuous. (σ_p is the symmetric group of order p acting on B_ε.)

The proof of Proposition 5.2 is also left to the reader (see also A. Bahri-J. M. Coron [3]).

We now turn to a local expansion of J.

Let u be in $V(p, \varepsilon)$; $\varepsilon < \varepsilon_0$. By Proposition 5.2, we write:

$$u = \sum_{i=1}^{p} \alpha_i P\delta_i + v \tag{5.14}$$

where v satisfies:

$$\int_{\mathbb{R}^n} \nabla v . \nabla P\delta_i = \int_{\mathbb{R}^n} \nabla v . \nabla \delta_i = 0 \tag{5.15}$$

$$\int_{\mathbb{R}^n} \nabla v . \nabla \frac{\partial}{\partial x_i} P\delta_i = \int_{\mathbb{R}^n} \nabla v . \frac{\nabla \partial \delta_i}{\partial x_i} = 0 \tag{5.16}$$

$$\int_{\mathbb{R}^n} \nabla v . \nabla \frac{\partial}{\partial \lambda_i} P\delta_i = \int_{\mathbb{R}^n} \nabla v . \frac{\nabla \partial \delta_i}{\partial \lambda_i} = 0 \tag{5.17}$$

Such a v thus satisfies (G0) and has support in $\overline{\Omega}$.

Let $H(x, y)$ be the regular part of the Green's function on Ω ; i.e.

$$\begin{cases} \Delta_x H(x,y) = 0 \\ H(x,y) = \dfrac{1}{|x-y|^{n-2}} \quad \text{on } \partial\Omega \end{cases} \tag{5.18}$$

We have:

<u>Proposition 5.3:</u>

$$J\left(\sum_{i=1}^{p} \alpha_i P\delta_i + v\right)$$

$$= \frac{(\Sigma \alpha_i^2)^{n/(n-2)}}{\Sigma \alpha_i^{2n/(n-2)} K(x_i)} \left(\int \delta^{\frac{2n}{n-2}}\right)^{\frac{2}{n-2}} \left\{ 1 - \left(\Sigma \alpha_i^{\frac{2n}{n-2}} K(x_j)\right)^{-1} c_7 \Sigma \alpha_i^{\frac{2n}{n-2}} \frac{\Delta K(x_i)}{\lambda_i^2} + o_K\left(\Sigma \frac{1}{\lambda_i^2}\right)\right.$$

$$- c_8\Big[\sum \frac{H(x_i, x_i)}{\lambda_i^{n-2}} \left(\frac{1}{2} \frac{\alpha_i^2}{\Sigma \alpha_j^2} - \alpha_i^2 \frac{\alpha_i^{4/(n-2)} K(x_i)}{\Sigma \alpha_j^2 (\alpha_j^{4/(n-2)} K(x_j))} \right)$$

$$+ \sum_{i \neq j} \left(\frac{H(x_i, x_j)}{(\lambda_i \lambda_j)^{(n-2)/2}} - \varepsilon_{ij} \right) \left(\frac{1}{2} \frac{\alpha_i \alpha_j}{\Sigma \alpha_k^2} - \frac{\alpha_i \alpha_j \alpha_i^{4/(n-2)} K(x_i)}{\Sigma \alpha_k^2 (\alpha_k^{4/(n-2)} K(x_k))} \right) \Big]$$

$$+ \frac{c_9}{\Sigma \alpha_i^2} \left(\int |\nabla v|^2 - \frac{n+2}{n-2} \Sigma \alpha_j^2 \, \Sigma \frac{\alpha_i^{4/(n-2)} K(x_i)}{\Sigma \alpha_j^2 (\alpha_j^{4/(n-2)} K(x_j))} \int \delta_i^{4/(n-2)} v^2 \right) \Big\}$$

$$+ (f, v)$$

$$+ 0 \left(\int |\nabla v|^2 \right)^{\frac{n}{n-2}} \text{ if } n \geq 6 + 0 \left(\int |\nabla v|^2 \left(\sum_{i \neq j} \varepsilon_{ij}^{\frac{2}{n-2}} (\operatorname{Log} \varepsilon_{ij}^{-1})^{2/n} \right) \right)$$

$$+ 0_K \left(\Sigma \frac{|DK(x_i)|}{\lambda_i} + \frac{(\operatorname{Log} \lambda_i)^{2/n}}{\lambda_i^2} \right) + \Sigma \frac{1}{\lambda_i^2 d_i^4} + \begin{cases} \Sigma \dfrac{1}{\lambda_j^2 d_j^{n-2}}, & n \geq 6 \\[2ex] \Sigma \dfrac{1}{\lambda_i^{(6-n)/2} \lambda_j^{(n-2)/2}} \dfrac{1}{d_j^{n-2}}, & n < 6 \end{cases}$$

$$+ 0\left(\left(\int |\nabla v|^2 \right)^{3/2} \right) \text{ if } n < 6$$

$$+ 0 \Bigg(\sum_{i \neq j} \varepsilon_{ij}^{n/(n-2)} + \Sigma \Bigg(\frac{1}{\lambda_i^{(n+2)/2} \lambda_j^{(n-2)/2}} \frac{\operatorname{Log} \lambda_i}{d_j^n} + \frac{1}{\lambda_i^{(n+2)/2}} \frac{1}{\lambda_j^{(n-2)/2}} \frac{1}{d_i^n}$$

$$+ \frac{1}{(\lambda_i d_i)^n} + \frac{1}{\lambda_i^2} \frac{1}{\lambda_j^{n-2}} \frac{1}{d_j^{2(n-2)}} \Bigg) \Bigg)$$

$$+ o_K(\varepsilon_{ij}) + (\text{if } n=3) \;\; 0\left(\int |\nabla v|^2 (\Sigma \varepsilon_{ij} \operatorname{Log} \varepsilon_{ij}^{-1}) \right),$$

where

$$(f, v) = \int_{\mathbb{R}^n} K(x) (\Sigma \alpha_i P\delta_i)^{\frac{n+2}{n-2}} v$$

$$= 0\left(\left(\int |\nabla v|^2\right)^{1/2}\left(\sum_{i\neq j} \varepsilon_{ij}^{\frac{2(n+2)}{n-2}} (\text{Log } \varepsilon_{ij}^{-1})^{\frac{n+2}{2n}} + \sum \frac{1}{\lambda_i^{(n+2)/2} d_i^{n+2}}\right.\right.$$

$$+ \sum \frac{1}{\lambda_i^2} \frac{1}{\lambda_j^{(n-2)/2}} \frac{1}{d_j^{n-2}} + 0_K\left(\sum \frac{|DK(x_i)|}{\lambda_i} + \frac{1}{\lambda_i^2}\right)$$

$$\left.\left. + (\text{if } n < 6) \sum_{i\neq j} \varepsilon_{ij} (\text{Log } \varepsilon_{ij}^{-1})^{\frac{n-2}{n}}\right)\right)$$

and

$$o_K(\varepsilon_{ij}) \leq \frac{c}{\lambda_i} 0(\varepsilon_{ij}((\text{Log } \varepsilon_{ij}^{-1})^{\frac{n-2}{2}}))\left(\frac{\text{Log } \lambda_i}{\lambda_i}^{2/n} + 1\right)\left[|DK(x_i)| + \frac{1}{\lambda_i^2}\right]$$

appears when K is not constant.

The previous expansion holds under the sole assumption that the ε_{ij} are small and v is small also. No assumption on the α_i's is required.

<u>Proof of Proposition 5.3</u>: We first expand:

$$\int_{\mathbb{R}^n} |\nabla(\sum_{i=1}^p \alpha_i P\delta_i + v)|^2 dx = \int_{\mathbb{R}^n} |\nabla v|^2 dx + \int_{\mathbb{R}^n} |\nabla(\sum \alpha_i P\delta_i)|^2 dx . \tag{5.19}$$

Now:

$$\int_{\mathbb{R}^n} |\nabla P\delta_i|^2 = \int_{\mathbb{R}^n} \nabla \delta_i \nabla P\delta_i dx = \int_\Omega \delta_i^{\frac{n+2}{n-2}} (\delta_i - \theta_i) dx. \tag{5.20}$$

We have:

$$\int_\Omega \delta_i^{\frac{2n}{n-2}} = \int_{\mathbb{R}^n} \delta_i^{\frac{2n}{n-2}} - \int_{\Omega^c} \delta_i^{\frac{2n}{n-2}} = \int_{\mathbb{R}^n} \delta_i^{\frac{2n}{n-2}} + 0\left(\frac{1}{(\lambda_i d_i)^n}\right) \tag{5.21}$$

We consider now the function θ_i defined in (65).

On $\partial\Omega$, we have:

$$\theta_i = c_0 \left(\frac{\lambda_i}{1+\lambda_i^2 |x-x_i|^2} \right)^{\frac{n-2}{2}} \quad , \quad x \in \partial\Omega; \tag{5.22}$$

hence

$$\left| \theta_i - \frac{c_0}{\lambda_i^{(n-2)/2}} \frac{1}{|x-x_i|^{n-2}} \right| \leq \frac{c}{\lambda_i^{(n+2)/2} d_i^n} \quad \text{on } \partial\Omega \tag{5.23}$$

while

$$\int \delta_i^{\frac{n+2}{n-2}} dx = 0 \left(\frac{1}{\lambda_i^{(n-2)/2}} \right) \tag{5.24}$$

Thus

$$\left| \theta_i - \frac{c_0}{\lambda_i^{(n-2)/2}} H(x, x_i) \right| = 0 \left(\frac{1}{\lambda_i^{(n+2)/2} d_i^n} \right) \quad \forall\, x \in \Omega, \tag{5.25}$$

and

$$\int_{\mathbb{R}^n} |\nabla P\delta_i|^2 = \int_{\mathbb{R}^n} \delta^{\frac{2n}{n-2}} - \int_\Omega \delta_i^{\frac{n+2}{n-2}} \frac{c_0}{\lambda_i^{(n-2)/2}} H(x, x_i)\, dx + 0 \left(\frac{1}{(\lambda_i d_i)^n} \right) \tag{5.26}$$

Now

$$\int_\Omega \delta_i^{\frac{n+2}{n-2}} H(x, x_i)\, dx = \int_{B_i} \delta_i^{\frac{n+2}{n-2}} H(x, x_i)\, dx + 0 \left(\frac{1}{\lambda_i^{(n+2)/2}} \right) \tag{5.27}$$

$$= \int_{B_i} \delta_i^{\frac{n+2}{n-2}} (H(x, x_i) - H(x_i, x_i))\, dx + H(x_i, x_i) \int_{B_i} \delta_i^{\frac{n+2}{n-2}} dx\, .$$

In

$$\int_{B_i} \delta_i^{\frac{n+2}{n-2}} (H(x, x_i) - H(x_i, x_i))\, dx\, ,$$

we expand $H(x, x_i)$ around x_i.

The first order term vanishes by oddness; the second order term vanishes because $\Delta H(x,x_i) = 0$; the third order term vanishes also by oddness and we are left with:

$$\int_{B_i} \delta_i^{\frac{n+2}{n-2}} (H(x,x_i) - H(x_i,x_i))\, dx = 0 \int_{B(0,\varepsilon)} \left(\frac{\lambda_i}{1+\lambda_i^2 r^2}\right)^{\frac{n+2}{2}} r^{n+3} dr \qquad (5.28)$$

Thus:

$$\int_{B_i} \delta_i^{\frac{n+2}{n-2}} H(x,x_i) = \frac{H(x_i,x_i)\,\bar{c}}{\lambda_i^{(n-2)/2}} + 0\left(\frac{1}{\lambda_i^{(n+2)/2}}\right) \qquad (5.29)$$

where

$$\bar{c} = c_0^{\frac{n+2}{n-2}} \int_{\mathbb{R}^n} \frac{1}{(1+|y|^2)^{(n+2)/2}} dy \,. \qquad (5.30)$$

Finally, summing up our estimates:

$$\int_{\mathbb{R}^n} |\nabla P\delta_i|^2 = \int_{\mathbb{R}^n} \delta^{\frac{2n}{n-2}} - c_0 \bar{c} \frac{H(x_i,x_i)}{\lambda_i^{n-2}} + 0\left(\frac{1}{(\lambda_i d_i)^n}\right) . \qquad (5.31)$$

We now consider:

$$\int_{\mathbb{R}^n} \nabla P\delta_i \nabla P\delta_j = \int_{\mathbb{R}^n} \delta_j^{\frac{n+2}{n-2}} (\delta_i - \theta_i) \qquad (5.32)$$

$$= \int_{\mathbb{R}^n} \delta_j^{\frac{n+2}{n-2}} \delta_i - \int_{\Omega^c} \delta_j^{\frac{n+2}{n-2}} \delta_i - \int_{\mathbb{R}^n} \delta_j^{\frac{n+2}{n-2}} \theta_i \,.$$

We have, by (E1):

$$\int_{\mathbb{R}^n} \delta_j^{\frac{n+2}{n-2}} \delta_i = c_0^{\frac{2n}{n-2}} c_1 \varepsilon_{ij} + 0(\varepsilon_{ij}^{\frac{n}{n-2}}) = c_0 \bar{c} \varepsilon_{ij} + 0(\varepsilon_{ij}^{\frac{n}{n-2}}). \qquad (5.33)$$

On the other hand

$$\int_{\Omega^c} \delta_j^{\frac{n+2}{n-2}} \delta_i \leq C\left[\int_{\Omega^c} \delta_j^{\frac{2n}{n-2}} + \int_{\Omega^c} \delta_i^{\frac{2n}{n-2}}\right] = 0\left(\frac{1}{(\lambda_i d_i)^n} + \frac{1}{(\lambda_j d_j)^n}\right) \tag{5.34}$$

Finally, by the same kind of estimates used for $\int \delta_i^{\frac{n+2}{n-2}} \theta_i$, we derive:

$$\int_{\mathbb{R}^n} \delta_j^{\frac{n+2}{n-2}} \theta_i = c_0^{\frac{2n}{n-2}} c_1 \frac{H(x_i, x_j)}{\lambda_i^{(n-2)/2} \lambda_j^{(n-2)/2}} + 0\left(\frac{1}{\lambda_j^{(n-2)/2}} \cdot \frac{1}{\lambda_i^{(n+2)/2}} \frac{1}{d_i^n}\right) \tag{5.35}$$

and thus:

$$\int_{\mathbb{R}^n} \nabla P\delta_i \nabla P\delta_j = c_0 \bar{c} \varepsilon_{ij} - c_0^{\frac{2n}{n-2}} c_1 \frac{H(x_i, x_j)}{(\lambda_i \lambda_j)^{(n-2)/2}} + \tag{5.36}$$

$$+0\left(\varepsilon_{ij}^{\frac{n}{n-2}} + \frac{1}{(\lambda_i d_i)^n} + \frac{1}{(\lambda_j d_j)^n}\right)$$

Thus, noticing that $c_0^{\frac{2n}{n-2}} c_1 = c_0 \bar{c}$,

$$\int_{\mathbb{R}^n} |\nabla(\Sigma \alpha_i P\delta_i + v)|^2 = \int_{\mathbb{R}^n} |\nabla v|^2 + \left(\int_{\mathbb{R}^n} \delta^{\frac{2n}{n-2}}\right) \Sigma \alpha_i^2 - c_0 \bar{c} \left\{ \sum_i \alpha_i^2 \frac{H(x_i, x_i)}{\lambda_i^{n-2}} \right.$$

$$\left. - \sum_{i \neq j} \left(\varepsilon_{ij} - \frac{H(x_i, x_j)}{(\lambda_i \lambda_j)^{(n-2)/2}}\right) \alpha_i \alpha_j \right\} + 0\left(\sum_i \frac{1}{(\lambda_i d_i)^n} + \sum_{i \neq j} \varepsilon_{ij}^{\frac{n}{n-2}}\right) \tag{5.37}$$

and

$$\left(\int_{\mathbb{R}^n} |\nabla(\Sigma \alpha_i P\delta_i + v)|^2\right)^{\frac{n}{n-2}} = (\Sigma \alpha_i^2)^{\frac{n}{n-2}} \left(\int_{\mathbb{R}^n} \delta^{\frac{2n}{n-2}}\right)^{\frac{n}{n-2}} \times \tag{5.38}$$

$$\times \left(1 - \frac{c_0 \bar{c}}{(\Sigma \alpha_i^2)\left(\int_{\mathbb{R}^n} \delta^{2n/(n-2)}\right)} \cdot \frac{n}{n-2} \left\{ \sum_i \alpha_i^2 \frac{H(x_i, x_i)}{\lambda_i^{n-2}} \right.\right.$$

$$- \sum_{i \neq j} \left(\varepsilon_{ij} - \frac{H(x_i, x_j)}{(\lambda_i \lambda_j)^{(n-2)/2}} \right) \alpha_i \alpha_j - \frac{\int |\nabla v|^2}{c_0 \bar{c}} \Bigg\}$$

$$+ 0 \left(\sum_i \frac{1}{(\lambda_i d_i)^n} + \sum_{i \neq j} \varepsilon_{ij}^{\frac{n}{n-2}} \right) .$$

We now estimate

$$\int_{\mathbb{R}^n} K(x) \left(\Sigma \alpha_i P\delta_i + v \right)^{\frac{2n}{n-2}} \tag{5.39}$$

We first remark that Estimate (G13) holds with δ_i replaced by $P\delta_i$. Indeed, throughout the proof of (G13) we used only upper bounds which still hold with $P\delta_i$ as $|P\delta_i(x)| \leq C|\delta_i(x)|$. Thus:

$$\int_{\mathbb{R}^n} K(x) \left(\Sigma \alpha_i P\delta_i + v \right)^{\frac{2n}{n-2}} - \int_{\mathbb{R}^n} K(x) \left(\Sigma \alpha_i P\delta_i \right)^{\frac{2n}{n-2}} \tag{5.40}$$

$$- \frac{2n}{n-2} \int_{\mathbb{R}^n} K(x) \left(\Sigma \alpha_i P\delta_i \right)^{\frac{n+2}{n-2}} v - \frac{n(n+2)}{(n-2)^2} \int_{\mathbb{R}^n} K(x) \left(\Sigma \alpha_i P\delta_i \right)^{\frac{4}{n-2}} v^2$$

$$= 0\left(\left(\int |\nabla v|^2 \right)^{\frac{n}{n-2}} \right) \text{ if } n \geq 6,$$

$$= 0\left(\left(\int |\nabla v|^2 \right)^{3/2} \right) \text{ if } n < 6.$$

Next we turn to

$$\int_{\mathbb{R}^n} K(x) \left(\Sigma \alpha_i P\delta_i \right)^{\frac{n+2}{n-2}} v = \int_{\mathbb{R}^n} K(x) \left(\Sigma \alpha_i \delta_i \right)^{\frac{n+2}{n-2}} v \tag{5.41}$$

$$+ \int_{\mathbb{R}^n} K(x) \left[\left(\Sigma \alpha_i (\delta_i - \theta_i) \right)^{\frac{n+2}{n-2}} - \left(\Sigma \alpha_i \delta_i \right)^{\frac{n+2}{n-2}} \right] v$$

By (G10), we have (we use here the fact that v satisfies (G0)):

$$\int_{\mathbb{R}^n} K(x)\,(\Sigma\,\alpha_i\delta_i)^{\frac{n+2}{n-2}}v = \begin{cases} 0\left((\int|\nabla v|^2)^{1/2}\left[(\sum_{i\neq j}\varepsilon_{ij}^{\frac{n+2}{2(n-2)}}(\mathrm{Log}\,\varepsilon_{ij}^{-1})^{\frac{n+2}{2n}} \right.\right. \\ \qquad \left.\left. + 0_K\ \ \Sigma\frac{|DK(x_i)|}{\lambda_i} + \frac{1}{\lambda_i^2}\right]\right. \text{if } n\geq 6, \\ 0\left((\int|\nabla v|^2)^{1/2}\left[(\sum_{i\neq j}\varepsilon_{ij}(\mathrm{Log}\,\varepsilon_{ij}^{-1})^{\frac{n-2}{n}})\right.\right. \\ \qquad \left.\left. + 0_K\left(\Sigma\frac{|DK(x_i)|}{\lambda_i} + \frac{1}{\lambda_i^2}\right)\right]\right. \text{if } n<6. \end{cases} \quad (5.42)$$

On the other hand:

$$\left|\int_{\mathbb{R}^n} K(x)\left[\left(\Sigma\,\alpha_i(\delta_i-\theta_i)\right)^{\frac{n+2}{n-2}} - (\Sigma\,\alpha_i\delta_i)^{\frac{n+2}{n-2}}\right]v\right| \quad (5.43)$$

$$\leq C\left[\int_{\mathbb{R}^n}\Sigma\,\alpha_i^{\frac{n+2}{n-2}}\theta_i^{\frac{n+2}{n-2}}|v| + \Sigma\,(\alpha_i\delta_i)^{\frac{4}{n-2}}\theta_j|v|\right]$$

$$\leq C(\int|\nabla v|^2)^{1/2}\left\{\Sigma\frac{1}{\lambda_i^{(n+2)/2}d_i^{n+2}} + \Sigma\frac{1}{\lambda_i^2}\frac{1}{\lambda_j^{(n-2)/2}}\frac{1}{d_j^{n-2}}\right\}$$

To obtain (5.43), we used estimates on θ_i and the Hölder inequality on Ω, as all integrals may be seen as integrals taken over Ω since v is zero outside Ω. Thus

$$\int_{\mathbb{R}^n} K(x)\,(\Sigma\,\alpha_i P\delta_i)^{\frac{n+2}{n-2}}v \text{ is a linear form on } v \text{ which is} \quad (5.44)$$

upperbounded by

$$0\left((\int|\nabla v|^2)^{1/2}\left(\sum_{i\neq j}\varepsilon_{ij}^{\frac{n+2}{2(n-2)}}(\mathrm{Log}\,\varepsilon_{ij}^{-1})^{\frac{n+2}{2n}} + \Sigma\frac{1}{\lambda_i^{(n+2)/2}}\frac{1}{d_i^{n+2}}\right.\right.$$

$$+ \Sigma \frac{1}{\lambda_i^2} \frac{1}{\lambda_j^{(n-2)/2}} \frac{1}{d_j^{n-2}} + 0_K \left(\Sigma \frac{|DK(x_i)|}{\lambda_i} + \frac{1}{\lambda_i^2} \right) \Bigg) \Bigg) \quad \text{if } n \geq 6,$$

and by

$$0 \left(\left(\int |\nabla v|^2 \right)^{1/2} \left(\sum_{i \neq j} \varepsilon_{ij} (\mathrm{Log}\, \varepsilon_{ij}^{-1})^{\frac{n-2}{n}} + \Sigma \frac{1}{\lambda_i^{(n+2)/2} d_i^{n+2}} \right.\right.$$

$$\left.\left. + \Sigma \frac{1}{\lambda_i^2} \frac{1}{\lambda_j^{(n-2)/2}} \frac{1}{d_j^{n-2}} + 0_K \left(\Sigma \frac{|DK(x_i)|}{\lambda_i} + \frac{1}{\lambda_i^2} \right) \right) \right) \quad \text{if } n < 6.$$

Next, we estimate

$$\int_{\mathbb{R}^n} K(x) (\Sigma\, \alpha_i P\delta_i)^{\frac{4}{n-2}} v^2 \tag{5.45}$$

$$= \int_{\mathbb{R}^n} K(x) (\Sigma\, \alpha_i \delta_i)^{\frac{4}{n-2}} v^2 + \int_{\mathbb{R}^n} K(x) \left((\Sigma\, \alpha_i P\delta_i)^{\frac{4}{n-2}} - (\Sigma\, \alpha_i \delta_i)^{\frac{4}{n-2}} \right) v^2$$

by (G11), we have:

$$\int_{\mathbb{R}^n} K(x) (\Sigma\, \alpha_i \delta_i)^{\frac{4}{n-2}} v^2 = \Sigma\, \alpha_i^{\frac{4}{n-2}} K(x_i) \int_{\mathbb{R}^n} \delta_i^{\frac{4}{n-2}} v^2 + 0 \left(\left(\int |\nabla v|^2 \right) \times \right. \tag{5.46}$$

$$\times \left[\sum_{i \neq j} (\alpha_i \alpha_j)^{\frac{2}{n-2}} \varepsilon_{ij}^{\frac{2}{n-2}} (\mathrm{Log}\, \varepsilon_{ij}^{-1})^{\frac{2}{n-2}} \quad \text{if } n > 3, \text{ or} \right.$$

$$\lambda \left(\sum_{i \neq j} \alpha_i^3 \alpha_j \varepsilon_{ij} \,\mathrm{Log}\, \varepsilon_{ij}^{-1} \right) \quad \text{if } n = 3;$$

$$\left. + 0_K \left(\Sigma \frac{|DK(x_i)|}{\lambda_i} + \frac{(\mathrm{Log}\, \lambda_i)^{2/n}}{\lambda_i^2} \right) \right]$$

On the other hand:

$$\left| \int_{\mathbb{R}^n} K(x) \left((\Sigma\, \alpha_i P\delta_i)^{\frac{4}{n-2}} - (\Sigma\, \alpha_i \delta_i)^{\frac{4}{n-2}} \right) v^2 \right| \tag{5.47}$$

$$\leq C\left\{\Sigma\, \alpha_i^{\frac{4}{n-2}} \int_{\mathbb{R}^n} \theta_i^{\frac{4}{n-2}} v^2 + \sum_j \int_{\mathbb{R}^n} (\Sigma\, \alpha_i \delta_i)^{\frac{4}{n-2}-1} \alpha_j \theta_j v^2\right\}$$

$$= 0\left(\int |\nabla v|^2 \left(\Sigma \frac{1}{\lambda_i^2 d_i^4} + \begin{cases} \sum_{i,j} [\int_{\Omega} (\alpha_j \delta_j)^{\frac{6-n}{n-2} \cdot \frac{n}{2}}]^{\frac{2}{n}} \operatorname{Sup} \theta_i & \text{if } n \geq 6 \\ \sum_{i,j} [\int_{\Omega} (\alpha_j \delta_j)^{\frac{6-n}{n-2} \frac{n}{2}}]^{\frac{2}{n}} \operatorname{Sup} \theta_i & \text{if } n < 6 \end{cases}\right)\right)$$

$$= 0\left(\int |\nabla v|^2 \left(\Sigma \frac{1}{\lambda_i^2 d_i^4} + \begin{cases} \Sigma \frac{1}{\lambda_j^2} \frac{1}{d_j^{n-2}} & \text{if } n \geq 6 \\ \Sigma \frac{1}{\lambda_i^{(6-n)/2}} \frac{1}{\lambda_j^{(n-2)/2}} \frac{1}{d_j^{n-2}} & \text{if } n < 6 \end{cases}\right)\right)$$

Here again, we used estimates on θ_i and the Hölder inequality on Ω. We thus have:

$$\int_{\mathbb{R}^n} K(x) (\Sigma\, \alpha_i P\delta_i)^{\frac{4}{n-2}} v^2 = \Sigma\, \alpha_i^{\frac{4}{n-2}} K(x_i) \int_{\mathbb{R}^n} \delta_i^{\frac{4}{n-2}} v^2 \tag{5.48}$$

$$+ 0\left(\int |\nabla v|^2 [\Sigma (\alpha_i \alpha_j)^{\frac{2}{n-2}} \varepsilon_{ij}^{\frac{2}{n-2}} (\operatorname{Log} \varepsilon_{ij}^{-1})^{2/n}]\right) \quad \text{for } n > 3$$

$$+ 0\left(\int |\nabla v|^2 [\sum_{i \neq j} \alpha_i^3 \alpha_j \varepsilon_{ij} \operatorname{Log} \varepsilon_{ij}^{-1}]\right) \quad \text{for } n = 3$$

$$+ 0_K\left(\left(\Sigma \frac{|DK(x_i)|}{\lambda_i} + \frac{\operatorname{Log} \lambda_i}{\lambda_i^2}\right)^{2/n}\right) + \Sigma \frac{1}{\lambda_i^2 d_i^4}$$

$$+ \Sigma \frac{1}{\lambda_j^2 d_j^{n-2}} \quad \text{if } n \geq 6,$$

$$+ \Sigma \frac{1}{\lambda_i^{(6-n)/2}} \frac{1}{\lambda_j^{(n-2)/2}} \frac{1}{d_j^{n-2}} \quad \text{if } n < 6.$$

Lastly, we estimate, extending θ_i by δ_i on Ω^c:

$$\int_{\mathbb{R}^n} K(x) (\Sigma \alpha_i P\delta_i)^{\frac{2n}{n-2}} = \int_{\mathbb{R}^n} K(x) (\Sigma \alpha_i \delta_i)^{\frac{2n}{n-2}} \tag{5.49}$$

$$- \frac{2n}{n-2} \int_{\mathbb{R}^n} K(x) (\Sigma \alpha_i \delta_i)^{\frac{n+2}{n-2}} (\Sigma \alpha_i \theta_i) + 0 \left(\Sigma \int \theta_i^{\frac{2n}{n-2}} + \Sigma \int \delta_i^{\frac{4}{n-2}} \theta_j^2 \right)$$

$$= \int_{\mathbb{R}^n} K(x) (\Sigma \alpha_i \delta_i)^{\frac{2n}{n-2}} - \frac{2n}{n-2} \int_{\mathbb{R}^n} K(x) \left(\Sigma \alpha_i^{\frac{n+2}{n-2}} \delta_i^{\frac{n+2}{n-2}} \right) (\Sigma \alpha_i \theta_i)$$

$$+ 0 \left(\Sigma \int_{\mathbb{R}^n} \theta_i^{\frac{2n}{n-2}} + \Sigma \int_{\mathbb{R}^n} \delta_i^{\frac{4}{n-2}} \theta_j^2 + \sum_{i \neq j} \int_{\mathbb{R}^n} (\alpha_i \delta_i)^{\frac{4}{n-2}} \operatorname{Inf}(\alpha_i \delta_i, \alpha_j \delta_j)^2 \right).$$

Using then Estimates (F8) and (F10) and (2.21), (2.22) (see also (2.64)), we derive:

$$\int_{\mathbb{R}^n} K(x) (\Sigma \alpha_i P\delta_i)^{\frac{2n}{n-2}} = \Sigma (\alpha_i c_0)^{\frac{2n}{n-2}} \left\{ K(x_i) \int \frac{r^{n-1} dr\, d\sigma}{(1+r^2)^n} \right. \tag{5.50}$$

$$\left. + \frac{\Delta K(x_i)}{\lambda_i^2} \int \frac{|x|^2 dx}{(1+|x|^2)^n} + o_K(\frac{1}{\lambda_i^2}) \right\} + \frac{2n}{n-2} \sum_{i \neq j} \alpha_i^{\frac{n+2}{n-2}} \alpha_j c_0^{\frac{2n}{n-2}} c_1 \varepsilon_{ij} K(x_i)$$

$$- \frac{2n}{n-2} c_0^{\frac{n+2}{n-2}} c_1 \Sigma \frac{K(x_i)\, \theta_j(x_i)}{\lambda_i^{(n-2)/2}} \alpha_i^{\frac{n+2}{n-2}} \alpha_j + 0 \left(\Sigma \frac{1}{\lambda_i^{(n+2)/2}} \frac{1}{\lambda_j^{(n-2)/2}} \frac{\operatorname{Log} \lambda_i}{d_j^n} \right.$$

$$\left. + \sum_{i \neq j} \varepsilon_{ij}^{\frac{n}{n-2}} \operatorname{Log} \varepsilon_{ij}^{-1} + \Sigma \frac{1}{(\lambda_i d_i^2)^n} + \Sigma \frac{1}{\lambda_i^2} \frac{1}{\lambda_j^{n-2}} \frac{1}{(d_j^2)^{n-2}} \right)$$

$$+ o_K(\varepsilon_{ij}) + (\text{if } n = 3) \;\; 0(\sum_{i \neq j} \varepsilon_{ij}^2 (\operatorname{Log} \varepsilon_{ij}^{-1})^{2/3})$$

where

$$o_K(\varepsilon_{ij}) \leq \frac{c}{\lambda_i} \, 0 \left(\varepsilon_{ij} \; (\operatorname{Log} \varepsilon_{ij}^{-1})^{\frac{n-2}{2}} \left(\left(\frac{\operatorname{Log} \lambda_i}{\lambda_i} \right)^{2/n} + 1 \right) [\, |DK(x_i)| + \frac{1}{\lambda_i}] \right) \tag{5.51}$$

appears when K is nonconstant.

Hence, using (5.25):

$$\int_{\mathbb{R}^n} K(x)\,(\Sigma\,\alpha_i P\delta_i)^{\frac{2n}{n-2}} = \Sigma\,(\alpha_i c_0)^{\frac{2n}{n-2}} \Bigg\{ K(x_i) \int \frac{r^{n-1}\,dr\,d\sigma}{(1+r^2)^n} \tag{5.52}$$

$$+ \frac{\Delta K(x_i)}{\lambda_i^2} \int \frac{|x|^2 dx}{(1+|x|^2)^n} + o_K\left(\frac{1}{\lambda_i^2}\right)\Bigg\} + \frac{2n}{n-2}\Bigg\{ \sum_{i\neq j} \alpha_i^{\frac{n+2}{n-2}} \alpha_j c_0^{\frac{2n}{n-2}} c_1 \varepsilon_{ij} K(x_i)$$

$$- c_0^{\frac{n+2}{n-2}} c_1\, \Sigma \frac{K(x_i)\,H(x_i, x_j)}{(\lambda_i \lambda_j)^{(n-2)/2}}\Bigg\} + 0\Bigg(\Sigma\Bigg[\frac{1}{\lambda_1^{(n+2)/2}}\,\frac{1}{\lambda_j^{(n-2)/2}}\,\frac{\operatorname{Log}\lambda_i}{d_j^n}$$

$$+ \frac{1}{\lambda_i^{(n+2)/2}}\,\frac{1}{\lambda_j^{(n-2)/2}}\,\frac{1}{d_i^n} + \frac{1}{\lambda_i^2}\,\frac{1}{\lambda_j^{n-2}}\,\frac{1}{(d_j^2)^{n-2}}\Bigg] + \sum_{i\neq j} \varepsilon_{ij}^{\frac{n}{n-2}}\Bigg]\Bigg)$$

$$+ o_K(\varepsilon_{ij}) + (\text{if } n = 3)\ 0\Big(\sum_{i\neq j} \varepsilon_{ij}^2 (\operatorname{Log} \varepsilon_{ij}^{-1})^{2/3}\Big)$$

where o_K satisfies (5.51).

Summing up (5.40), (5.44), (5.48) and (5.52), we obtain:

$$\int_{\mathbb{R}^n} K(x)\,(\Sigma\,\alpha_i P\delta_i + v)^{\frac{2n}{n-2}} = \Sigma\ \alpha_i c_0^{\frac{2n}{n-2}} \Bigg\{ K(x_i) \int \frac{r^{n-1}\,dr\,d\sigma}{(1+r^2)^n} \tag{5.53}$$

$$+ \frac{\Delta K(x_i)}{\lambda_i^2} \int \frac{|x|^2 dx}{(1+|x|^2)^n} + o_K\left(\frac{1}{\lambda_i^2}\right)\Bigg\} + \frac{2n}{n-2}\Bigg\{ \sum_{i\neq j} \alpha_i^{\frac{n+2}{n-2}} \alpha_j c_0^{\frac{2n}{n-2}} c_1 \varepsilon_{ij} K(x_i)$$

$$- \alpha_i^{\frac{n+2}{n-2}} \alpha_j c_0^{\frac{2n}{n-2}} c_1\, \Sigma \frac{K(x_i)\,H(x_i, x_j)}{(\lambda_i \lambda_j)^{(n-2)/2}}\Bigg\} + \frac{n(n+2)}{(n-2)^2}\, \Sigma\, \alpha_i^{\frac{4}{n-2}} K(x_i) \int_{\mathbb{R}^n} \delta_i^{\frac{4}{n-2}} v^2 + (f, v)$$

$$+ \begin{cases} 0\Big(\big(\int |\nabla v|^2\big)^{\frac{n}{n-2}}\Big) & \text{if } n \geq 6 \\ 0\Big(\big(\int |\nabla v|^2\big)^{3/2}\Big) & \text{if } n < 6 \end{cases}$$

$$+ 0\left(\sum_{i\neq j} \varepsilon_{ij}^{\frac{n}{n-2}} + \Sigma\left(\frac{1}{\lambda_i^{(n+2)/2}\lambda_j^{(n-2)/2}} \frac{\operatorname{Log}\lambda_i}{d_j^n} + \frac{1}{\lambda_i^{(n+2)/2}} \frac{1}{\lambda_j^{(n-2)/2}} \frac{1}{d_i^n}\right.\right.$$

$$\left.\left. + \frac{1}{\lambda_i^2} \frac{1}{\lambda_j^{n-2}} \frac{1}{d_j^{2(n-2)}}\right)\right) + o_K(\varepsilon_{ij}) + 0\left(\int|\nabla v|^2\left[\sum_{i\neq j}(\alpha_i\alpha_j)^{\frac{2}{n-2}} \varepsilon_{ij}^{\frac{2}{n-2}} (\operatorname{Log}\varepsilon_{ij}^{-1})^{2/n}\right.\right.$$

$$+ 0_K\left(\Sigma \frac{|DK(x_i)|}{\lambda_i} + \frac{(\operatorname{Log}\lambda_i)^{2/n}}{\lambda_i^2}\right) + \Sigma \frac{1}{\lambda_i^2 d_i^4}$$

$$+ \left\{\begin{array}{ll} \Sigma \dfrac{1}{\lambda_j^2 d_j^{n-2}}, & n \geq 6 \\ & \\ \Sigma \dfrac{1}{\lambda_i^{(6-n)/2}\lambda_j^{(n-2)/2}} \dfrac{1}{d_j^{n-2}}, & n < 6 \end{array}\right.\left.\vphantom{\dfrac{1}{1}}\right] + (\text{if } n=3)\ 0\Big(\sum_{i\neq j} \varepsilon_{ij}^2 (\operatorname{Log}\varepsilon_{ij}^{-1})^{2/3}$$

$$+ 0((\int|\nabla v|^2)(\Sigma\, \varepsilon_{ij} \operatorname{Log}\varepsilon_{ij}^{-1}))$$

where:

$$(f, v) = \int_{\mathbb{R}^n} K(x)(\Sigma\, \alpha_i P\delta_i)^{\frac{n+2}{n-2}} v \tag{5.54}$$

$$= 0\left((\int|\nabla v|^2)^{1/2}\left(\sum_{i\neq j} \varepsilon_{ij}^{\frac{n+2}{2(n-2)}} (\operatorname{Log}\varepsilon_{ij}^{-1})^{\frac{n+2}{2n}}\right.\right.$$

$$+ \Sigma \frac{1}{\lambda_i^{(n+2)/2}} \frac{1}{d_i^{n+2}} + \Sigma \frac{1}{\lambda_i^2} \frac{1}{\lambda_j^{(n-2)/2}} \frac{1}{d_j^{n-2}} + 0_K\left(\frac{|DK(x_i)|}{\lambda_i} + \frac{1}{\lambda_i^2}\right)$$

$$\left.\left. + (\text{if } n < 6) \quad \sum_{i\neq j} \varepsilon_{ij} (\operatorname{Log}\varepsilon_{ij}^{-1})^{\frac{n-2}{2}}\right)\right).$$

Noticing again that $c_0 \bar{c} = c_0^{\frac{2n}{n-2}} c_1$, we derive the following expansion for $J(\Sigma\, \alpha_i P\delta_i + v)$,

$$J(\Sigma \alpha_i P\delta_i + v) = \frac{(\Sigma \alpha_i^2)^{n/(n-2)}}{\Sigma \alpha_i^{2n/(n-2)} K(x_i)} \left(\int \delta^{\frac{2n}{n-2}}\right)^{\frac{2}{n-2}} \left\{ 1 - \frac{(\Sigma \alpha_j^{2n/(n-2)} K(x_j))^{-1}}{\int \delta^{2n/(n-2)}} \times \right.$$

$$\cdot \int \frac{|x|^2 dx}{(1+|x|^2)^n} \sum_i \alpha_i^{\frac{2n}{n-2}} \frac{\Delta K(x_i)}{\lambda_i^2} + o_K\left(\Sigma \frac{1}{\lambda_i^2}\right) \tag{5.55}$$

$$- c_0 \bar{c}^{\frac{2n}{n-2}} \left(\Sigma \frac{H(x_i, x_i)}{\lambda_i^{n-2}} \left[\frac{1}{2} \frac{\alpha_i^2}{(\Sigma \alpha_j^2)(\int_{\mathbb{R}^n} \delta^{2n/(n-2)})}\right.\right.$$

$$\left. - \frac{\alpha_i^{2n/(n-2)} K(x_i)}{\Sigma \alpha_j^{2n/(n-2)} K(x_j)} \cdot \frac{1}{\int \delta^{2n/(n-2)}}\right] + \sum_{i \neq j} \left(\frac{H(x_i, x_j)}{(\lambda_i \lambda_j)^{(n-2)/2}} - \varepsilon_{ij}\right)$$

$$\left. \cdot \left[\frac{1}{2} \frac{\alpha_i \alpha_j}{(\Sigma \alpha_k^2)(\int_{\mathbb{R}^n} \delta^{2n/(n-2)})} - \frac{\alpha_i^{(n+2)/(n-2)} \alpha_j K(x_i)}{\Sigma \alpha_j^{2n/(n-2)} K(x_j)} \cdot \frac{1}{\int \delta^{2n/(n-2)}}\right]\right)$$

$$\left. + \frac{n}{n-2} \frac{1}{\int \delta^{2n/(n-2)}} \left(\frac{\int |\nabla v|^2}{\Sigma \alpha_i^2} - \frac{n+2}{n-2} \Sigma \frac{\alpha_i^{4/(n-2)} K(x_i)}{\Sigma \alpha_j^{2n/(n-2)} K(x_j)} \int \delta_i^{\frac{4}{n-2}} v^2\right) \right\}$$

$$+ (f, v) + \begin{cases} 0((\int |\nabla v|^2)^{\frac{n}{n-2}}) & \text{if } n \geq 6 \\ 0((\int |\nabla v|^2)^{3/2}) & \text{if } n < 6 \end{cases}$$

$$+ 0\left(\sum_{i \neq j} \varepsilon_{ij}^{\frac{n}{n-2}} + \Sigma\left(\frac{1}{\lambda_i^{(n+2)/2} \lambda_j^{(n-2)/2}} \frac{\text{Log } \lambda_i}{d_j^n} + \frac{1}{\lambda_i^{(n+2)/2}} \frac{1}{\lambda_j^{(n-2)/2}} \frac{1}{d_i^n}\right.\right.$$

$$\left.\left. + \frac{1}{(\lambda_i d_i)^n} + \frac{1}{\lambda_i^2} \frac{1}{\lambda_j^{n-2}} \frac{1}{d_j^{2(n-2)}}\right)\right) + o_K(\varepsilon_{ij})$$

$$+ 0\left((\int |\nabla v|^2)\left[\sum_{i \neq j} \varepsilon_{ij}^{2/(n-2)} (\text{Log } \varepsilon_{ij}^{-1})^{2/n} + 0_K\left(\Sigma \frac{|DK(x_i)|}{\lambda_i} + \frac{(\text{Log } \lambda_i)^{2/n}}{\lambda_i^2}\right)\right.\right.$$

$$+\Sigma \frac{1}{\lambda_i^2 d_i^4} + \begin{cases} \Sigma \dfrac{1}{\lambda_j^2 d_j^{n-2}}, & n \geq 6 \\ \Sigma \dfrac{1}{\lambda_i^{(6-n)/2}} \dfrac{1}{\lambda_j^{(n-2)/2}} \dfrac{1}{d_j^{n-2}}, & n < 6 \end{cases} \Bigg]\Bigg)$$

$$+ (\text{if } n = 3) \quad 0((\int |\nabla v|^2)(\Sigma \varepsilon_{ij} \operatorname{Log} \varepsilon_{ij}^{-1}))$$

where (f, v) satisfies (5. 54).

Hence the proof of Proposition 5. 3.

From the expansion given in Proposition 5. 3, we draw now a certain number of first conclusions which give an insight into the phenomenon at infinity. We first take care of the v-part in the expansion given by Proposition 5. 3. This v-part looks at first order like:

$$(f, v) + Q(v) + 0((\int |\nabla v|^2)^{1+\alpha}) \tag{5. 56}$$

where Q is a quadratic form of the type

$$\int |\nabla v|^2 - \frac{n+2}{n-2} \Sigma \alpha_j^2 \Sigma \frac{\alpha_i^{4/(n-2)} K(x_i)}{\Sigma \alpha_j^2 (\alpha_j^{4/(n-2)} K(x_j))} \int \delta_i^{\frac{4}{n-2}} v^2 + o(1) \int |\nabla v|^2 \tag{5. 57}$$

and v is in:

$$v \in H_1 = \{w: \Omega \subset \mathbb{R}^n \to \mathbb{R}; \int_\Omega |\nabla w|^2 < +\infty ; \int_\Omega |w|^{\frac{2n}{n-2}} < +\infty \tag{5. 58}$$

$$\int_{\mathbb{R}^n} \nabla \delta_i \nabla w = 0 ; \int_{\mathbb{R}} \nabla \frac{\partial \delta_i}{\partial x_i} \nabla w = 0 ; \int_{\mathbb{R}^n} \nabla \frac{\partial \delta_i}{\partial \lambda_i} \nabla w = 0,$$

$$i = 1, \ldots, p; \quad w = 0|_{\partial\Omega} \quad \text{if } \Omega \Subset \mathbb{R}^n \}$$

We assume u to be in $V(p, \varepsilon) \cap \Sigma^+$ with ε small.

$$0 < \varepsilon \ll 1 \tag{5. 59}$$

The assumption over ε implies

$$J(\sum_{i=1}^{p} \alpha_i \delta_i)\, \alpha_i^{4/(n-2)} K(x_i) = 1 + o(1) \tag{5.60}$$

Thus, Q is in fact close to:

$$\int |\nabla v|^2 - \frac{n+2}{n-2} \sum_i \int \delta_i^{4/(n-2)} v^2 + o(1) \int |\nabla v|^2. \tag{5.61}$$

On the other hand, the remainder term $0((\int|\nabla v|^2)^{1+\alpha})$ is, as the functional J is, twice differentiable in v. Its differential at zero, thus:

$$D0((\int|\nabla v|^2)^{1+\alpha}) = 0((\int|\nabla v^2|)^{1/2+\alpha}), \quad \alpha > 0. \tag{5.62}$$

From this and from the estimate on the linear term given in Proposition 5.3, we derive:

Proposition 5.4: For any given set of functions $(\delta_1, \ldots, \delta_p)$ with λ_i large, ε_{ij} small for $i \neq j$ and $(\alpha_1, \ldots, \alpha_p)$ satisfying (5.60), there exists an optimal $\bar{v}$ minimizing $J(\Sigma\, \alpha_i P\delta_i + v)$ on the set of v's belonging to a small neighbourhood of zero in H_1.

There is, in fact, a retraction by deformation of $V(p, \varepsilon)$ onto $\{ u \in V(p, \varepsilon)\ ; \quad u = \sum_{i=1}^{p} \alpha_i P\delta_i + \bar{v};\ \bar{v}$ optimal with respect to (α_i, δ_i); $i = 1, \ldots, p\}$, along which the functional decreases.

Finally, we have the estimate:

$$(f, \bar{v}) + Q(\bar{v}) + 0\left((\int|\nabla \bar{v}|^2)^{1+\alpha}\right) \leq C|\bar{v}|^2 \leq 0\Big(\sum_{i\neq j} \varepsilon_{ij}^{\frac{n+2}{n-2}} (\text{Log}\, \varepsilon_{ij}^{-1})^{\frac{n+2}{n}} \tag{5.63}$$

$$+ \sum \frac{1}{\lambda_i^{n+2}} \frac{1}{d_i^{2(n+2)}} + \sum \frac{1}{\lambda_i^4} \frac{1}{\lambda_j^{n-2}} \frac{1}{d_j^{2(n-2)}} + 0_K\left(\sum \frac{|DK(x_i)|^2}{\lambda_i^2} + \frac{1}{\lambda_i^4}\right)$$

$$+ (\text{if } n < 6) \sum_{i\neq j} \varepsilon_{ij}^2 (\text{Log}\, \varepsilon_{ij}^{-1})^{\frac{2(n-2)}{n}} \qquad .$$

Proof: The existence of $\bar{v}$ follows from the implicit function theorem. Indeed, by Proposition 5.1 and (5.61), Q is definite positive, lower-bounded by $\bar{\alpha}_{0/4} \int |\nabla v|^2$ on H_1; thus represented with respect to the H_1-scalar product by a self-adjoint, continuous, invertible and positive operator A:

$$(Av, v)_{H_1} = Q(v) ; \quad \alpha \, \mathrm{Id} \leq A \leq \beta \, \mathrm{Id}, \quad \alpha > 0. \tag{5.64}$$

On the other hand, f is small. Indeed by the estimate on (f, v) given in Proposition 5.3, we have:

$$|f|^2_H \leq 0\Bigg(\sum_{i \neq j} \varepsilon_{ij}^{\frac{n+2}{n-2}} (\mathrm{Log}\, \varepsilon_{ij}^{-1})^{\frac{n+2}{n}} + \Sigma \frac{1}{\lambda_i^{n+2}} \frac{1}{d_i^{2(n+2)}} \tag{5.65}$$

$$+ \Sigma \frac{1}{\lambda_i^4} \frac{1}{\lambda_j^{n-2}} \frac{1}{d_j^{2(n-2)}}$$

$$+ 0_K\Big(\Sigma \frac{|DK(x_i)|^2}{\lambda_i^2} + \frac{1}{\lambda_i^4}\Big) + (\text{if } n < 6) \sum_{i \neq j} \varepsilon_{ij}^2 (\mathrm{Log}\, \varepsilon_{ij}^{-1})^{\frac{2(n-2)}{n}} \Bigg)$$

Using then (5.62), we see that the operator:

$$A + D0((\int |\nabla v|^2)^{1+\alpha}) \tag{5.66}$$

is invertible in a neighbourhood of the origin (depending only on the smallness of $\int |\nabla v|^2$) with an inverse A^{-1} satisfying

$$A^{-1} \geq 1/2\beta \; \mathrm{Id}. \tag{5.67}$$

From (5.65), the existence of $\bar{v}$ follows. From the positiveness of A + D0, the existence of the retraction by deformation decreasing J follows. The estimate (5.63) is then an easy corollary of (5.65). Hence the proof of Proposition 5.4.

We turn now to the justification of the expansions in ε_{ij} for $i \neq j$. Namely, we prove that, if ε_{ij} is not small (which might occur if two points are close enough with respect to the concentrations λ_i's of the functions δ_i's), then the energy $J(\sum_{i=1}^{p} \alpha_i P\delta_i + v)$ is down from the critical level for p masses concentrated at points $(x_1, \dots, x_p)$ at infinity, i.e.

$$J(\sum_{i=1}^{p} \alpha_i P\delta_i + v) < (\int \delta^{\frac{2n}{n-2}})^{\frac{2}{n-2}} \left(\Sigma \frac{1}{(K(x_i))^{(n-2)/2}} \right)^{\frac{2}{n-2}}. \tag{5.68}$$

Down from this critical level, the p-masses $(\delta_1(x_1, \infty), \dots, \delta_p(x_p, +\infty))$ cannot build a critical point at infinity, i.e. the gradient flow will necessarily lead to other points and other waves of strictly less energy, if it is to go to infinity.

This is justified in the following

Proposition 5.5: Assume one of the ε_{ij} $(i \neq j)$ is larger than $\varepsilon_0 > 0$. Then

$$J(\sum_{i=1}^{p} \alpha_i P\delta_i + v) < (\int \delta^{\frac{2n}{n-2}})^{\frac{2}{n-2}} \left(\Sigma \frac{1}{(K(x_i))^{(n-2)/2}} \right)^{\frac{2}{n-2}} - \gamma_1(\varepsilon_0)$$

as soon as $|v|^2 < \phi(\varepsilon_0)$ where $\gamma_1, \phi : \mathbb{R}^{+\star} \to \mathbb{R}^{+\star}$ are strictly positive continuous functions. This holds under the assumptions $\frac{1}{\lambda_i} < \varepsilon_1$ and $\frac{1}{\sqrt{\lambda_i d_i}} < M$, where M is a fixed constant and $\varepsilon_1 > 0$ is small enough.

Proof of Proposition 5.5: Assuming that the concentrations λ_i are large enough $(\sum_{i=1}^{p} \alpha_i P\delta_i + v \in V(p, \varepsilon_1))$, $\varepsilon_{i_0 j_0}$ $(i_0 \neq j_0)$ cannot be larger than $\varepsilon_0 > 0$ if the two points x_{i_0} and x_{j_0} are not close enough. In fact, for such an i and j, we have:

$$\frac{\lambda_{i_0}}{\lambda_{j_0}} + \frac{\lambda_{j_0}}{\lambda_{i_0}} + \lambda_{i_0}\lambda_{j_0}|x_{i_0} - x_{j_0}|^2 < \frac{1}{\varepsilon_0^{(n-2)/2}}. \tag{5.69}$$

Let us first assume that K is constant. We will adjoint the argument for the general case later on.

Let

$$\bar{a}_i = \frac{\alpha_i P\delta_i}{\Sigma \alpha_j P\delta_j} \ ; \quad a_i = \frac{\alpha_i \delta_i}{\Sigma \alpha_j \delta_j} \ . \tag{5.70}$$

We have:

$$\left(\frac{\Sigma\, \alpha_i (\delta_i)^{(n+2)/(n-2)}}{\Sigma\, \alpha_i \delta_i} \right)^{\frac{2n}{n+2}} \leq \frac{1}{\Sigma\, \alpha_i \delta_i} \ \Sigma\, \alpha_i (\delta_i)^{\frac{8n}{n^2-4}} \ ; \tag{5.72}$$

hence

$$\left(\Sigma\, \alpha_i (\delta_i)^{\frac{n+2}{n-2}} \right)^{\frac{2n}{n+2}} \leq (\Sigma\, \alpha_i \delta_i)^{\frac{2n}{n+2}} \ \Sigma\, a_i (\delta_i)^{\frac{8n}{n^2-4}} \tag{5.70}$$

hence

$$(\Sigma\, \alpha_i (\delta_i)^{\frac{n+2}{n-2}})(\Sigma\, \alpha_i \delta_i) \leq (\Sigma\, \alpha_i \delta_i)^2 [\Sigma a_i (\delta_i)^{\frac{8n}{n^2-4}}]^{\frac{n+2}{2n}} \tag{5.73}$$

using Hõlder inequality and the convexity of $x \to |x|^{\frac{n+2}{4}}$. This yields

$$\left(\int (\Sigma\, \alpha_i (\delta_i)^{\frac{n+2}{n-2}}) (\Sigma\, \alpha_i \delta_i) \right)^{\frac{n}{n-2}} \leq \left(\int \Sigma a_i (\delta_i)^{\frac{2n}{n-2}} \right)^{\frac{2}{n-2}} \left(\int (\Sigma\, \alpha_i \delta_i)^{\frac{2n}{n-2}} \right) \tag{5.74}$$

Now

$$|P\delta_i - \delta_i|(x) \leq \frac{c}{\lambda_i^{(n-2)/2} d_i^{n-2}} < C M^{n-2} \tag{5.75}$$

Thus

$$\left| \int (\Sigma\, \alpha_i (P\delta_i)^{\frac{n+2}{n-2}}) (\Sigma\, \alpha_i P\delta_i) - \int (\Sigma\, \alpha_i \delta_i^{\frac{n+2}{n-2}}) (\Sigma\, \alpha_i \delta_i) \right| \tag{5.76}$$

$$\leq C M^{n-2} \sum_{i,j} \int \delta_i^{\frac{4}{n-2}} \delta_j \leq C M^{n-2} \sum_i \frac{1}{\lambda_i^{(n-2)/2}}$$

While

$$\left| \int (\Sigma\, \alpha_i \delta_i)^{\frac{2n}{n-2}} - \int (\Sigma\, \alpha_i P\delta_i)^{\frac{2n}{n-2}} \right| \leq C M^{n-2} \Sigma \int \delta_i^{\frac{n+2}{n-2}} \tag{5.77}$$

$$\leq C M^{n-2} \Sigma \frac{1}{\lambda_i^{(n-2)/2}}$$

(5.74), (5.76) and (5.77) yield:

$$\left(\int (\Sigma\, \alpha_i \delta_i^{\frac{n+2}{n-2}}) (\Sigma\, \alpha_i P\delta_i) \right)^{\frac{n}{n-2}} \leq \left(\int \Sigma\, a_i \delta_i^{\frac{2n}{n-2}} \right)^{\frac{2}{n-2}} \int (\Sigma\, \alpha_i P\delta_i)^{\frac{2n}{n-2}} \qquad (5.78)$$

Thus

$$\left(\int |\nabla (\Sigma\, \alpha_i P\delta_i)|^2 \right)^{\frac{n}{n-2}} \leq \int (\Sigma\, \alpha_i P\delta_i)^{\frac{2n}{n-2}} \left(\int \Sigma a_i \delta_i^{\frac{2n}{n-2}} \right)^{\frac{2}{n-2}} \qquad (5.79)$$

$$+ 0\left(\sum_i \frac{1}{\lambda_i^{(n-2)/2}} \right)$$

This yields:

$$J\left(\sum_{i=1}^{p} \alpha_i P\delta_i + v \right) \leq \left(\int \sum_{i=1}^{p} a_i \delta_i^{\frac{2n}{n-2}} \right)^{\frac{2}{n-2}} + 0\left(\left(\int |\nabla v|^2 \right)^{1/2} + \sum_i \frac{1}{\lambda_i^{(n-2)/2}} \right) \qquad (5.80)$$

Assuming now (5.69) on one of the ε_{ij}, say ε_{i_0, j_0}, we have:

$$\left(\int \sum_{i=1}^{p} a_i \delta_i^{\frac{2n}{n-2}} \right)^{\frac{2}{n-2}} < (p - \psi(\varepsilon_0))^{\frac{2}{n-2}} \left(\int \delta^{\frac{2n}{n-2}} \right)^{\frac{2}{n-2}} \qquad (5.81)$$

where ψ is a strictly positive continuous function. Hence the proof of Proposition 5.5 in the case K is constant.

In the case when K nonconstant, we proceed as follows:

Let $1 > \varepsilon > 0$ be given. Assume $i_0 = 1$, $j_0 = 2$. We consider the following sequence of subsets of $\{1, \ldots, p\}$,

$$A_1 = \{1\}; \quad A_2 = \{i \neq 1;\ \text{such that } \varepsilon_{1i} > \varepsilon\, \varepsilon_0\} \ldots \qquad (5.82)$$

$$A_n = \{i \in \bigcup_{i=1}^{n-1} A_i \ \text{ such that } \exists k/\varepsilon_{ki} > \varepsilon\, \varepsilon_0 \ \text{ with } k \in A_{n-1}\} \ldots$$

The sequence A_n is empty for $n \geq n_0$. Let

$$B_0 = \bigcup_{i=1}^{n_0} A_i; \quad B_0 \supset \{1,2\} \text{ as } \varepsilon_{12} > \varepsilon_0. \tag{5.83}$$

We may assume that

$$B_0 = \{1,2,\ldots,k\} \subset \{1,\ldots,p\}. \tag{5.84}$$

By construction, we have, if $i \in B_0$

$$|x_i - x_1| \leq C_\varepsilon \sum_{i=1}^{k} \frac{1}{\lambda_j} . \tag{5.85}$$

Indeed, any $i \in B_0$ is related to 1 by a sequence

$$i = j_1; \; j_2, \ldots, j_r = 1 \text{ or } 2; \quad j_{r+1} = 1 \tag{5.86}$$

with the property:

$$\varepsilon_{j_m j_{m+1}} > \varepsilon \, \varepsilon_0, \tag{5.87}$$

hence

$$\frac{1}{(\varepsilon \, \varepsilon_0)^{(n-2)/2}} > \lambda_{j_m} \lambda_{j_{m+1}} |x_{j_m} - x_{j_{m+1}}|^2 . \tag{5.88}$$

From (5.88), (5.85) follows. On the other hand, we have on $B_0^c = \{k+1,\ldots,p\}$

$$\varepsilon_{ik} < \varepsilon \, \varepsilon_0 \; \forall \, i \in B_0^c \; \forall \, k \in B_0. \tag{5.89}$$

Now, on B_0^c, we may pick up the index $k+1$ and define a sequence of the same type as in (5.82). By this procedure, we split $\{1,\ldots,p\}$ into subsets $B_0, \ldots, B_r$ having the following property.

$$\forall \, i,j \in B_m, \quad |x_i - x_j| < C_\varepsilon \sum_{s \in B_m} \frac{1}{\lambda_s} ; \tag{5.90}$$

$\forall\, i \in B_m$ and $j \in B_m^c$, $\varepsilon_{ij} < \varepsilon\varepsilon_0$.

When computing $J(\Sigma\, \alpha_i P\delta_i + v)$, there is a corresponding splitting which arises in a first expansion.

Indeed, for each block B_k, we have:

$$\int K(x)\Big(\sum_{i\in B_k} \alpha_i \delta_i\Big)^{\frac{2n}{n-2}} = K(x_{l_k}) \int\Big(\sum_{i\in B_k} \alpha_i \delta_i\Big)^{\frac{2n}{n-2}} + 0\left(\sum_{i\in B_i} \frac{1}{\lambda_i}\right) \tag{5.91}$$

where $l_k \in B_k$, while

$$\int K(x)\big(\Sigma\, \alpha_i P\delta_i\big)^{\frac{2n}{n-2}} - \sum_k \int K(x)\Big(\sum_{i\in B_k} \alpha_i P\delta_i\Big)^{\frac{2n}{n-2}} = 0(\varepsilon) \tag{5.92}$$

Similarly:

$$\int \big|\nabla(\Sigma\, \alpha_i P\delta_i)\big|^2 = \sum_k \int \Big|\nabla\Big(\sum_{i\in B_k} \alpha_i P\delta_i\Big)\Big|^2 + 0(\varepsilon)\,. \tag{5.93}$$

Thus

$$J(\Sigma\, \alpha_i P\delta_i + v) \tag{5.94}$$

$$= \Big(\sum_k \Big(\int \Big|\nabla\Big(\sum_{i\in B_k} \alpha_i P\delta_i\Big)\Big|^2\Big)\Big)^{\frac{n}{n-2}} \Big/ \Big(\sum_k K(x_{l_k}) \int\Big(\sum_{i\in B_k} \alpha_i P\delta_i\Big)^{\frac{2n}{n-2}}$$

$$+ 0\big(\varepsilon + \big(\textstyle\int |\nabla v|^2\big)^{1/2} + \Sigma\, 1/\lambda_i\big).$$

Using now (5.79), we derive:

$$J(\Sigma\, \alpha_i P\delta_i + v) \le \frac{1}{\sum_k K(x_k) \int\Big(\sum_{i\in B_k} \alpha_i P\delta_i\Big)^{2n/(n-2)}} \tag{5.95}$$

$$\times \Big[\sum_k \left[\left(\int \sum_{i\in B_k} a_i \delta_i^{\frac{2n}{n-2}}\right)^{2/n} \left[\int\Big(\sum_{i\in B_k} \alpha_i P\delta_i\Big)^{\frac{2n}{n-2}}\right]^{\frac{n-2}{n}}\right]^{\frac{n}{n-2}} +$$

$$+ 0(\varepsilon + (\int |\nabla v|^2 + \Sigma \frac{1}{\lambda_i}).$$

On B_0, we have $\varepsilon_{12} > \varepsilon_0$. Thus:

$$(\int \sum_{i \in B_0} a_i \delta_i^{\frac{2n}{n-2}})^{2/n} < (\text{card } B_0)^{2/n} (\int \delta^{\frac{2n}{n-2}})^{2/n} - \gamma(\varepsilon_0) , \tag{5.96}$$

$\gamma : \mathbb{R}^{+\star} \to \mathbb{R}^{+\star}$, a continuous function. Thus:

$$J(\Sigma \alpha_i P\delta_i + v) \tag{5.97}$$

$$< (\int \delta^{\frac{2n}{n-2}})^{\frac{2}{n-2}} \left[\sum_k (\text{card } B_k)^{2/n} \left[\int (\sum_{i \in B_k} \alpha_i P\delta_i)^{\frac{2n}{n-2}} \right]^{\frac{n-2}{n}} \right]^{\frac{n}{n-2}} /$$

$$/ (\sum_k K(x_k) \int (\sum_{i \in B_k} \alpha_i P\delta_i)^{\frac{2n}{n-2}}) - \gamma_1(\varepsilon) + 0(\varepsilon + |v| + \Sigma \frac{1}{\lambda_i}),$$

$\gamma_1 : \mathbb{R}^{+\star} \to \mathbb{R}^{+\star}$, a continuous function. We now have:

$$\Sigma v_k b_k \le (\Sigma v_k^{n/2})^{2/n} (\Sigma b_k^{n/(n-2)})^{\frac{n-2}{n}} . \tag{5.98}$$

Hence

$$J(\Sigma \alpha_i P\delta_i + v) < (\int \delta^{\frac{2n}{n-2}})^{\frac{2}{n-2}} (\Sigma \frac{\text{card } B_k}{K(x_k)^{(n-2)/2}})^{\frac{2}{n-2}} \tag{5.99}$$

$$- \gamma_1(\varepsilon_0) + 0(\varepsilon + |v| + \Sigma \frac{1}{\lambda_i})$$

$$< (\int \delta^{\frac{2n}{n-2}})^{\frac{2}{n-2}} \left[\sum_{i=1}^{p} \frac{1}{(K(x_i))^{(n-2)/2}} \right]^{\frac{2}{n-2}} - \gamma_1(\varepsilon_0) + 0(\varepsilon + |v| + \Sigma \frac{1}{\lambda_i})$$

(5.99) yields the result.

Before concluding this chapter of the book, we derive, taking into account Proposition 5.4, an expansion of $J(\Sigma \alpha_i P\delta_i + \bar{v})$, where $\bar{v}$ is the optimal v Proposition 5.4 provides. We have:

Proposition 5.6: $J(\Sigma \alpha_i P\delta_i + \bar{v})$

$$= \frac{(\Sigma \alpha_i^2)^{n/(n-2)}}{\Sigma \alpha_i^{2n/(n-2)} K(x_i)} \left(\int \delta^{\frac{2n}{n-2}}\right)^{\frac{2}{n-2}} \left\{ 1 - \left(\Sigma \alpha_j^{\frac{2n}{n-2}} K(x_j)\right)^{-1} c_7 \sum_i \alpha_i^{\frac{2n}{n-2}} \frac{\Delta K(x_i)}{\lambda_i^2} \right.$$

$$+ o_K\left(\Sigma \frac{1}{\lambda_i^2}\right) - c_8 \left(\Sigma \frac{H(x_i, x_i)}{\lambda_i^{n-2}} \left(\frac{1}{2} \frac{\alpha_i^2}{\Sigma \alpha_j^2} - \alpha_i^2 \frac{\alpha_i^{4/(n-2)} K(x_i)}{\Sigma \alpha_j^2 (\alpha_j^{4/(n-2)} K(x_j)} \right) \right.$$

$$\left. \left. + \sum_{i \neq j} \left(\frac{H(x_i, x_j)}{(\lambda_i \lambda_j)^{(n-2)/2}} - \varepsilon_{ij} \right) \left(\frac{1}{2} \frac{\alpha_i \alpha_j}{\Sigma \alpha_k^2} - \frac{\alpha_i \alpha_j \alpha_i^{4/(n-2)} K(x_i)}{\Sigma \alpha_k^2 (\alpha_k^{4/(n-2)} K(x_k))} \right) \right) \right\}$$

$$+ o_K(\varepsilon_{ij}) + 0\left(\sum_{i \neq j} \varepsilon_{ij}^{n/(n-2)} + \Sigma \left(\frac{1}{\lambda_i^{(n+2)/2} \lambda_j^{(n-2)/2}} \frac{\operatorname{Log} \lambda_i}{d_j^n} \right.\right.$$

$$\left. + \frac{1}{\lambda_i^{(n+2)/2}} \frac{1}{\lambda_j^{(n-2)/2}} \frac{1}{d_i^n} + \frac{1}{(\lambda_i d_i)^n} + \frac{1}{\lambda_i^2} \frac{1}{\lambda_j^{n-2}} \frac{1}{d_j^{2(n-2)}} \right)$$

$$+ 0\left(\Sigma \frac{1}{\lambda_i^{n+2}} \frac{1}{d_i^{2(n+2)}} + \Sigma \frac{1}{\lambda_i^4} \frac{1}{\lambda_j^{n-2}} \frac{1}{d_j^{2(n-2)}} \right.$$

$$\left. + (\text{if } n < 6) \sum_{i \neq j} \varepsilon_{ij}^2 (\operatorname{Log} \varepsilon_{ij}^{-1})^{\frac{2(n-2)}{n}} \right) + 0_K\left(\Sigma \frac{|DK(x_i)|^2}{\lambda_i^2} + \frac{1}{\lambda_i^4} \right).$$

Proof: Straightforward from Propositions 5.3 and 5.4.

5.2 The situation when K is constant, $\Omega \in \mathbb{R}^n$, along compact sets

We assume now that K is a constant equal to 1 and $\Omega \in \mathbb{R}^n$.

$$K = 1, \quad \Omega \in \mathbb{R}^n. \tag{5.100}$$

Let

$$W_p = \left\{ u \in \Sigma^+ \mid J(u) < (p+1)^{\frac{2}{n-2}} \left(\int \delta^{\frac{2n}{n-2}}\right)^{\frac{2}{n-2}} \right\} \tag{5.101}$$

Let

$$A \text{ is a compact contained in } \Omega. \tag{5.102}$$

Let

$$\Delta_p \text{ is the standard p-simplex} = \{(t_1, \ldots, t_{p+1});\ t_i \geq 0;\ \sum_i^{p+1} t_i = 1\} \tag{5.103}$$

Proposition 5.7: There exist $\lambda_0(A,p)$ such that, for any $(x_1, \ldots, x_p) \in A^p$ and $(\alpha_1, \ldots, \alpha_p) \in \Delta_{p-1}$ and $(\lambda_1, \ldots, \lambda_p) \in]\lambda_0, +\infty[^p$, we have:

$$\sum_{i=1}^{p} \alpha_i P\delta(x_i, \lambda_i) \in W_p.$$

Proposition 5.8: Let $\gamma > 1$ be given. Assume $\lambda_i/\lambda_j \leq \gamma$. There exists an $\eta_p > 0$ and a $\tilde{\lambda}_0(A,p)$ such that for $\lambda_1 > \tilde{\lambda}_0$,

$$\sum_{i=1}^{p} \alpha_i P\delta(x_i, \lambda_i) \in W_{p-1} \text{ if } \inf_{i \neq j} |x_i - x_j| < \eta_p.$$

Proof of Propositions 5.7 and 5.8: As A is compact in Ω, we may always assume that λ_0 and $\tilde{\lambda}_0$ are large enough, so that the conditions of Proposition 5.5 are fulfilled. Then, if one of the ε_{ij} $(i \neq j)$ is not small, Proposition 5.5 holds and yields

$$\sum_{i=1}^{p} \alpha_i P\delta_i \in W_{p-1}. \tag{5.104}$$

Thus, Propositions 5.7 and 5.8 are proven in this situation.

We are thus left with the case:

$$\varepsilon_{ij} \text{ small } \forall i \neq j. \tag{5.105}$$

Then the expansion of Proposition 5.6 holds; and yields Proposition 5.7 immediately. For Proposition 5.8, with K a constant equal to 1, we have, with $x_i \in A$, for any i:

$$J(\Sigma \alpha_i P\delta_i) \tag{5.106}$$

$$= \frac{(\Sigma \alpha_i^2)^{n/(n-2)}}{\Sigma \alpha_i^{2n/(n-2)}} (\int \delta^{\frac{2n}{n-2}})^{\frac{2}{n-2}} \left\{ 1 - c_8 \left[\Sigma \frac{H(x_i, x_i)}{\lambda_i^{n-2}} \left(\frac{1}{2} \frac{\alpha_i^2}{\Sigma \alpha_j^2} - \frac{\alpha_i^{2n/(n-2)}}{\Sigma \alpha_j^{2n/(n-2)}} \right) \right.\right.$$

$$\left.\left. + \sum_{i \neq j} \left(\frac{H(x_i, x_j)}{(\lambda_i \lambda_j)^{(n-2)/2}} - \varepsilon_{ij} \right) \left(\frac{1}{2} \frac{\alpha_i \alpha_j}{\Sigma \alpha_k^2} - \frac{\alpha_i^{(n+2)/(n-2)} \alpha_j}{\Sigma \alpha_k^{2n/(n-2)}} \right) \right] + \sum_{\substack{i,j \\ i \neq j}} o\left(\varepsilon_{ij} + \frac{1}{\lambda_i^{n-1}} \right) \right\}$$

As we assume $\lambda_i / \lambda_j \leq \gamma \ \forall i \neq j$, we have:

$$\varepsilon_{ij} \sim \frac{1}{(\lambda_i \lambda_j)^{(n-2)/2} |x_i - x_j|^{n-2}} \tag{5.107}$$

and thus

$$J(\Sigma \alpha_i P\delta_i) = \frac{(\Sigma \alpha_i^2)^{n/(n-2)}}{\Sigma \alpha_i^{2n/(n-2)}} (\int \delta^{\frac{2n}{n-2}})^{\frac{2}{n-2}} \tag{5.108}$$

$$\times \left\{ 1 - c_8 \left[(\frac{\Sigma H(x_i x_i)}{\lambda_i^{n-2}} \left(\frac{1}{2} \frac{\alpha_i^2}{\Sigma \alpha_j^2} - \frac{\alpha_i^{2n/(n-2)}}{\Sigma \alpha_k^{2n/(n-2)}} \right) \right.\right.$$

$$+ \sum_{i \neq j} \left(\frac{H(x_i, x_j)}{(\lambda_i \lambda_j)^{(n-2)/2}} - \frac{1}{(\lambda_i \lambda_j)^{(n-2)/2}} \frac{1}{|x_i - x_j|^{n-2}} \right)$$

$$\left.\left. \times \left(\frac{1}{2} \frac{\alpha_i \alpha_j}{\Sigma \alpha_k^2} - \frac{\alpha_i^{(n+2)/(n-2)} \alpha_j}{\Sigma \alpha_k^{2n/(n-2)}} \right) \right\} + o(\Sigma \frac{1}{\lambda_i^{n-2}})$$

The conclusion follows now from the fact that $\lambda_i/\lambda_j \leq \gamma$ and $H(x_i,x_j)$ is uniformly bounded for x_i and x_j in A. Indeed, while all the terms involving $H(x_i,x_j)$ are bounded by $c(\gamma)/\lambda_1^{n-2}$, we have, under the assumption:

$$|x_{i_0}-x_{j_0}| = \min_{i\neq j} |x_i-x_j| < \eta_p, \tag{5.109}$$

$$J(\Sigma\, \alpha_i P\delta_i) \leq \frac{(\Sigma\, \alpha_i^2)^{n/(n-2)}}{\Sigma\, \alpha_i^{2n/(n-2)}} \left(\int \delta^{\frac{2n}{n-2}}\right)^{\frac{n}{n-2}} \left\{1 + \frac{c(\gamma)}{\lambda_1^{n-2}}\right. \tag{5.110}$$

$$\left. + c_8 \left(\frac{1}{2} \frac{\alpha_{i_0}\alpha_{j_0}}{\Sigma\, \alpha_k^2} - \frac{\alpha_{i_0}^{(n+2)/(n-2)}\alpha_{j_0}}{\Sigma\, \alpha_k^{2n/(n-2)}}\right) \frac{1}{\eta_p^{n-2}} \times \frac{1}{\gamma^{n-2}\lambda_1^{n-2}} + o\left(\frac{1}{\lambda_1^{n-2}}\right)\right\}.$$

We then have two cases:

Either the α_i's are almost equal. Then:

$$\frac{1}{2} \frac{\alpha_{i_0}\alpha_{j_0}}{\Sigma\, \alpha_k^2} - \frac{\alpha_{i_0}^{(n+2)/(n-2)}\alpha_{j_0}}{\Sigma\, \alpha_k^{2n/(n-2)}} \text{ is almost equal to } -\frac{1}{2p} \tag{5.111}$$

and thus

$$J(\Sigma\, \alpha_i P\delta_i) < p^{\frac{2}{n-2}} \left(\int \delta^{\frac{2n}{n-2}}\right)^{\frac{n}{n-2}} \text{ if } \eta_p \text{ is small enough.} \tag{5.112}$$

Otherwise, if the α_i's are "far" from being equal, we have:

$$\frac{(\Sigma\, \alpha_i^2)^{n/(n-2)}}{\Sigma\, \alpha_i^{2n/(n-2)}} < p^{\frac{2}{n-2}} - \varepsilon_4; \quad \varepsilon_4 > 0 \tag{5.113}$$

The result follows as $\lambda_1 > \tilde{\lambda}_0 \gg 1$. Hence the proof of Propositions 5.7 and 5.8.

We introduce now for $(a_1, \ldots, a_p)$ in A^p the following matrix:

$$M(x_1, \dots, x_p) = \begin{bmatrix} & -G(x_i, x_j) \\ & \\ & H(x_i, x_i) & \end{bmatrix}, \tag{5.114}$$

i.e. the coefficient of M at line i and column j is:

$$\begin{cases} -G(x_i, x_j) = H(x_i, x_j) - \dfrac{1}{|x_i - x_j|^{n-2}} & \text{if } i \neq j \\ \\ H(x_i, x_i) & \text{otherwise.} \end{cases} \tag{5.115}$$

As $H(x_i, x_i)$ is positive (strictly) and $-G(x_i, x_j)$ is strictly negative, the matrix M has a unique least eigenvalue, corresponding to a unique eigenvector, with positive (strictly) coordinates.

Definition 5.9: Let $\rho(x_1, \dots, x_p)$ be the least eigenvalue of $M(x_1, \dots, x_p)$ and let $(\Lambda_1, \dots, \Lambda_p)$ be the corresponding eigenvector such that $\Sigma \Lambda_i^2 = 1$. We have:

$$\Lambda_i > 0. \tag{5.116}$$

We define now a map γ_p as follows:

Let λ be a large positive number (which might depend on p); and $\varepsilon > 0$, small enough; λ will be in the sequel made as large as necessary so that all statements hold. (5.117)

By Proposition 5.8, there exists $\eta_p > 0$ and $\tilde{\lambda}_0(A, p)$ such that for $(x_1, \dots, x_p) \in A^p$ with

$$\operatorname*{Min}_{i \neq j} |x_i - x_j| < \eta_p ; \quad \lambda > \tilde{\lambda}_0(A, p) , \tag{5.118}$$

we have:

$$\sum_{i=1}^{p} \alpha_i P(x_i, \lambda) \in W_{p-1} \quad \forall (\alpha_1, \dots, \alpha_p) \in \Delta_{p-1}. \tag{5.119}$$

We choose λ large enough so that:

$$\lambda \eta_p \gg 1. \tag{5.120}$$

Let

$$A_1^p = \{(x_1, \dots, x_p) \in A^p \text{ such that } \min_{i \neq j} |x_i - x_j| \geq \eta_p\} \tag{5.121}$$

$$A_2^p = \{(x_1, \dots, x_p) \in A^p \text{ such that } \min_{i \neq j} |x_i - x_j| \leq \eta_p/2\} \tag{5.122}$$

$$A_3^p = A^p/(A_1^p \cup A_2^p). \tag{5.123}$$

On A_1^p, we define γ_p as follows: Let

$$\bar{\lambda}_i = \frac{\lambda}{p^{1/(n-2)}} \frac{1}{\Lambda_i^{2/(n-2)}}; \quad i = 1, \dots, p, \tag{5.124}$$

where $(\Lambda_1, \dots, \Lambda_p)$ has been defined in Definition 5.9 at $(x_1, \dots, x_p)$. Notice that the vector:

$$\frac{1}{\bar{\lambda}_1^{(n-2)/2}}, \dots, \frac{1}{\bar{\lambda}_p^{(n-2)/2}} \tag{5.125}$$

is positively collinear to $(\Lambda_1, \dots, \Lambda_p)$; and we have, as $\sum_{i=1}^p \Lambda_i^2 = 1$;

$$\sum_{i=1}^p \frac{1}{\bar{\lambda}_i^{n-2}} = \frac{p}{\lambda^{n-2}} \tag{5.126}$$

then, on $A_1^p \times \Delta_{p-1}$, we set:

$$\gamma_p(x_1, \dots, x_q, \alpha_1, \dots, \alpha_p) = \sum_{i=1}^p \alpha_i P\delta(x_i, \bar{\lambda}_i). \tag{5.127}$$

On $A_2^p \times \Delta_{p-1}$, we set:

$$\gamma_p(x_1, \dots, x_p, \alpha_1, \dots, \alpha_p) = \sum_{i=1}^p \alpha_i P\delta(x_i, \lambda) \tag{5.128}$$

Finally, on $A_3^p \times \Delta_{p-1}$, we set:

$$\gamma_p(x_1,\dots,x_p,\ \alpha_1,\dots,\alpha_p) = \sum_{i=1}^{p} \alpha_i P\delta_i(x_i,\omega_i) \tag{5.129}$$

with

$$\frac{1}{\omega_i^{(n-2)/2}} = \frac{2}{\eta_p}\left\{ \frac{\underset{i\neq j}{\mathrm{Min}}\ |x_i-x_j| - \eta_{p/2}}{\bar{\lambda}_i^{(n-2)/2}} + \frac{\eta_p - \underset{i\neq j}{\mathrm{Min}}\ |x_i-x_j|}{\lambda^{(n-2)/2}} \right\}; \tag{5.130}$$

$\omega_i > 0.$

We then have:

Proposition 5.10: γ_p is an equivariant map (with respect to the action of σ_p), which sends:

$$\gamma_p: (A^p \times_{\sigma_p} \Delta_{p-1},\ A^p \times_{\sigma_p} \partial\Delta_{p-1}) \to (W_p, W_{p-1}) \tag{5.131}$$

for p large enough, γ_p sends $A^p \times_{\sigma_p} \Delta_{p-1}$ into W_{p-1}. (5.132)

for any p, if η_p is small enough, γ_p sends (5.133)

$(A_2^p \cup A_3^p) \times \Delta_{p-1}$ into W_{p-1}.

Consider the set of $x \in A^p$ such that $\underset{i\neq j}{\mathrm{Min}}\ |x_i-x_j| \leq \eta_p$ and the (5.134)

least eigenvalue $\rho(x)$ of $M(x) = M(x_1,\dots,x_p)$ is less than

a small strictly negative number $-\varepsilon_p$. Then, γ_p can be

chosen so that any such (x, α), $\alpha \in \Delta_{p-1}$ is sent into W_{p-1}.

Proof of Proposition 5.10: The fact that γ_p is equivariant follows immediately from its very definition. Next, observe that for any $(x, \alpha) \in A^p \times \Delta_{p-1}$, the function:

$$\gamma_p(x, \alpha) = \Sigma\ \alpha_i P(x_i, \lambda_i) \tag{5.135}$$

satisfies:

$$\lambda_i/\lambda_j \le \gamma \; ; \quad \gamma \text{ fixed positive constant} \tag{5.136}$$

independent of the size of the λ_i's (i.e. independent of $\lambda_i > \tilde{\lambda}_0(A, p)$).

This is due to the following:

On A_1^p, the matrix $M(x_1, \ldots, x_p)$ is bounded and so is its least eigenvalue $\rho(x_1, \ldots, x_p)$ and least corresponding eigenvector $(\Lambda_1, \ldots, \Lambda_p)$, which vary continuously with respect to $(x_1, \ldots, x_p)$. In fact, $(\Lambda_1, \ldots, \Lambda_p)$ are strictly positive, bounded away from zero, and bounded from above. Thus (5.136) holds on A_1^p. (5.136) holds as well on A_2^p as, on this set, $\lambda_i = \lambda$ independent of i. Finally, on A_3^p, the λ_i's are (non) -linear combination of their definition on A_1^p and their definition on A_2^p and (5.136) is easily seen to hold by a direct computation.

As (5.136) holds, (5.107) and (5.108) hold on $\gamma_p(x, \alpha) = \Sigma\, \alpha_i P(x_i, \lambda_i)$; $(x, \alpha) \in A^p \times \Delta_{p-1}$; as soon as ε_{ij} is small for any (i, j), $i \neq j$. Thus:

$$J\Big(\sum_{i=1}^{p} \alpha_i P\delta_i(x_i, \lambda_i)\Big) = J(\gamma_p(x, \alpha)) \tag{5.137}$$

$$= \frac{(\Sigma \alpha_i^2)^{n/(n-2)}}{\Sigma \alpha_i^{2n/(n-2)}} \left(\int \delta^{\frac{2n}{n-2}}\right)^{\frac{2}{n-2}} \left\{ 1 - c_8 \left[\left(\Sigma \frac{H(x_i, x_i)}{\lambda_i^{n-2}} \; \frac{1}{2} \; \frac{\alpha_i^2}{\Sigma \alpha_j^2} - \right.\right.\right.$$

$$\left. - \frac{\alpha_i^{2n/(n-2)}}{\Sigma \alpha_k^{2n/(n-2)}}\right) + \sum_{i \neq j} \left(\frac{H(x_i, x_j)}{(\lambda_i \lambda_j)^{(n-2)/2}} - \frac{1}{(\lambda_i \lambda_j)^{(n-2)/2}} \; \frac{1}{|x_i - x_j|^{n-2}} \right)$$

$$\left. \left. \cdot \left(\frac{1}{2} \; \frac{\alpha_i \alpha_j}{\Sigma \alpha_k^2} - \frac{\alpha_i^{(n+2)/(n-2)} \alpha_j}{\Sigma \alpha_k^{2n/(n-2)}} \right) \right] + o\left(\Sigma \frac{1}{\lambda_i^{n-2}} \right) \right\}$$

By Proposition 5.7, γ_p sends $A^p \times_{\sigma_p} \Delta_{p-1}$ into W_p. Now, by (5.137), on $A^p \times_{\sigma_p} \delta\Delta_{p-1}$, assuming the ε_{ij} to be small, we have:

$$J\Big(\sum_{i=1}^{p} \alpha_i P\delta_i(x_i, \lambda_i)\Big) = \frac{(\Sigma \alpha_i^2)^{n/(n-2)}}{\Sigma \alpha_i^{2n/(n-2)}} \left(\int \delta^{\frac{2n}{n-2}}\right)^{\frac{2}{n-2}} + 0 \left(\Sigma \frac{1}{\lambda_i^{n-2}} \right). \tag{5.138}$$

If one of the α_i is zero, we have:

$$\frac{(\Sigma \alpha_i^2)^{n/(n-2)}}{\Sigma \alpha_i^{2n/(n-2)}} \leq (p-1)^{\frac{2}{n-2}} \tag{5.139}$$

and thus:

$$J(\sum_{i=1}^{p} \alpha_i P\delta_i(x_i, \lambda_i)) \leq (p-1)^{\frac{2}{n-2}} (\int \delta^{\frac{2n}{n-2}})^{\frac{2}{n-2}} + 0\left(\Sigma \frac{1}{\lambda_i^{n-2}}\right), \tag{5.140}$$

for $(x, \alpha) \in A^p \times_{\sigma_p} \partial\Delta_{p-1}$ and ε_{ij} small. Thus, $\gamma_p(x, \alpha)$ belongs to W_{p-1} under these hypotheses. If ε_{ij} is not small, $\gamma_p(x, \alpha)$ belongs to W_{p-1} by Proposition 5.5. Thus:

$$\gamma_p(x, \alpha) \in W_{p-1} \ \forall (x, \alpha) \in A^p \times_{\sigma_p} \partial\Delta_{p-1}, \tag{5.141}$$

and (5.131) is proved.

From now on, we will assume we are dealing with x's in A^p such that the ε_{ij} corresponding to $\gamma_p(x, \alpha)$ are small. Otherwise, the conditions of Proposition 5.10 follow immediately from Proposition 5.5.

Notice that, by (5.138), we have:

$$J(\sum_{i=1}^{p} \alpha_i P\delta_i(x_i, \lambda_i)) < (p \int \delta^{\frac{2n}{n-2}})^{\frac{2}{n-2}} \tag{5.142}$$

as soon as

$$\sum_{i=1}^{p} \left| \frac{\alpha_i^2}{\sum_{j=1}^{p} \alpha_j^2} - \frac{1}{p} \right| \text{ is not } o(1). \tag{5.143}$$

Indeed, under (5.143), i.e. if the α_i's are not all nearly equal, $(\Sigma \alpha_i^2)^{\frac{n}{n-2}} / \Sigma \alpha_i^{\frac{2n}{n-2}}$ is away (below) from $p^{2/(n-2)}$ and (5.142) follows from (5.138) as soon as $\tilde{\lambda}_0(A, p)$ is large enough. Thus Proposition 5.10 holds if (5.143) holds.

Consequently, we may assume that α is such that:

$$\sum_{i=1}^{p} \left| \frac{\alpha_i^2}{\sum_{j=1}^{p} \alpha_j^2} - \frac{1}{p} \right| = o(1). \tag{5.144}$$

(5.144) and (5.137) imply,

$$J(\gamma_p(x,\alpha)) \le p^{\frac{2}{n-2}} (\int \delta^{\frac{2n}{n-2}})^{\frac{2}{n-2}} \left(1 + \frac{c_9}{p}\left[\frac{1}{\lambda_1^{(n-2)/2}}, \ldots, \frac{1}{\lambda_p^{(n-2)/2}}\right] \right.$$
$$\left. \cdot\, M(x_1, \ldots, x_p) \begin{bmatrix} \frac{1}{\lambda_1^{(n-2)/2}} \\ \vdots \\ \frac{1}{\lambda_p^{(n-2)/2}} \end{bmatrix} + o\left(\Sigma \frac{1}{\lambda_i^{n-2}}\right)\right) \tag{5.145}$$

where

$$c_9 = \frac{c_8}{2}. \tag{5.146}$$

Indeed, (5.144) implies:

$$|\alpha_i/\alpha_j - 1| = o(1), \tag{5.147}$$

hence (5.145) from (5.137) and (5.147).

Thus, from now on, ε_{ij} is small and α satisfies (5.147) so that (5.145) holds. Otherwise, Proposition 5.10 has been seen to hold.

We prove now (5.132), i.e. that γ_p sends $A^p \times_{\sigma_p} \Delta_{p-1}$ into W_{p-1} for p large enough. If x belongs to A_1^p, the argument is as follows: For such an x, $(\bar{\lambda}_1, \ldots, \bar{\lambda}_p)$ is by definition collinear to $\frac{1}{\Lambda_1^{2/(n-2)}}, \ldots, \frac{1}{\Lambda_p^{2/(n-2)}}$, the collinearity coefficient being given through (5.126), where $(\Lambda_1, \ldots, \Lambda_p)$ is the normalized eigenvector of least eigenvalue ρ of $M(x_1, \ldots, x_p)$. Such a $(\lambda_1, \ldots, \lambda_p)$ has been denoted $(\bar{\lambda}_1, \ldots, \bar{\lambda}_p)$. Thus, for such an x, we have, using (5.145):

$$J(\gamma_p(x,\alpha)) \leq p^{\frac{2}{n-2}} \left(\int \delta^{\frac{2n}{n-2}}\right)^{\frac{2}{n-2}} \left(1 + \frac{c_9}{p} \rho(x_1,\ldots,x_p) \cdot \frac{p}{\lambda^{n-2}}\right)$$

$$+ o\left(\Sigma \frac{1}{\bar{\lambda}_i^{n-2}}\right). \tag{5.148}$$

We used here (5.124) which tells us that $\sum_{i=1}^{p} \frac{1}{\bar{\lambda}_i^{n-2}} = \frac{p}{\lambda^{n-2}}$. This in fact yields:

$$J(\gamma_p(x,\alpha)) \leq p^{\frac{2}{n-2}} \left(\int \delta^{\frac{2n}{n-2}}\right)^{\frac{2}{n-2}} \left(1 + c_9 \frac{\rho(x_1,\ldots,x_p)}{\lambda^{n-2}} \tag{5.149}$$

$$+ o\left(\frac{1}{\lambda^{n-2}}\right)\right) .$$

Now, as A is compact, there exists a $p_0 \in N$ such that for $p \geq p_0$, we have:

$$\rho(x_1,\ldots,x_p) < 0, \quad \forall\, (x_1,\ldots,x_p) \in A^p;\ p \geq p_0 . \tag{5.150}$$

Indeed, for p large enough, two of the points $(x_1,\ldots,x_p)$ at least have to come close; e.g.

$$|x_1 - x_2| < \varepsilon_3. \tag{5.151}$$

Then:

$$-G(x_1,x_2) \text{ is very large and negative.} \tag{5.152}$$

Indeed, $-G(x_1,x_2) = H(x_1,x_2) - \frac{1}{|x_1-x_2|^{n-2}}$; and $H(x_1,x_2)$ is uniformly bounded for x_1 and x_2 in A, as A is compact. Then, the matrix:

$$\begin{bmatrix} H(x_1,x_1) & -G(x_1,x_2) \\ -G(x_1,x_2) & H(x_2,x_2) \end{bmatrix} \tag{5.153}$$

is negative definite; and so is $M(x_1,\ldots,x_p)$; thus (5.150) holds. We, in fact, proved more than (5.150). Namely, we have proved:

$$\text{for any } \varepsilon_4 > 0, \text{ there exists a } p_0 \in \mathbb{N} \text{ such that for } p \geq p_0, \quad \rho(x_1,\dots,x_p) < -\varepsilon_4 \text{ for any } (x_1,\dots,x_p) \in A^p. \tag{5.154}$$

Indeed, the control of the negativeness of ρ can be achieved using the matrix (5.153) and its least eigenvalue which upper bounds $\rho(x_1,\dots,x_p)$.

Thus (5.154) holds on A_1^p for $p \geq p_0$. Consequently, using (5.149), we derive:

$$J(\gamma_p(x,\alpha)) \leq p^{\frac{2}{n-2}} (\int \delta^{\frac{2n}{n-2}})^{\frac{2}{n-2}} \left(1 - \frac{c_9 \varepsilon_4}{\lambda^{n-2}} + o\left(\frac{1}{\lambda^{n-2}}\right)\right) \tag{5.155}$$

$$< p^{\frac{2}{n-2}} (\int \delta^{\frac{2n}{n-2}})^{\frac{2}{n-2}} \quad \text{for } \lambda \text{ large enough.}$$

Thus $\gamma_p(x,\alpha)$ belongs to W_{p-1} and (5.132) is proved in case x is an A_1^p.

We now turn to the x's belonging to A_2^p, for which the $\bar{\lambda}_i$'s are equal to λ. For such an x, (5.145) yields:

$$J(\gamma_p(x,\alpha)) \leq p^{\frac{2}{n-2}} (\int \delta^{\frac{2n}{n-2}})^{\frac{2}{n-2}} \times \tag{5.156}$$

$$\cdot \left(1 + \frac{c_9}{p\lambda^{n-2}} [1,\dots,1] M(x_1,\dots,x_p) \begin{bmatrix} 1 \\ \vdots \\ 1 \end{bmatrix} + o\left(\frac{1}{\lambda^{n-2}}\right)\right).$$

Consider now a strictly positive number C such that, for any (x,y) in A^2, we have:

$$H(x,x) \leq C; \qquad G(x,y) \geq \frac{1}{C}. \tag{5.157}$$

By compactness of A, such a number exists. Then:

$$[1,\dots,1] M \begin{bmatrix} 1 \\ \vdots \\ 1 \end{bmatrix} \leq pC - (p-1)^2 \frac{1}{C}. \tag{5.158}$$

Thus:

$$J(\gamma_p(x,\alpha)) \le p^{\frac{2}{n-2}} (\int \delta^{\frac{2n}{n-2}})^{\frac{2}{n-2}} \left[1 + \frac{c_9 C}{\lambda^{n-2}} \right. \tag{5.159}$$

$$\left. - \frac{c_9}{C} \frac{1}{\lambda^{n-2}} \frac{(p-1)^2}{p} + o\left(\frac{1}{\lambda^{n-2}}\right) \right]$$

Clearly, (5.159) implies, for p large enough, that $\gamma_p(x,\alpha)$ belongs to W_{p-1}. Hence (5.132) for x in A_2^p.

We are left with the case where x belongs to A_3^p. For such x's, the argument combines the two ways we have been arguing previously, when x was in A_1^p and when x was in A_2^p respectively. Indeed, for such an x, we have:

$$\gamma_p(x,\alpha) = \sum_{i=1}^{p} \alpha_0 p \delta_i(x_i, \omega_i) \tag{5.160}$$

where the ω_i's satisfy:

$$\frac{1}{\omega_i^{(n-2)/2}} = \frac{2}{\eta_p} \left\{ \frac{\operatorname{Min}_{i\neq j} |x_i - x_j| - \eta_{p/2}}{\bar{\lambda}_i^{(n-2)/2}} + \frac{\eta_p - \operatorname{Min}_{i\neq j} |x_i - x_j|}{\lambda^{(n-2)/2}} \right\} \tag{5.161}$$

Then, denoting by d the quantity $\operatorname{Min}_{i\neq j} |x_i - x_j|$:

$$\left[\frac{1}{\omega_1^{(n-2)/2}}, \dots, \frac{1}{\omega_p^{(n-2)/2}} \right] M(x_1, \dots, x_p) \begin{bmatrix} \frac{1}{\omega_1^{(n-2)/2}} \\ \vdots \\ \frac{1}{\omega_p^{(n-2)/2}} \end{bmatrix} = \tag{5.162}$$

$$= \frac{4}{\eta_p^2} (d - \eta_p)^2 \rho \Sigma \frac{1}{\bar{\lambda}_i^{n-2}} + 2\rho \frac{4}{\eta_p^2} (d - \eta_{p/2})(\eta_p - d) \left(\Sigma \frac{1}{\bar{\lambda}_i^{(n-2)/2}} \right) \frac{1}{\lambda^{(n-2)/2}}$$

$$+ \frac{4}{\eta_p^2} \frac{(\eta_p - d)^2}{\lambda^{n-2}} [1, \dots, 1] M \begin{bmatrix} 1 \\ \vdots \\ 1 \end{bmatrix} .$$

Now, for $p \ge p_0$, (5.154) holds with a certain $\varepsilon_4 > 0$ the choice of which is at our disposal and we also have (5.158) (which does not depend on p and always holds). Thus:

$$\left[\frac{1}{\omega_1^{(n-2)/2}}, \ldots, \frac{1}{\omega_p^{(n-2)/2}}\right] M \begin{bmatrix} 1/\omega_1^{(n-2)/2} \\ \\ 1/\omega_p^{(n-2)/2} \end{bmatrix} \tag{5.163}$$

$$\leq -\frac{4}{\eta_p^2}(d-\eta_{p/2})^2 \frac{\rho\varepsilon_4}{\lambda^{n-2}} + \frac{4}{\eta_p^2} \frac{(\eta_p-d)^2}{\lambda^{n-2}} \left(pC - \frac{(p-1)^2}{C}\right)$$

$$- 2\varepsilon_4(d-\eta_{p/2})(\eta_p-d) \frac{1}{\lambda^{(n-2)/2}} \left(\Sigma \frac{1}{\bar{\lambda}_i^{(n-2)/2}}\right) .$$

(We also used here (5.126), i.e. $\sum_{i=1}^{p} \frac{1}{\bar{\lambda}_i^{n-2}} = \frac{p}{\lambda^{n-2}}$.) Now, by (5.126), we have:

$$\frac{c_{10}}{\lambda^{(n-2)/2}} \leq \sum_{i=1}^{p} \frac{1}{\bar{\lambda}_i^{(n-2)/2}} \leq \frac{c_{11}}{\lambda^{(n-2)/2}} , \tag{5.164}$$

where c_{10} and c_{11} are fixed positive constants (independent of λ, although they depend on p), and

$$\left|d - \frac{\eta_p}{2}\right| + \left|\eta_p - d\right| \geq \eta_p/10 . \tag{5.165}$$

Finally,

$$pC - \frac{(p-1)^2}{C} \leq -c_{12}p^2 ; \quad c_{12} \text{ fixed constant independent of } p, \tag{5.166}$$

for $p \geq p_0$. Thus:

$$\left[\frac{1}{\omega_1^{(n-2)/2}}, \ldots, \frac{1}{\omega_p^{(n-2)/2}}\right] M \begin{bmatrix} \frac{1}{\omega_1^{(n-2)/2}} \\ \vdots \\ \frac{1}{\omega_p^{(n-2)/2}} \end{bmatrix} \tag{5.167}$$

$$\leq -\frac{4}{\eta_p^2}\left[c_{12}p^2(\eta_p-d)^2 + (d-\eta_p/2)^2 p\varepsilon_4 + \varepsilon_4(d-\eta_{p/2})(\eta_p-d)c_{10}\right] \frac{1}{\lambda^{n-2}} .$$

On A_3^p, we have:

$$\eta_{p/2} < d < \eta_p . \tag{5.168}$$

Thus, using (5.165),

$$\frac{4}{\eta_p^2} [c_{12} p^2 (\eta_p - d)^2 + (d - \eta_{p/2})^2 p \varepsilon_4 + (d - \eta_{p/2})(\eta_p - d) c_{10}] \geq c_{13} > 0; \tag{5.169}$$

c_{13} fixed positive constant independent of x in A_3^p.

Thus:

$$\left[\frac{1}{\omega_1^{(n-2)/2}}, \dots, \frac{1}{\omega_p^{(n-2)/2}}\right] M \begin{bmatrix} \frac{1}{\omega_1^{(n-2)/2}} \\ \vdots \\ \frac{1}{\omega_p^{(n-2)/2}} \end{bmatrix} \leq -\frac{c_{13}}{\lambda^{n-2}} . \tag{5.170}$$

This implies, using (5.145) and the fact that $o\left(\Sigma \frac{1}{\omega_i^{n-2}}\right) = o\left(\frac{1}{\lambda^{n-2}}\right)$,

$$J(\gamma_p(x, \alpha)) \leq p^{\frac{2}{n-2}} \left(\int \delta^{\frac{2n}{n-2}}\right)^{\frac{2}{n-2}} \left(1 - \frac{c_9 c_{13}}{p \lambda^{n-2}}\right) + o\left(\frac{1}{\lambda^{n-2}}\right) . \tag{5.171}$$

Hence, here also, $\gamma_p(x, \alpha)$ belongs to W_{p-1}, for $p \geq p_0$ and x in A_3^p. (5.132) is thus proved.

We are left with (5.133) and (5.134). (5.133) is implied by (5.132) for $p \geq p_0$. We are thus left with a finite number of p's; $p \leq p_0$. We are considering x's belonging to $A_2^p \cup A_3^p$; thus:

$$d = \min_{i \neq j} |x_i - x_j| \leq \eta_p . \tag{5.172}$$

For η_p small enough, we have:

$$\rho(x_1, \dots, x_p) < -\varepsilon_4; \quad \varepsilon_4 > 0 \tag{5.173}$$

arbitrarily fixed; for any x in $A_2^p \cup A_3^p$.

$$[1,\dots,1]\, M(x_1,\dots,x_p)\begin{bmatrix}1\\ \vdots\\ 1\end{bmatrix} < -\varepsilon_4 \; ; \tag{5.174}$$

$\varepsilon_4 > 0$ arbitrarily fixed; for any x in $A_1^p \cup A_3^p$.

Indeed, if $d \le \eta_p$, with η_p small enough, we are in the situation where an inequality of the type (5.151) holds and the subsequent arguments already stated, which led to (5.154), lead here to an analogous statement, namely (5.173). To avoid redundancies, we do not repeat the argument here. Thus (5.173) holds.

For (5.174), we have:

$$[1,\dots,1] M \begin{bmatrix}1\\ \vdots\\ 1\end{bmatrix} = \sum_{i=1}^{p} H(x_i,x_i) - 2 \sum_{i=1}^{p} G(x_i,x_j). \tag{5.175}$$

We know that $H(x_i,x_i)$ is bounded, as well as $H(x_i,x_j)$ for x_i and x_j in A.

$$H(x,y) \le C \;\; \forall\, (x,y) \in A^2. \tag{5.176}$$

$$[1,\dots,1] M \begin{bmatrix}1\\ \vdots\\ 1\end{bmatrix} \le 3pC - \sum_{i=1}^{p} \frac{1}{|x_i \; x_j|^{n-2}} \le 3pC - \frac{1}{\eta_p^{n-2}} \tag{5.177}$$

if x is in $A_2^p \cup A_3^p$. (5.174) follows for η_p small enough.

(5.156), (5.145) and (5.167) imply then (5.133) using (5.173) and (5.174). As for (5.134), one uses the expansion (5.145) and this yields the result through arguments very similar to those already stated. Proposition 5.10 is thereby proved.

5.3 A local type deformation argument: the case K = 1

Let $p \in N^\star$ be fixed; θ_p and $\bar{\varepsilon}_p$ are two strictly positive real numbers; we will first choose θ_p large enough and the $\bar{\varepsilon}_p$ small enough.

Let μ be a function in $C^\infty([0,+\infty[\,;R^+)$ such that

$$\begin{cases} \mu(0) = \bar{\varepsilon}_p \\ -\frac{2}{\theta_p} \leq \mu' \leq 0 \\ \mu(r) = 0 \text{ for } r \text{ in } [\theta_p \bar{\varepsilon}_p; +\infty[. \end{cases} \quad (5.178)$$

Let:

$R_p : \Sigma \to R$ be defined by:

$$\begin{cases} R_p(u) = J(u) - \mu(|\partial J(u)|^2) \text{ for } J(u) \leq ((p + \frac{1}{2}) S)^{2/n-2} \\ R_p(u) = J(u) \text{ elsewhere.} \end{cases} \quad (5.179)$$

R_p is defined on Σ, the unit sphere of H_0^1, and is C^1. Let

$$\boldsymbol{l}(u) = |\partial J(u)|^2.$$

Clearly, there exists a constant C such that

$$|(\partial J(v), \partial \boldsymbol{l}(v))| \leq M |\partial J(v)|^2 \ \forall v \in \Sigma \quad (5.180)$$

such that $J(v) \leq b_p = ((p+1) S)^{2/n-2}$.

Let θ_p be such that:

$$\theta_p > 2M. \quad (5.181)$$

We then have:

$$\partial R_p(v) . \partial J(v) > 0 \ \forall v \in \Sigma^+ \text{ such that } J(v) \leq b_p. \quad (5.182)$$

This implies:

Proposition 5.11: The pair $(R_p^{b_{p-1}} \cap \Sigma^+, W_{p-1})$ is a retract by deformation of (W_p, W_{p-1}). Given $\varepsilon_1 > 0$, one can choose $\bar{\varepsilon}_p$ small enough so that

$$\overline{\Sigma^+ \cap (R_p^{b_{p-1}} \backslash W_{p-1})} \subset V(p, \varepsilon_1).$$

Proof: Consider the differential equation:

$$\frac{\partial u}{\partial s} = -\partial J(u) \; ; \eta(s, \cdot) \quad \text{its flow.} \tag{5.183}$$

Using Proposition 5.1, we have:

$$\{s \geq 0 \text{ s.t. } R_p(\eta(s, u_0)) \leq b_{p-1}\} \neq \emptyset, \; \forall u_0 \in W_p. \tag{5.184}$$

Let:

$$s(u) = \min \{s \geq 0 \text{ s.t. } R_p(\eta(s, u)) \leq b_{p-1}\}. \tag{5.185}$$

Observe that:

$$s(u) = 0 \text{ if } u \in W_{p-1} \tag{5.186}$$

The map:

$$\begin{aligned} [0,1] \times W_p &\to W_p \\ (\tau, u) &\to \eta(\tau s(u), u) \end{aligned} \tag{5.187}$$

is our retraction by deformation.

Observe now that:

$$\forall u \in \overline{\Sigma^+ \cap (R_p^{b_{p-1}} \backslash W_{p-1})}, \quad |\partial J(u)|^2 \leq \theta_p \bar{\varepsilon}_p \tag{5.188}$$

Hence, by Proposition 5.1, $\overline{\Sigma_+ \cap (R_p^{b_{p-1}} \backslash W_{p-1})}$ is in $V(p, \varepsilon_1)$ if $\theta_p \bar{\varepsilon}_p$ is small enough. This completes the proof of Proposition 5.11.

6 Topological arguments (existence proof)

Consider the map:

$$\gamma_p : (A^p \times_{\sigma_p} \Delta_{p-1} A^p \times_{\sigma_p} \partial\Delta_{p-1} \rightarrow (W_p, W_{p-1}). \tag{6.1}$$

Using a filtration of Ω by compact sets, we may assume $(\gamma_p)_\star$ to be defined as well, as far as homology or homotopy classes are involved, on:

$$\gamma_{p\star} : H_\star(\Omega^p \times_{\sigma_p} \Delta_{p-1}, \Omega^p \times_{\sigma_p} \partial\Delta_{p-1} \rightarrow H_\star(W_p, W_{p-1}). \tag{6.2}$$

6.1: <u>The rational homology case</u>

Let us first deal with the case where there is a nonzero odd dimensional generator in the rational homology of Ω. Then, we may as well, because of the nontriviality of the arrow:

$$H_\star(\Omega^p \times \Delta_{p-1}, \Omega^p \times \partial\Delta_{p-1}) \rightarrow H_\star(\Omega^p \times_{\sigma_p} \partial\Delta_{p-1}, \Omega^p \times_{\sigma_p} \partial\Delta_{p-1}) \tag{6.3}$$

on these classes in $H_\star(\Omega^p \times \Delta_{p-1}, \Omega^p \times \partial\Delta_{p-1}) = \bigoplus_{\cdot \leq \star} H_\cdot(\Omega^p) \otimes$ $H_{\star - \cdot}(\Delta_{p-1}, \partial\Delta_{p-1})$ which come from odd dimensional generators (see G. Bredon [4] cf.), consider the map:

$$\tilde{\gamma}_p : (\Omega^p \times \Delta_{p-1}, \Omega^p \times \partial\Delta_{p-1}) \rightarrow (W_p, W_{p-1}) \tag{6.4}$$

hence

$$\tilde{\gamma}_{p\star} : H_\star(\Omega^p \times \Delta_{p-1}, \Omega^p \times \partial\Delta_{p-1}) \rightarrow H_\star(W_p, W_{p-1}). \tag{6.5}$$

Let:

$$e_{p-1} \text{ generator of } H_{p-1}(\Delta_{p-1}, \partial\Delta_{p-1}; \mathbb{Q}). \tag{6.6}$$

From now until further notice, the homology is taken with rational coefficients. Let

$$w \in H_{\star -p+1}(\Omega^p). \tag{6.7}$$

Then

$$w \otimes e_{p-1} \in H_\star(\Omega^p \times \Delta_{p-1}, \ \Omega^p \times \partial\Delta_{p-1}). \tag{6.8}$$

We then define:

$$\delta_\star : H_\star(\Omega^p \times \Delta_{p-1}, \Omega^p \times \partial\Delta_{p-1}) \to H_{\star -1}(\Omega^{p-1} \times \Delta_{p-2}, \Omega^{p-1} \times \partial\Delta_{p-2}) \tag{6.9}$$

by

$$\delta_\star(w \otimes e_{p-1}) = (-1)^{\star -p+1} \left(\sum_{i=1}^{p} (-1)^{i-1} p_{i\star}(w) \right) \otimes e_{p-2}, \tag{6.10}$$

where p_i stands for the projection:

$$p_i : \Omega^p \to \Omega^{p-1} \tag{6.11}$$
$$(x_1, \ldots, x_p) \to (x_1, \ldots, \overset{x_i}{\Lambda}, \ldots, x_p).$$

Let ∂ be the connecting homomorphism (see Dold cf. [13]):

$$H_\star(W_p, W_{p-1}) \xrightarrow{\partial} H_{\star -1}(W_{p-1}, W_{p-2}). \tag{6.12}$$

We then have, in homology, the following diagram:

$$\begin{array}{ccc} H_\star(\Omega^p \times \Delta_{p-1}, \Omega^p \times \partial\Delta_{p-1}) & \xrightarrow{\tilde{\gamma}_{p\star}} & H_\star(W_p, W_{p-1}) \\ \Big| \delta & & \Big| \partial \\ H_{\star -1}(\Omega^{p-1} \times \Delta_{p-2}, \Omega^{p-1} \times \partial\Delta_{p-2}) & \xrightarrow{\tilde{\gamma}_{p-1\star}} & H_\star(W_{p-1}, W_{p-2}) \end{array} \tag{6.13}$$

This diagram is commutative as, for each i and each $A \subset \Omega^p$, the following one is commutative up to homotopy (see the end of 6.1 for the proof).

$$\begin{array}{ccccc} A^p \times \Delta_{p-2} & \xrightarrow{g_i} & A^p \times \Delta_{p-1} & \xrightarrow{\tilde{\gamma}_p} & W_p \\ \downarrow & & & \nearrow & \\ A^{p-1} \times \Delta_{p-2} & \xrightarrow{\tilde{\gamma}_{p-1}} & W_{p-1} & & \end{array} \tag{6.14}$$

where

$$g_i : A^p \times \Delta_{p-2} \to A^p \times \Delta_{p-1} \tag{6.15}$$

corresponds to sending Δ_{p-2} as the i-th side of Δ_{p-1}.

We then have the cap-product:

$$H^{\star}(\Omega^p \times \Delta_{p-1}) \otimes H_{\star+i}(\Omega^p \times \Delta_{p-1}, \Omega^p \times \partial \Delta_{p-1}) \to H_i(\Omega^p \times \Delta_{p-1}, \Omega^p \times \partial \Delta_{p-1}), \tag{6.16}$$

which provides $H_\star(\Omega^p \times \Delta_{p-1}, \Omega^p \times \partial \Delta_{p-1})$ with a structure of module over $H_\star(\Omega^p)$. Via the natural map:

$$\tau : \Omega^p \to \Omega^p/\sigma_p; \quad \text{hence} \quad \tau^\star : H^\star(\Omega^p) \to H^\star(\Omega^p/\sigma_p) \tag{6.17}$$

$H_\star(\Omega^p \times \Delta_{p-1}, \Omega^p \times \partial \Delta_{p-1})$ is equipped with a structure of module over $H^\star(\Omega^p/\sigma_p)$. On the other hand, we consider:

$$H_\star(W_p, W_{p-1}) = H_\star(\overline{W}_p, \overline{W}_{p-1}). \tag{6.18}$$

(Observe that, as we assume there is no critical point in W_p, the points (W_p, W_{p-1}) and $(\overline{W}_p, \overline{W}_{p-1})$ are homotopy equivalent.) This, by Proposition 5.11 is equal as well to:

$$H(R_p^{b_{p-1}}, \overline{W}_{p-1}) = H(R_p^{b_{p-1}} \cap V(p, \varepsilon_1), \overline{W}_{p-1} \cap V(p, \varepsilon_1)). \tag{6.19}$$

Thus, $H_\star(W_p, W_{p-1})$ may as well be considered as a module over $H^\star(R_p^{b_{p-1}} \cap V(p, \varepsilon_0))$. In Proposition 5.2, we define a continuous map:

$$V(p, \varepsilon_0) \to \Omega^p/\sigma_p. \tag{6.20}$$

Hence, $H_\star(W_p, W_{p-1})$ is a module over $H^\star(\Omega^p/\sigma_p)$.

We now claim that the map $\tilde{\gamma}_{p\star}$ in (6.13) is $H^\star(\Omega^p/\sigma_p)$-linear. Indeed the following diagram commutes (we again cite the end of 6.1):

$$\begin{array}{ccc} (A^p \times \Delta_{p-1}, A^p \times \partial\Delta_{p-1}) & \xrightarrow{\beta_p} (R_p^{b_{p-1}} \cap & V(p, \varepsilon_1; \overline{W}_{p-1} \cap V(p, \varepsilon_1)) \\ \downarrow & & \downarrow \\ A^p & \longrightarrow & \Omega^p/\sigma_p \end{array} \tag{6.21}$$

where β_p is homotopic to γ_p relatively to $(\overline{W}_p, \overline{W}_{p-1})$. Hence the claim.

Now let

$$v \in H_{2k+1}(\Omega); \quad v \neq 0. \tag{6.22}$$

Let

$$u \in H^{2k+1}(\Omega) \text{ such that } \langle u, v \rangle = 1; \text{ where } \langle u, v \rangle \text{ is the Poincaré duality.} \tag{6.23}$$

We use now an induction argument to show:

$$\tilde{\gamma}_{p\star}(v \otimes \ldots \otimes v \otimes e_{p-1}) \text{ is nonzero.} \tag{6.24}$$

Proposition 6.1: As long as there is no critical point to J ($K = 1$; $\Omega \subset \mathbb{R}^n$) in W_p, $\tilde{\gamma}_{p\star}(v \otimes \ldots \otimes v \otimes e_{p-1})$ is nonzero.

Corollary 6.2: The equation $\begin{cases} \Delta u + u^{\frac{n+2}{n-2}} \\ u = 0 \big|_{\partial\Omega}; \quad u > 0 \end{cases}$ has a solution.

Proof of Proposition 6.1: For $p = 1$, (6.24) holds by (6.22). Consider now

$$u_p = u \otimes 1 \otimes \ldots \otimes 1 + 1 \otimes u \otimes 1 \otimes \ldots \otimes 1 + \ldots + 1 \otimes 1 \ldots \otimes 1 \otimes u \qquad (6.25)$$

in $H^\star(\Omega^p)$.

τ is defined in (6.17).

There is an element (see G. Bredon [4]) denoted $\mathrm{tr}\, u_p$ such that:

$$\tau^\star(\mathrm{tr}\, u) = u_p, \qquad (6.26)$$

and, in fact, tr u is uniquely defined through (6.26) as, in rational homology, there is an isomorphism between $H_\star(\Omega^p/\sigma_p)$ and the invariants under the action of σ_p in $H_\star(\Omega^p)$.

We now have:

$$\partial(\mathrm{tr}\, u \cdot \tilde{\gamma}_{p\star}(v \otimes v \ldots \otimes v \otimes e_{p-1})) = (-1)^p p \tilde{\gamma}_{p-1\star}(v \otimes \ldots \otimes v \otimes e_{p-2}). \quad (6.27)$$

Hence the proof of the proposition by induction.

Corollary 6.2 follows now using (5.132), Proposition 5.10.

Proof of claims (6.19) and (6.21): Observe that, as we defined $\tilde{\gamma}_p$ and $\tilde{\gamma}_{p-1}$, there are associated maps:

$$d_p: \quad A^p \to (\lambda_0, +\infty)^p$$
$$d_{p-1}: \quad A^{p-1} \to (\lambda_0, +\infty)^{p-1}$$

such that

$$\tilde{\gamma}_p(x, t) = \sum_{i=1}^{p} t_i P\delta_i(x_i, \lambda_i); \quad \text{when} \quad (\lambda_1, \ldots, \lambda_p) = d_p(x)$$

$$\tilde{\gamma}_{p-1}(x, t) = \sum_{i=1}^{p-1} t_i P\delta_i(x_i, \lambda_i').$$

Observe also that, given any other function:

$$d_p': A^p \to (\lambda_0, +\infty)^p$$
$$x \to d_p'(x) = (\mu_1, \ldots, \mu_p)$$

the map:

$$\tilde{\gamma}'_p(x,t) = \sum_{i=1}^{p} t_i P\delta_i(x_i,\mu_i)$$

is homotopic to $\tilde{\gamma}_p$ with values in (W_p, W_{p-1}). The homotopy is

$$[0,1]\times(A^p\times\Delta_{p-1},\ A^p\times\partial\Delta_{p-1}) \to \sum_{i=1}^{p} t_i P\delta_i(x_i,\ (1-s)\lambda_i+s\mu_i).$$

Indeed, $(1-s)\lambda_i + s\mu_i$ belongs to $(\lambda_0, +\infty)$. By the expansion provided by Proposition 5.10, which holds if the ε_{ij} are small we have, assuming $(\mu'_1,\ldots,\mu'_p)$ maps to $(\lambda_0,+\infty)^p$,

$$J\left(\sum_{i=1}^{p} p_i P\delta_i(x_i,\mu'_i)\right) = \frac{\left(\sum_{i=1}^{p} t_i^2\right)^{n/(n-2)}}{\sum_{i=1}^{p} t_i^{2n/(n-2)}} \left(\int \delta^{\frac{2n}{n-2}}\right)^{\frac{2}{n-2}} \left(1+0\left(\frac{1}{\lambda_0^{n-2}}\right)\right)$$

$$\leq (p+\tfrac{1}{2})^{\frac{2}{n-2}}\left(\int \delta^{\frac{2n}{n-2}}\right)^{\frac{2}{n-2}} < b_p$$

as soon as λ_0 is large enough.

If one t_i is zero, this inequality improves, yielding

$$J\left(\sum_{i=1}^{p} t_i P\delta_i(x_i,\mu_i)\right) \leq (p-\tfrac{1}{2})^{\frac{2}{n-2}}\left(\int \delta^{\frac{2n}{n-2}}\right)^{\frac{2}{n-2}} < b_{p-1}.$$

If now one of the ε_{ij} is not small, then Proposition 5.5 guarantees that $J\left(\sum_{i=1}^{p} t_i P\delta_i(x_i,\mu'_i)\right) < b_p$; and if one t_i is zero, this inequality improves, yielding $J\left(\sum_{i=1}^{p} t_i P\delta_i(x_i,\mu'_i)\right) < b_{p-1}$. Thus, we may make homotopies through the concentrations: Having noticed this, the commutativity up to homotopy of (6.14) follows readily by adjusting the concentrations of $\tilde{\gamma}_p \circ g_i$ and $\tilde{\gamma}_{p-1} \circ p_1$ where p_1 is the projection of A^p onto A^{p-1}.

For (6.21), we proceed as follows:

We first homotop $\tilde{\gamma}_p$ to $\tilde{\gamma}'_p$ defined as follows:

$$\tilde{\gamma}'_p(x,t) = \sum_{i=1}^{p} t_i P\delta_i(x_i, (\lambda_0+1)^i)$$

i.e. we replace the concentrations $(\lambda_1, \dots, \lambda_p)$ by $(\lambda_0+1, (\lambda_0+1)^2, \dots, (\lambda_0+1)^p)$. This is a homotopy of $\tilde{\gamma}_p$ relatively to $(\overline{W}_p, \overline{W}_{p-1})$. Observe now that denoting ε'_{ij} the correspondent of ε_{ij} for $\tilde{\gamma}'_p(x,t)$, we have:

$$\varepsilon'_{ij} = \left(\frac{(\lambda_0+1)^i}{(\lambda_0+1)^j} + \frac{(\lambda_0+1)^j}{(\lambda_0+1)^i} + (\lambda_0+1)^i (\lambda_0+1)^j |x_i - x_j|^2 \right)^{\frac{-2}{n-2}}$$

$$< \operatorname{Inf}((\lambda_0+1)^{i-j}, (\lambda_0+1)^{j-i}),$$

hence the ε'_{ij} are small and Proposition 5.10 holds, yielding:

$$J\left(\sum_{i=1}^{p} t_i P\delta_i(x_i, (\lambda_0+1)^i)\right) = \frac{\left(\sum_{i=1}^{p} t_i^2\right)^{n/(n-2)}}{\sum_{i=1}^{p} t_i^{2n/(n-2)}} \left(\int \delta^{\frac{2n}{n-2}}\right)^{\frac{2}{n-2}} \left(1 + 0\left(\frac{1}{\lambda_0^{n-2}}\right)\right).$$

Next, we define the following homotopy of $\tilde{\gamma}'_p$: consider $\varepsilon > 0$ large enough: define:

$$[\varepsilon, 1] \times (A^p \times \Delta_{p-1}, A^p \times \partial\Delta_{p-1}) \to (W_p, W_{p-1})$$

$$(s, x, t) \to \sum_{i=1}^{p} \left(st_i + (1-s)\frac{1}{p}\right) P\delta_i(x_i, (\lambda_0+1)^i)$$

This homotopy is relative to (W_p, W_{p-1}) using the expansion of Proposition 5.10, which holds since the ε_{ij} remain small along the homotopy (they are equal to ε'_{ij} defined for $\tilde{\gamma}'_p$ all along the homotopy). Indeed, using this proposition, we have:

$$J\left(\sum_{i=1}^{p} \left(st_i + (1-s)\frac{1}{p}\right) P\delta_i(x_i, (\lambda_0+1)^i)\right)$$

$$= \frac{\left(\sum_{i=1}^{p} (st_i+(1-s)\frac{1}{p})^2\right)^{n/(n-2)}}{\sum_{i=1}^{p} (st_i+(1-s)\frac{1}{p})^{2n/(n-2)}} \left(\int \delta^{\frac{2n}{n-2}}\right)^{\frac{2}{n-2}} + 0\left(\frac{1}{\lambda_0^{n-2}}\right)$$

this is always upper bounded, provided λ_0 is large enough, by

$$(p+\tfrac{1}{2})^{\frac{2}{n-2}} \left(\int \delta^{\frac{2n}{n-2}}\right)^{\frac{2}{n-2}}$$

(hence the homotopy takes place in W_p).

Next, if one t_i is zero, say t_1, then:

$$(1-s)\frac{1}{p} = st_1 + (1-s)\frac{1}{p} < \frac{\varepsilon}{p-1} + \frac{1}{p} < \frac{1}{p-1} \sum_{i=2}^{p} (st_i + (1-s)\frac{1}{p})$$

$$= \frac{s}{p-1} \sum_{i=1}^{p} t_i + \frac{1-s}{p-1} = \frac{s}{p-1} + \frac{1-s}{p-1} = \frac{1}{p-1}$$

thus the value of the first coefficient remains bounded strictly above by a quantity which is upper bounded by the average of the other coefficients, which implies, using the expansion that, if λ_0 is large enough, we have:

$$J\left(\sum_{i=1}^{p} (st_i+(1-s)\frac{1}{p})\, P\delta_i(x_i, (\lambda_0+1)^i)\right) \le p^{\frac{2}{n-2}} \left(\int \delta^{\frac{2n}{n-2}}\right)^{\frac{2}{n-2}} - \phi(\varepsilon),$$

where ϕ is a strictly positive continuous function from $\mathbb{R}^{+\star}$ into $\mathbb{R}^{+\star}$. Hence the homotopy maps $A^p \times \partial\Delta_{p-1}$ into W_{p-1}. We are now left with a map β_p, homotopic to $\tilde{\gamma}_p$, such that: on $\beta_p(x,t)$, the ε_{ij} are small and the coefficients $\varepsilon t_i + (1-\varepsilon)\frac{1}{p}$ are close to $1/p$. Hence $\beta_p(x,t)$ belongs to $V(p,\varepsilon')$, with ε' arbitrarily small depending on ε small. After extending β_p to all of Ω^p, we have thus: $\tilde{\beta}_p : (\Omega^p \times \Delta_{p-1}, \Omega^p \times \partial\Delta_{p-1}) \to (W_p \cap V(p,\varepsilon), W_{p-1} \cap V(p,\varepsilon))$.

We now have:

$$R_p(\tilde{\beta}_p(x,t)) = J(\tilde{\beta}(x,t)) - \mu(|\partial J(\tilde{\beta}_p(x,t)|^2).$$

When ε goes to zero, we have:

$$\lim_{\varepsilon\to 0} \sup_{u\in V(p,\varepsilon)} J(u) = b_{p-1}$$

$$\lim_{\varepsilon\to 0} \sup_{u\in V(p,\varepsilon)} |\partial J(u)|^2 = 0$$

Thus

$$\lim_{\substack{\lambda_0\to+\infty\\ \varepsilon\to 0}} \sup_{(x,t)\in\Omega^p\times\Delta_{p-1}} J(\tilde{\beta}_p(x,t)) = b_{p-1},$$

$$\lim_{\substack{\lambda_0\to+\infty\\ \varepsilon\to 0}} \sup_{(x,t)\in\Omega^p\times\Delta_{p-1}} \mu(|\partial J(\tilde{\beta}_p(x,t)|^2) = \bar{\varepsilon}_p \quad (\mu(0) = \bar{\varepsilon}_p > 0).$$

Hence

$$\lim_{\substack{\lambda_0\to+\infty\\ \varepsilon\to 0}} \sup_{(x,t)\in\Omega^p\times\Delta_{p-1}} R_p(\tilde{\beta}_p(x,t)) = b_{p-1} - \bar{\varepsilon}_p < b_{p-1}.$$

Thus, the map $\tilde{\beta}_p$ is a map into $(R_p^{b_{p-1}} \cap V(p,\varepsilon), \overline{W}_p \cap V(p,\varepsilon))$.

We now take $\varepsilon = \varepsilon_1$ of Proposition 5.11. We consider the diagram:

$$\begin{array}{ccc}
(\Omega^p\times\Delta_{p-1}, \Omega^p\times\partial\Delta_{p-1}) & \xrightarrow{\tilde{\beta}_p} & (R_p^{b_{p-1}} \cap V(p,\varepsilon), \overline{W}_p \cap V(p_1\varepsilon_1)) \\
\downarrow & & \downarrow \\
\Omega^p/\sigma_p & \rightarrow & \Omega^p/\sigma_p
\end{array}$$

where the right vertical arrow is provided by solving problem (5.12) which can be done by Proposition 5.2.

This diagram commutes as the solution of (5.12) with $u = \tilde{\beta}_p(x,t) =$

$\sum_{i=1}^{p} \alpha_i P\delta_i(x_i, (\lambda_0+1)^i)$ is evidently $(\alpha_1, \ldots, \alpha_p, x_1, \ldots, x_p, (\lambda_0+1), \ldots, (\lambda_0+1)^p)$.
Hence the right vertical arrow maps (x, t) into x. Hence the result on (6.21).

6.2 The general case: some facts in algebraic topology

We need first some preliminaries. Observe that the argument of 6.1 breaks down in case $H_{2k+1}(\Omega, \mathbb{Q}) = 0 \ \forall k \in N$, (6.3) is zero on even dimensional generators.

We assume now:

$$H_k(\Omega;\mathbb{Z}_2) \neq 0 \text{ for } k \neq 0 \ (k \text{ given}). \tag{6.28}$$

All homology groups will be taken in 6.2 with $\mathbb{Z}_2$-coefficients. This section is devoted to the proof of a nontrivial homological arrow: Consider Ω^p with the action of σ_p on it. This action degenerates on a subset F_p of Ω^p:

$$F_p = \{(x_1, \ldots, x_p) \in \Omega^p \text{ such that } \exists(i,j),\ i \neq j \text{ with } x_i = x_j\}. \tag{6.29}$$

Let

$$V_p \text{ be a tubular neighbourhood of } F_p \text{ invariant under } \sigma_p, \tag{6.30}$$

and

$$(\Omega^p)\star = \overline{\Omega^p / V_p} \ (\text{closure in } \Omega^p), \tag{6.31}$$

$$\partial V_p \text{ be the boundary of this tubular neighbourhood.} \tag{6.32}$$

The pair $((\Omega^p)\star, \partial V_p)$ is homologically equivalent to (Ω^p, V_p), hence to (Ω^p, F_p); and the action of σ_p is free on it.

Observe that we have, by excision and deformation:

$$H_\star((\Omega^p)\star/\sigma_p, \partial V_p/\sigma_p) = H_\star(\Omega^p/\sigma_p, V_p/\sigma_p), \tag{6.33}$$

$$H_\star((\Omega^p)\star \times_{\sigma_p} \partial\Delta_{p-1}, \partial V_p \times_{\sigma_p} \partial\Delta_{p-1}) = H_\star(\Omega^p \times_{\sigma_p} \partial\Delta_{p-1},$$

$$V_p \times_{\sigma_p} \partial\Delta_{p-1}), \tag{6.34}$$

$$H_\star(\Omega^p \times_{\sigma_p} \Delta_{p-1}, V_p \times \Delta_{p-1} \cup_{\sigma_p} \Omega^p \times \partial\Delta_{p-1}) \tag{6.35}$$

$$= H_\star((\Omega^p)\star \times_{\sigma_p} \Delta_{p-1}, \partial V_p \times \Delta_{p-1} \cup_{\sigma_p} (\Omega^p)\star \times \partial\Delta_{p-1})$$

$$= H_\star(\Omega^p \times_{\sigma_p} \Delta_{p-1}, F_p \times \Delta_{p-1} \cup_{\sigma_p} \Omega^p \times \partial\Delta_{p-1}).$$

Observe also that we may write:

$$(\Omega^p)\star \times_{\sigma_p} \Delta_{p-1}/\partial V_p \times \Delta_{p-1} \cup_{\sigma_p} (\Omega^p)\star \times \partial\Delta_{p-1} \tag{6.36}$$

$$= ((\Omega^p)\star \times_{\sigma_p} \Delta_{p-1}/(\Omega^p)\star \times_{\sigma_p} \partial\Delta_{p-1})/(\partial V_p \times_{\sigma_p} \Delta_{p-1}/\partial V_p \times_{\sigma_p} \partial\Delta_{p-1})$$

[A/B is the quotient of A by B].

We thus have the following exact sequence:

$$H_\star((\Omega^p)\star \times_{\sigma_p} \Delta_{p-1}, (\Omega^p)\star \times_{\sigma_p} \partial\Delta_{p-1}) \tag{6.37}$$

$$\to H_\star((\Omega^p)\star \times_{\sigma_p} \Delta_{p-1}, (\Omega^p)\star \times \partial\Delta_{p-1} \cup_{\sigma_p} \partial V_p \times \Delta_{p-1})$$

$$\to H_{\star-1}(\partial V_p \times_{\sigma_p} \Delta_{p-1}, \partial V_p \times_{\sigma_p} \partial\Delta_{p-1}) \to \cdots$$

On the other hand, we introduce, choosing a point $\hat{x}$ in ∂V_p,

$$j_{\hat{x}}: (\Delta_{p-1}, \partial\Delta_{p-1}) \to (\partial V_p \times_{\sigma_p} \Delta_{p-1}, \partial V_p \times_{\sigma_p} \partial\Delta_{p-1}) \tag{6.38}$$

$$(t_1, \ldots, t_p) \to [(x_1, \ldots, x_p, t_1, \ldots, t_p)]; \text{ where } \hat{x} = (x_1, \ldots, x_p),$$

$\partial V_p \times_{\sigma_p} \Delta_{p-1}$ is a disc-bundle over $\partial V_p/\sigma_p$. Then, there is a Thom isomorphism from $H_\star(\partial V_p \times_{\sigma_p} \Delta_{p-1}, \partial V_p \times_{\sigma_p} \partial\Delta_{p-1}) \to H_{\star-p+1}(\partial V_p/\sigma_p)$. This isomorphism, which we denote $\phi_{1\star}$ is given (see Switzer [14]) by taking the cap-product of any element in $H_\star(\partial V_p \times_{\sigma_p} \Delta_{p-1}, \partial V_p \times_{\sigma_p} \partial\Delta_{p-1})$ with the fundamental class u_1^p of this bundle, i.e.:

$$u_1^p \in H^{p-1}(\partial V_p \times_{\sigma_p} \Delta_{p-1}, \partial V_p \times_{\sigma_p} \partial\Delta_{p-1}) \quad \text{such that:} \tag{6.39}$$

$$j_{\hat{x}}^{\star}(u_1^p) \text{ generates } H^{p-1}(\Delta_{p-1}, \partial\Delta_{p-1}), \text{ for any } \hat{x} \text{ in } \partial V_p. \tag{6.40}$$

Thus:

$$\phi_{1\star} : H_\star(\partial V_p \times_{\sigma_p} \Delta_{p-1}, \partial V_p \times_{\sigma_p} \partial\Delta_{p-1}) \underset{\simeq}{\overset{u_1 \cap}{\rightarrow}} H_{\star-p+1}(\partial V_p/\sigma_p) \tag{6.41}$$

(6.39), (6.40) and (6.41) hold also for $((\Omega^p)\star \times_{\sigma_p} \Delta_{p-1}, (\Omega^p)\star \times_{\sigma_p} \partial\Delta_{p-1})$ and we thus have:

$$u_2^p \in H^{p-1}((\Omega^p)\star \times_{\sigma_p} \Delta_{p-1}, (\Omega^p)\star \times_{\sigma_p} \partial\Delta_{p-1}) \quad \text{such that} \tag{6.42}$$

$$\tilde{j}_{\hat{x}}^{\star}(u_2^p) \text{ generates } H^{p-1}(\Delta_{p-1}, \partial\Delta_{p-1}) \text{ for any } \hat{x} \text{ in } (\Omega^p)\star, \tag{6.43}$$

where $\tilde{j}_{\hat{x}} : (\Delta_{p-1}, \partial\Delta_{p-1}) \rightarrow ((\Omega^p)\star \times_{\sigma_p} \Delta_{p-1}, (\Omega^p)\star \times_{\sigma_p} \partial\Delta_{p-1})$ is:

$$(t_1, \ldots, t_p) \rightarrow [(x_1, \ldots, x_p, t_1, \ldots, t_p)]; \; \hat{x} = (x_1, \ldots, x_p)$$

and we have:

$$\phi_{2\star} : H_\star((\Omega^p)\star \times_{\sigma_p} \Delta_{p-1}, (\Omega^p)\star \times_{\sigma_p} \partial\Delta_{p-1}) \underset{\simeq}{\overset{u_2^p \cap}{\rightarrow}} H_{\star-p+1}((\Omega^p)\star/\sigma_p). \tag{6.44}$$

Observe now that, denoting:

$$k : (\partial V_p \times_{\sigma_p} \Delta_{p-1}, \partial V_p \times_{\sigma_p} \partial\Delta_{p-1}) \rightarrow ((\Omega^p)\star \times_{\sigma_p} \Delta_{p-1}, (\Omega^p)\star \times_{\sigma_p} \partial\Delta_{p-1}) \tag{6.45}$$

the inclusion, we have, for any $\hat{x}$ in ∂V_p:

$$\tilde{j}_{\hat{x}} = k \circ j_{\hat{x}}. \tag{6.46}$$

We may thus choose u_1^p as:

$$u_1^p = k^\star(u_2^p) \ . \tag{6.47}$$

We then have the commutative diagram:

$$\begin{array}{ccc} H_\star(\partial V_p \times_{\sigma_p} \Delta_{p-1},\ \partial V_p \times_{\sigma_p} \partial\Delta_{p-1}) & \xrightarrow{k_\star} & H_\star((\Omega^p)^\star \times_{\sigma_p} \Delta_{p-1}, (\Omega^p)^\star \times_{\sigma_p} \partial\Delta_{p-1}) \\ \Big\downarrow u_1^p\cap & & \Big\downarrow u_2^p\cap \\ H_{\star-p+1}(\partial V_p/\sigma_p) & \xrightarrow{l_\star} & H_{\star-p+1}((\Omega^p)^\star/\sigma_p) \end{array} \tag{6.48}$$

where l is the inclusion:

$$l : \partial V_p/\sigma_p \longleftrightarrow (\Omega^p)^\star/\sigma_p \ . \tag{6.49}$$

Finally, observe that, via formula (6.36), u_2^p acts by cap-product on $H_\star((\Omega^p)^\star \times_{\sigma_p} \Delta_{p-1},\ \partial V_p \times \Delta_{p-1} \cup_{\sigma_p} (\Omega^p)^\star \times \partial\Delta_{p-1})$ with values in $H_{\star-p+1}((\Omega^p)^\star/\sigma_p,\ \partial V_p/\sigma_p)$. Indeed:

$$u_2^p \text{ belongs to } H^{p-1}((\Omega^p)^\star \times_{\sigma_p} \Delta_{p-1},\ (\Omega^p)^\star \times_{\sigma_p} \partial\Delta_{p-1}) \ , \tag{6.50}$$

thus the cap-product $u_2^p\cap$ maps:

$$H_\star((\Omega^p)^\star \times_{\sigma_p} \Delta_{p-1},\ \partial V_p \times \Delta_{p-1} \cup_{\sigma_p} (\Omega^p)^\star \times \partial\Delta_{p-1}) \tag{6.51}$$

$$\xrightarrow{u_2^p\cap\cdot} H_{\star-p+1}((\Omega^p)^\star \times_{\sigma_p} \Delta_{p-1},\ \partial V_p \times_{\sigma_p} \Delta_{p-1}) \ .$$

Now, the pair $((\Omega^p)^\star \times_{\sigma_p} \Delta_{p-1},\ \partial V_p \times_{\sigma_p} \Delta_{p-1})$ is homotopy equivalent to the pair $((\Omega^p)^\star \times_{\sigma_p} \Delta_{p-1},\ \partial V_p \times_{\sigma_p} \Delta_{p-1})$ and the result follows.

We then have:

Proposition 6.3: (1) The map $u_2^p : H_*((\Omega^p)^* \times_{\sigma_p} \Delta_{p-1}, \partial V_p \times \Delta_{p-1} \cup_{\sigma_p} (\Omega^p)^* \times \partial \Delta_{p-1}) \to H_{*-p+1}((\Omega^p)^*/\sigma_p, \partial V_p/\sigma_p)$ is an isomorphism.

(2) $H^*(\Omega^p/\sigma_p)$ acts on both $H_*((\Omega^p)^* \times_{\sigma_p} \Delta_{p-1}, \partial V_p \times \Delta_{p-1} \cup_{\sigma_p} (\Omega^p)^* \times \partial \Delta_{p-1})$ and $H_*((\Omega^p)^*/\sigma_p, \partial V_p/\sigma_p)$ by cap-product also. These two groups are thus $H^*(\Omega^p)/\sigma_p)$-modules and we have the following commutative diagram: Let ξ belong to $H^{l p}(\Omega^p/\sigma_p)$:

$$\begin{array}{ccc} H^*((\Omega^p)^* \times_{\sigma_p} \Delta_{p-1}, \partial V_p \times \Delta_{p-1} \cup_{\sigma_p} (\Omega^p)^* \times \partial \Delta_{p-1}) & \xrightarrow{\xi \cap \cdot} & H_{*-l}((\Omega^p)^* \times_{\sigma_p} \Delta_{p-1}, \partial V_p \times \Delta_{p-1} \cup_{\sigma_p} (\Omega^p)^* \times \partial \Delta_{p-1}) \\ \downarrow & & \downarrow \\ H_{*-p+1}((\Omega^p)^*/\sigma_p, \partial V_p/\sigma_p) & \xrightarrow{z \cap \cdot} & H_{*-(p+l)+1}((\Omega^p)^*/\sigma_p, \partial V_p/\sigma_p) \end{array}$$

Proof of Proposition 6.3: For (1), observe that ϕ_{1*} and ϕ_{2*} may be identified as cap-products respectively from $H_*(\partial V_p \times_{\sigma_p} \Delta_{p-1}, \partial V_p \times_{\sigma_p} \partial \Delta_{p-1})$ into $H_{*-p+1}(\partial V_p \times_{\sigma_p} \Delta_{p-1})$ and from $H_*((\Omega^p)^* \times_{\sigma_p} \Delta_{p-1}, (\Omega^p)^* \times_{\sigma_p} \partial \Delta_{p-1})$ into $H_{*-p+1}((\Omega^p)^* \times_{\sigma_p} \Delta_{p-1})$. In this way, these maps are the usual cap-products of elements of $H^{p-1}(X, A_2)$ with elements of $H_*(X, A_1 \cup A_2)$, yielding elements of $H_{*-p+1}(X, A_1)$. For ϕ_{1*}, we have:

$$X = \partial V_p \times_{\sigma_p} \Delta_{p-1} \ ; \ A_2 = \partial V_p \times_{\sigma_p} \partial \Delta_{p-1} \ ; \ A_1 = \phi . \tag{6.52}$$

For ϕ_{2*}, we have:

$$X = (\Omega^p)^* \times_{\sigma_p} \Delta_{p-1}; \ A_2 = (\Omega^p)^* \times_{\sigma_p} \Delta_{p-1}; \ A_1 = \phi \tag{6.53}$$

and in fact, as $u_1^p = k^*(u_2^p)$, the two actions agree, yielding the commutative diagram (6.48).

They also agree with the action of u_2 on $H_\star((\Omega^p)\star \times_{\sigma_p} \Delta_{p-1},$ $\partial V_p \times \Delta_{p-1} \cup_{\sigma_p} (\Omega^p)\star \times \partial\Delta_{p-1})$. Here:

$$X = (\Omega^p)\star \times_{\sigma_p} \Delta_{p-1};\ A_2 = (\Omega^p)\star \times_{\sigma_p} \partial\Delta_{p-1}; \tag{6.54}$$

$$A_1 = \partial V_p \times_{\sigma_p} \Delta_{p-1}$$

From the naturality of the cap-product, we have:

$$k_\star(u_1 \cap v) = u_2^p \cap k_\star(v) \quad \text{as} \quad u_2^p = k^\star(u_1^p) \tag{6.55}$$

for any v in $H_\star(\partial V_p \times_{\sigma_p} \Delta_{p-1}, \partial V_p \times_{\sigma_p} \partial\Delta_{p-1})$,

where k has to be considered as the map of triads:

$$k: (\partial V_p \times_{\sigma_p} \Delta_{p-1};\ \partial V_p \times_{\sigma_p} \partial\Delta_{p-1};\phi) \to ((\Omega^p)\star \times_{\sigma_p} \Delta_{p-1}; \tag{6.56}$$
$$(\Omega^p)\star \times_{\sigma_p} \partial\Delta_{p-1};\phi).$$

Observe that $k_\star = \boldsymbol{l}_\star$ from $H_\star(\partial V_p/\sigma_p)$ into $H_\star((\Omega^p)\star/\sigma_p)$. Hence, the justification of the commutativity of (6.48).

Next, denoting m the map of triads:

$$m: ((\Omega^p)\star \times_{\sigma_p} \Delta_{p-1}; (\Omega^p)\star \times_{\sigma_p} \partial\Delta_{p-1};\phi) \tag{6.57}$$

$$\to ((\Omega^p)\star \times_{\sigma_p} \Delta_{p-1};(\Omega^p)\star \times_{\sigma_p} \partial\Delta_{p-1};\ \partial V_p \times_{\sigma_p} \Delta_{p-1})$$

$$(X; A_2; A_1) \to (X';\ A_2';\ A_1')$$

we have:

$$m_\star(u_2^p \cap v) = u_2^p \cap m_\star(v) \quad \text{for any } v \text{ in} \tag{6.58}$$

$$H_\star((\Omega^p)\star \times_{\sigma_p} \Delta_{p-1}, (\Omega^p)\star \times_{\sigma_p} \partial\Delta_{p-1})$$

and then the following commutative diagram:

$$
\begin{array}{ccc}
H_{\star}((\Omega^p)\star\times_{\sigma_p}\Delta_{p-1};(\Omega^p)\star\times_{\sigma_p}\partial\Delta_{p-1}) & \xrightarrow{m_\star} & H_{\star}((\Omega^p)\star\times_{\sigma_p}\Delta_{p-1},(\Omega^p)\star \times\partial\Delta_{p-1}\cup_{\sigma_p}\partial V_p\times\Delta_{p-1}) \\
\downarrow u_2^p\cap\cdot & & \downarrow u_2^p\cap\cdot \\
H_{\star-p+1}((\Omega^p)\star\times_{\sigma_p}\Delta_{p-1}) & \xrightarrow{m_\star} & H_{\star-p+1}((\Omega^p)\star\times_{\sigma_p}\Delta_{p-1},\partial V_p\times_{\sigma_p}\Delta_{p-1})
\end{array}
\tag{6.59}
$$

Finally, by stability of the cap-product (see Dold [13]), we have the following commutative diagram:

$$
\begin{array}{ccc}
H_{\star}((\Omega^p)\star\times_{\sigma_p}\Delta_{p-1},(\Omega^p)\star\times\partial\Delta_{p-1}\cup_{\sigma_p}\partial V_p\times\Delta_{p-1}) & \xrightarrow{\partial} & H_{\star-1}((\Omega^p)\star\times\partial\Delta_{p-1}\cup_{\sigma_p}\partial V_p\times\Delta_{p-1},(\Omega^p)\star\times_{\sigma_p}\partial\Delta_{p-1}) \\
 & & \simeq H_{\star-1}(\partial V_p\times_{\sigma_p}\Delta_{p-1},\partial V_p\times_{\sigma_p}\partial\Delta_{p-1}) \\
\downarrow u_2^p\cap\cdot & & \downarrow u_1^p\cap\cdot \\
H_{\star-p+1}((\Omega^p)\star\times_{\sigma_p}\Delta_{p-1},\partial V_p\times_{\sigma_p}\Delta_{p-1}) & \xrightarrow{\partial} & H_{\star-p+1}(\partial V_p\times_{\sigma_p}\Delta_{p-1})
\end{array}
\tag{6.60}
$$

Summing up (6.48), (6.59) and (6.60) yields:

$$
\begin{array}{ccccccccc}
A & \xrightarrow{k_\star} & B & \xrightarrow{m_\star} & C & \xrightarrow{\partial} & D & \longrightarrow & E \\
\downarrow u_1^p\cap\cdot & & \downarrow u_2^p\cap\cdot & & \downarrow u_2^p\cap\cdot & & \downarrow u_1^p\cap\cdot & & \downarrow u_2^p \\
F & \xrightarrow{k_\star} & G & \xrightarrow{m_\star} & \underline{H} & \xrightarrow{\partial} & I & \longrightarrow & J
\end{array}
\tag{6.61}
$$

where

$$A = H_{\star}(\partial V_p \times_{\sigma_p} \Delta_{p-1}, \partial V_p \times_{\sigma_p} \partial\Delta_{p-1}),$$

$$B = H_{\star}((\Omega^p)\star \times_{\sigma_p} \Delta_{p-1}, (\Omega^p)\star \times_{\sigma_p} \partial\Delta_{p-1}),$$

$$C = H_{\star}((\Omega^p)\star \times_{\sigma_p} \Delta_{p-1}, (\Omega^p)\star \times \partial\Delta_{p-1} \cup_{\sigma_p} \partial V_p \times \Delta_{p-1}),$$

$$D = H_{\star-1}(\partial V_p \times_{\sigma_p} \Delta_{p-1}, \partial V_p \times_{\sigma_p} \partial\Delta_{p-1}),$$

$$E = H_{\star-1}((\Omega^p)\star \times_{\sigma_p} \Delta_{p-1}, (\Omega^p)\star \times_{\sigma_p} \partial\Delta_{p-1}),$$

$$F = H_{\star-p+1}(\partial V_p \times_{\sigma_p} \Delta_{p-1}),$$

$$G = H_{\star-p+1}((\Omega^p)\star \times_{\sigma_p} \Delta_{p-1}),$$

$$\underline{H} = H_{\star-p+1}((\Omega^p)\star \times_{\sigma_p} \Delta_{p-1}, \partial V_p \times_{\sigma_p} \Delta_{p-1}),$$

$$I = H_{\star-p}(\partial V_p \times_{\sigma_p} \Delta_{p-1}),$$

$$J = H_{\star-p}((\Omega^p)\star \times_{\sigma_p} \Delta_{p-1}),$$

where the two left and the two right vertical arrows are Thom isomorphisms. Hence the middle arrow is also an isomorphism and (1) of Proposition 6.3 is thereby proved.

For (2) of this proposition, observe that, by excision:

$$H_{\star}((\Omega^p)\star/\sigma_p, \partial V_p/\sigma_p) \simeq H_{\star}(\Omega^p/\sigma_p, V_p/\sigma_p) \tag{6.62}$$

$$H_{\star}((\Omega^p)\star \times_{\sigma_p} \Delta_{p-1} \partial V_p \times \Delta_{p-1} \cup_{\sigma_p} (\Omega^p)\star \times \partial\Delta_{p-1}) \tag{6.63}$$

$$\simeq H_{\star}(\Omega^p \times_{\sigma_p} \Delta_{p-1}, V_p \times \Delta_{p-1} \cup_{\sigma_p} \Omega^p \times \partial\Delta_{p-1}).$$

Thus, both groups are modules over $H^{\star}(\Omega^p/\sigma_p)$ acting by cap-product. Let

$$f: ((\Omega^p)\star \times_{\sigma_p} \Delta_{p-1}, \partial V_p \times_{\sigma_p} \Delta_{p-1}) \to (\Omega^p \times_{\sigma_p} \Delta_{p-1}, V_p \times_{\sigma_p} \Delta_{p-1}) \tag{6.64}$$

$$g: ((\Omega^p)^\star \times_{\sigma_p} \Delta_{p-1}, \partial V_p \times \Delta_{p-1} \cup_{\sigma_p} (\Omega^p)^\star \times \partial\Delta_{p-1}) \tag{6.65}$$

$$\to (\Omega^p \times_{\sigma_p} \Delta_{p-1}, V_p \times \Delta_{p-1} \cup_{\sigma_p} \Omega^p \times \partial\Delta_{p-1})$$

be the injections, which are isomorphisms in homology. Let

$$\xi \text{ be in } H^l(\Omega^p/\sigma_p). \tag{6.66}$$

The action of ξ on the left-side groups in (6.62), (6.63) is via excision, i.e. if w belongs to $H_\star((\Omega^p)^\star/\sigma_p, \partial V_p/\sigma_p)$, we have:

$$\xi \cap w = f_\star^{-1}(\xi \cap f_\star(w)) \tag{6.67}$$

if v belongs to $H_\star((\Omega^p)^\star \times_{\sigma_p} \Delta_{p-1}, \partial V_p \times \Delta_{p-1} \cup_{\sigma_p} (\Omega^p)^\star \times \partial\Delta_{p-1})$:

$$\xi \cap v = g_\star^{-1}(\xi \cap g_\star(v)). \tag{6.68}$$

Thus:

$$u_2^p \cap (\xi \cap v) = u_2^p \cap (g_\star^{-1}(\xi \cap g_\star(v))). \tag{6.69}$$

On the other hand:

$$\xi \cap (u_2^p \cap v) = f_\star^{-1}(\xi \cap f_\star(u_2^p \cap v)) \tag{6.70}$$

Consider f as a map of triads:

$$f: ((\Omega^p)^\star \times_{\sigma_p} \Delta_{p-1}; \partial V_p \times_{\sigma_p} \Delta_{p-1}; \phi) \to (\Omega^p \times_{\sigma_p} \Delta_{p-1}; \tag{6.71}$$

$$V_p \times_{\sigma_p} \Delta_{p-1}; \phi).$$

Then:

$$\xi \cap (u_2^p \cap v) = f_\star^{-1}(\xi \cap f_\star(u_2^p \cap v)) = f_\star^{-1}(f_\star(f^\star(\xi) \cap (u_2^p \cap v))) \tag{6.72}$$

$$= f^\star(\xi) \cap (u_2^p \cap v) \quad .$$

With $\mathbb{Z}_2$ coefficients, we then have:

$$\xi \cap (u_2^p \cap v) = f^\star(\xi) \cap (u_2^p \cap v) = (f^\star(\xi) \cup u_2^p) \cap v \tag{6.73}$$

$$= (u_2^p \cup f^\star(\xi)) \cap v = u_2^p \cap (f^\star(\xi) \cap v).$$

If we consider now g as a map of triads:

$$g: ((\Omega^p)^\star \times_{\sigma_p} \Delta_{p-1}; \partial V_p \times \Delta_{p-1} \cup_{\sigma_p} (\Omega^p)^\star \times \partial\Delta_{p-1}; \phi) \tag{6.74}$$

$$\to (\Omega^p \times_{\sigma_p} \Delta_{p-1}; V_p \times \Delta_{p-1} \cup_{\sigma_p} \Omega^p \times \partial\Delta_{p-1}, \phi).$$

Then:

$$g_\star(\xi) = f_\star(\xi) \quad \text{as} \quad \xi = H^l(\Omega^p \times_{\sigma_p} \Delta_{p-1}; \phi). \tag{6.75}$$

Thus:

$$\xi \cap (u_2^p \cap v) = u_2^p \cap (g^\star(\xi) \cap v) = u_2^p \cap g_\star^{-1}(g_\star(g^\star(\xi) \cap v)) \tag{6.75'}$$

$$= u_2^p \cap g_\star^{-1}(\xi \cap g_\star(v))$$

Hence by (6.69):

$$\xi \cap (u_2^p \cap v) = u_2^p \cap (\xi \cap v) \tag{6.76}$$

and (2) of Proposition 6.3 is thereby proved.

The proof of Proposition 6.3 is thereby complete.

6.3 The existence argument in the general case

Lemma 6.4: There exists a commutative diagram:

$$\begin{array}{ccc} H_\star(\Omega^p \times_{\sigma_p} \Delta_{p-1}, \Omega^p \times \partial\Delta_{p-1} \cup_{\sigma_p} F_p \times \Delta_{p-1}) & \to & H_\star(W_p, W_{p-1}) \\ \downarrow & & \downarrow \\ H_{\star-1}(\Omega^{p-1} \times_{\sigma_{p-1}} \Delta_{p-2}, \Omega^{p-1} \times \partial\Delta_{p-2} \cup_{\sigma_{p-1}} F_{p-1} \times \Delta_{p-2}) & \to & H_\star(W_{p-1}, W_{p-2}) \end{array}$$

The horizontal arrows come from the maps γ_p and γ_{p-1} defined in Chapter 5. The right vertical arrow is equal to the connecting homomorphism of the pair (W_p, W_{p-1}) composed with the projection map $W_{p-1} \to W_{p-1}/W_{p-2}$. The left vertical arrow is described thereafter.

Description of arrows:

The horizontal arrows come from γ_p and γ_{p-1}: each homology class in the set considered is realized by a chain with simplices contained in a compact set A of Ω. Hence γ_p is well defined, as well as γ_{p-1}.

To make things precise, we also choose the η_i involved in the definitions of the maps γ_i small enough and uniform for all i less than p (p given), so that all the definitions of the γ_i's agree on the considered homological classes which will be controlled by a fixed compact set A in Ω (i.e. the simplices of the chains at order p will be contained in A^p, A fixed compact set in Ω). By Proposition 5.10, γ_p then sends $\Omega^p \times_{\sigma_p} \partial\Delta_{p-1} \cup_{\sigma_p} F_p \times \Delta_{p-1}$ into W_{p-1}. Indeed, by (5.131) of this proposition, for a fixed compact A, γ_p may be chosen so that $\gamma_p(A^p \times_{\sigma_p} \partial\Delta_{p-1}) \subset W_{p-1}$. Hence, by the previous remarks, the map $\gamma_{p\star}$ will send $H_\star(\Omega^p \times_{\sigma_p} \Delta_{p-1}, \Omega^p \times_{\sigma_p} \partial\Delta_{p-1})$ into $H_\star(W_p, W_{p-1})$.

By (5.134) of Proposition 5.10, γ_p sends every x such that $\operatorname{Min}_{i\neq j}|x_i - x_j| \leq \eta$ into W_{p-1}. If x is in F_p, $\operatorname{Min}_{i\neq j}|x_i - x_j| = 0$. Thus $\gamma_p(F_p) \subset W_{p-1}$. The horizontal arrows are thus well defined. The right vertical arrow is clear. We have to state now the left vertical arrow:

Consider the connecting homomorphism δ of the pair $(\Omega^p \times_{\sigma_p} \Delta_{p-1}, \Omega^p \times_{\sigma_p} \partial\Delta_{p-1} \cup_{\sigma_p} F \times \Delta_{p-1})$,

$$H_\star(\Omega^p \times_{\sigma_p} \Delta_{p-1}, \Omega^p \times \partial\Delta_{p-1} \cup_{\sigma_p} F_p \times \Delta_{p-1}) \xrightarrow{\partial} H_{\star-1}(\Omega^p \times \partial\Delta_{p-1} \cup_{\sigma_p} F_p \times \Delta_{p-}$$

(6.77)

Δ_{p-1} is the standard p-simplex, i.e.

$$\Delta_{p-1} = \{(t_1, \ldots, t_p) ; t_i \geq 0; \sum_{i=1}^{p} t_i = 1\} . \qquad (6.78)$$

The boundary of Δ_{p-1}, $\partial\Delta_{p-1}$, may be thought of as a union of (p-1)-simplices Δ_{p-2}; or either of elements $(t_1, \ldots, t_p)$ such that one of the t_i's at least is zero.

Consequently, we may define a continuous map:

$$\theta_p : \Omega^p \times \partial\Delta_{p-1} \cup_{\sigma_p} F_p \times \Delta_{p-1} \to \Omega^{p-1} \times_{\sigma_p} \Delta_{p-2} / \Omega^{p-1} \times \partial\Delta_{p-2} \cup_{\sigma_{p-1}} F_p \times \Delta_{p-2}$$

(6.79)

as follows:

if $(x_1, \ldots, x_p, t_1, \ldots, t_p)$ is a representative of an element of $\Omega^p \times \partial\Delta_{p-1} \cup_{\sigma_p} F_p \times \Delta_{p-1}$ which is in $\Omega^p \times \partial\Delta_{p-1}$, (6.80)

then there is a t_i in $(t_1, \ldots, t_p)$ which is zero. We consider $(x_1, \ldots, x_{i-1}, x_{i+1}, \ldots, x_p, t_1, \ldots, t_{i-1}, t_{i+1}, \ldots, t_p)$, i.e. we drop from $(x_1, \ldots, x_p, t_1, \ldots, t_p)$ a couple (x_i, t_i), with $t_i = 0$. We then have an element y of $\Omega^{p-1} \times \Delta_{p-2}$.

We then define $\theta_p(x_1, \ldots, x_p, t_1, \ldots, t_p)$ to be $[y]$ = the class of y in $\Omega^{p-1} \times_{\sigma_{p-1}} \Delta_{p-2} / \Omega^{p-1} \times \partial\Delta_{p-2} \cup_{\sigma_{p-1}} F_{p-1} \times \Delta_{p-2}$.

if $(x_1, \ldots, x_p, t_1, \ldots, t_p)$ is a representative of an element of $\Omega^p \times \partial\Delta_{p-1} \cup_{\sigma_p} F_p \times \Delta_{p-1}$ which is in $F_p \times \Delta_{p-1}$, then there are two distinct indices i and j such that $x_i = x_j$. (6.81)

We then consider $y = (x_1, \ldots, x_{i-1}, x_{i+1}, \ldots, x_j, \ldots, x_p, t_1, \ldots, t_{i-1}, t_{i+1}, \ldots, t_i+t_j, \ldots, t_p)$, i.e. we drop x_i and t_i (or x_j and t_j); we replace t_j by t_i+t_j (or t_i with t_i+t_j); and we again define $\theta_p((x_1, \ldots, x_p, t_1, \ldots, t_p))$ to by $[y]$.

At first, it might seem that θ_p depends on the choice of a representative of $(x, t) = (x_1, \ldots, x_p, t_1, \ldots, t_p)$. Nevertheless, if $(x_1, \ldots, x_p)$ is not in F_p and if $(t_1, \ldots, t_p)$ has only one zero component, the definition of $\theta_p(x, t)$ is unambiguous and invariant under the action of σ_p, with values in the target space as defined in (6.79). On this set, θ_p is then clearly continuous. If, now $(x_1, \ldots, x_p)$ is not in F_p and $(t_1, \ldots, t_p)$ has more than one zero component, y belongs, for any representative of (x, t) and any subsequent dropped (x_i, t_i), to $\Omega^{p-1} \times \partial\Delta_{p-2}$; as one of its components in the t_j's is zero. Hence $[y] = 0$ and the definition of θ_p is unambiguous.

The continuity of θ_p follows also: assuming x_n converges to x not in F_p and $(t_1^n, \ldots, t_p^n)$ converge to $(t_1, \ldots, t_p)$ with more than one zero component, the corresponding components in $(t_1^n, \ldots, t_p^n)$ tend to zero with n going to $+\infty$. As only one of them is dropped, $\theta_p([y^n])$ (with obvious notations) enters, when $n \to +\infty$, any prescribed neighbourhood of $\Omega^{p-1} \times \partial\Delta_{p-1}$. Thus $[y^n]$ tends to zero and θ_p is continuous at such points.

We are thus left now with (x, t) such that x is in F_p. If t belongs to $\overset{\circ}{\Delta}_{p-1} = \Delta_{p-1}/\partial\Delta_{p-1}$, then θ_p is again well defined and continuous at such points. Indeed, either on such an (x, t), x has only two equal components. Then any neighbouring point in $F_p \times (\Delta_{p-1}, \partial\Delta_{p-1})$ has only two equal components. On such a neighbourhood, the corresponding y's of (6.81) are all, for a given (x', t'), in the same orbit under the action of σ_{p-1}. θ_p is thus well defined and continuous.

If now x has more than two equal components, y has at least two equal components and $[y] = 0$. Again θ_p is well defined and its continuity follows by standard arguments.

We are thus left now with (x, t) such that x is in F_p and t is in $\partial\Delta_{p-1}$. We then have two cases:

Either $x_i = x_j$ $(i \neq j)$ and t_i as well as t_j is nonzero. Then, according to (6.80), y belongs to $F_{p-1} \times \Delta_{p-2}$ and $[y] = 0$ while according to (6.81), y belongs to $\Omega^{p-1} \times \partial\Delta_{p-2}$ and $[y] = 0$. Thus, in this case, both definitions of θ_p agree; and θ_p is again continuous at those points.

Or $x_i = x_j$ $(i \neq j)$ and $t_i = 0$ (or $t_j = 0$). If, besides t_i, there is another

$t_k = 0$, then $[y] = 0$ for (6. 80) as well as (6. 81) and θ_p is continuous at such points. Thus, we may assume that $x_i = x_j$ $(i \neq j)$ and $t_i = 0$ while $t_k \neq 0$ for $k \neq i$.

According to (6. 80), we then have:

$$[y] = [(x_1,\ldots,x_{i-1},x_{i+1},\ldots,x_p,t_1,\ldots,t_{i-1},t_{i+1},\ldots,t_p)] \tag{6.82}$$

According to (6. 81):

$$\begin{aligned}[y] &= [(x_1,\ldots,x_{i-1},x_{i+1},\ldots,x_p,t_1,\ldots,t_{i-1},t_{i+1},\ldots,t_{j-1},t_i+t_j,\\ &\qquad t_{j+1},\ldots,t_p)]\\ &= [(x_1,\ldots,x_{i-1},x_{i+1},\ldots,x_p,t_1,\ldots,t_{i-1},t_{i+1},\ldots,t_j,\ldots,t_p)]\end{aligned} \tag{6.83}$$

as $t_i = 0$

or

$$\begin{aligned}[y] &= [(x_1,\ldots,x_{j-1},x_{j+1},\ldots,x_p,t_1,\ldots,t_{i-1},t_i+t_j,\\ &\qquad t_{i+1},\ldots,t_{j-1},t_{j+1},\ldots,t_p)]\\ &= [(x_1,\ldots,x_{j-1},x_{j+1},\ldots,x_p,t_1,\ldots,t_{i-1},t_{i+1},\ldots,t_{j-1},\\ &\qquad t_{j+1},\ldots,t_p)]\end{aligned} \tag{6.84}$$

as $t_i = 0$,

and (6. 83) and (6. 34) agree as the elements which classes in the target space considered in these formulae are in the same σ_{p-1}-orbit. Finally, (6. 32) and (6. 33) obviously agree. θ_p has as such a unique definition at such points and is continuous.

By definition, the left vertical arrow in the diagram of Lemma 6. 4 will be:

$$\begin{aligned}\theta_{p\star}\circ\delta : H_\star(\Omega^p \times_{\sigma_p} \Delta_{p-1}, \;&\Omega^p \times \partial\Delta_{p-1} \cup_{\sigma_p} F_p \times \Delta_{p-1})\\ &\downarrow\\ H_{\star-1}(\Omega^{p-1}\times_{\sigma_{p-1}} \Delta_{p-2}, \;&\Omega^{p-1}\times\partial\Delta_{p-2} \cup_{\sigma_{p-1}} F_{p-1}\times\Delta_{p-1}).\end{aligned} \tag{6.85}$$

We turn now to the proof of Lemma 6. 4.

Proof of Lemma 6.4: We have, using a filtration of Ω by compact sets A and using γ_p, a map of pairs:

$$\tilde{\gamma}_p : (\Omega^p \times_{\sigma_p} \Delta_{p-1}, \Omega^p \times \partial\Delta_{p-1} \cup_{\sigma_p} F_p \times \Delta_{p-1}) \to (W_p, W_{p-1}) \tag{6.86}$$

which agrees with γ_p homologically and may be taken to be equal to γ_p on an arbitrary given compact set A in Ω.

$$\tilde{\gamma}_{p\star} = \gamma_{p\star}, \quad \tilde{\gamma}_p = \gamma_p \quad \text{on A given.} \tag{6.87}$$

Thus, the diagram

$$\begin{array}{ccc} H_\star(\Omega^p \times_{\sigma_p} \Delta_{p-1}, \Omega^p \times \partial\Delta_{p-1} \cup_{\sigma_p} F_p \times \Delta_{p-1}) & \to & H_\star(W_p, W_{p-1}) \\ \delta \downarrow & & \downarrow \delta \\ H_{\star-1}(\Omega^p \times \partial\Delta_{p-1} \cup_{\sigma_p} F_p \times \Delta_{p-1}) & \to & H_{\star-1}(W_{p-1}) \end{array} \tag{6.88}$$

commutes; and due to the definition of the arrows in Lemma 6.4, the proof of Lemma 6.4 reduces to the fact that the following diagram is homotopically commutative:

$$\begin{array}{ccc} \Omega^p \times \partial\Delta_{p-1} \cup_{\sigma_p} F_p \times \Delta_{p-1} & \xrightarrow{\tilde{\gamma}_p} & W_{p-1} \\ \theta_p \downarrow & & \downarrow l_p \\ \Omega^{p-1} \times_{\sigma_{p-1}} \Delta_{p-2} / \Omega^{p-1} \times \partial\Delta_{p-1} \cup_{\sigma_{p-1}} F_p \times \Delta_{p-2} & \xrightarrow{\tilde{\gamma}_{p-1}} & W_{p-1}/W_{p-2} \end{array} \tag{6.89}$$

In (6.89), the map $W_{p-1} \xrightarrow{l_p} W_{p-1}/W_{p-2}$ is passage to the quotient.

We first claim that

$$\tilde{\gamma}_{p-1} \circ \theta_p = l_p \circ \tilde{\gamma}_p \quad \text{on a neighbourhood of } \theta_p^{-1}(\cdot), \tag{6.90}$$

where $\cdot$ is the image of $\Omega^{p-1} \times \partial\Delta_{p-2} \cup_{\sigma_{p-1}} \times \Delta_{p-1}$ under the quotient map:

$$\Omega^{p-1} \times_{\sigma_{p-1}} \Delta_{p-2} \to \Omega^{p-1} \times_{\sigma_{p-1}} \Delta_{p-2} / \Omega^{p-1} \times \partial\Delta_{p-2} \cup_{\sigma_{p-1}} F_{p-1} \times \Delta_{p-2}.$$

Indeed, we have:

$$\tilde{\gamma}_p(\theta_p^{-1}(\cdot)) \subset W_{p-2}, \tag{6.91}$$

as

$$\theta_p^{-1}(\cdot) = \{(x, t) \text{ s.t. } x \text{ has at least three equal components or } t \text{ has at least two zero components}\}. \tag{6.92}$$

Now:

$$\tilde{\gamma}_p(x, t) = \sum_{i=1}^{p} t_i P\delta(x_i, \lambda_i). \tag{6.93}$$

Following the definition of $\tilde{\gamma}_p$ which assigns equal concentrations to the $P\delta(x_i, \lambda_i)$ if two points are close enough in $(x_1, \ldots, x_p)$, and assuming for example that $x_1 = x_2 = x_3$, $\tilde{\gamma}_p(x, t) = \sum_{i=3}^{p} t_i P\delta(x_i, \lambda) + (t_1+t_2+t_3) P\delta(x_1, \lambda)$ which belongs to W_{p-2} by Proposition 5.7.

Thus (6.91) is proved.

By continuity of $\tilde{\gamma}_p$ and as W_{p-2} is open, the inclusion (6.91) holds in fact on a whole neighbourhood of $\theta_p^{-1}(\cdot)$. Consequently, there exists a neighbourhood V_1 of $\theta_p^{-1}(\cdot)$ such that:

$$l_p \circ \tilde{\gamma}_p(\theta_1) = \star' \text{ where } \star' \text{ is the image of } W_{p-2} \text{ under the quotient } W_{p-1} \to W_{p-1}/W_{p-2}. \tag{6.94}$$

On the other hand, $\tilde{\gamma}_{p-1}$ maps $\Omega^{p-1} \times \partial\Delta_{p-2} \cup_{\sigma_{p-1}} F_p \times \Delta_{p-2}$ into W_{p-2}. Thus, again by continuity of $\tilde{\gamma}_{p-1}$ and of openness of W_{p-2}, there exists a neighbourhood V_2 of $\theta_p^{-1}(\cdot)$ such that:

$$\tilde{\gamma}_{p-1} \circ \theta_p(V_2) = \star' . \tag{6.95}$$

Hence, (6.95) holds on $V_1 \cap V_2$. Let now:

$$U \text{ be the complement of } V_1 \cap V_2 \text{ in } \Omega^p \times \partial\Delta_{p-1} \cup_{\sigma_p} F_p \times \Delta_{p-1}. \tag{6.96}$$

Let

$$B_{p-1} = \Omega^{p-1} \times (\Lambda_0, +\infty)^{p-1} \times \Delta_{p-2}/\sigma_{p-1}; \text{ with obvious action of } \sigma_{p-1} \text{ on } \Omega^{p-1} \times (\Lambda_0, +\infty)^{p-1}. \tag{6.96'}$$

Λ_0 is a large positive number.

By Proposition 5.7, there is a well defined map ϕ_{p-1} from B_{p-1} into W_{p-1}/W_{p-2}, namely:

$$\phi_{p-1}((x_1, \dots, x_{p-1}, \lambda_1, \dots, \lambda_{p-1}, \alpha_1, \dots, \alpha_{p-1})) = \sum_{i=1}^{p-1} \alpha_i P(x_i, \lambda_i). \tag{6.97}$$

We claim now that both maps $\tilde{\gamma}_{p-1} \circ \theta_p$ and $l_p \circ \tilde{\gamma}_p$, restricted to U, factorize through ϕ_{p-1}, i.e. there exist two continuous maps:

$$g, g' : U \to B_{p-1} \tag{6.98}$$

such that

$$\tilde{\gamma}_{p-1} \circ \theta_p\big|_U = \phi_{p-1} \circ g' \; ; \; l_p \circ \tilde{\gamma}_p\big|_U = \phi_{p-1} \circ g . \tag{6.99}$$

Furthermore, we have:

$$n \circ g = n \circ g' \text{ when } n \text{ is the projection map:} \tag{6.100}$$

$$B_{p-1} \to \Omega^{p-1} \times_{\sigma_{p-1}} \Delta_{p-2} .$$

Once (6.98), (6.99) and (6.100) are proven, the homotopy between $\tilde{\gamma}_{p-1} \circ \theta_p$ and $l_p \circ \tilde{\gamma}_p$ follows readily: namely, on U, as these two maps are represented by $\phi_{p-1} \circ g'$ and $\phi_{p-1} \circ g$, it suffices to homotope g' to g through a homotopy which is constant whenever $g((x,t)) = g'((x,t))$. This homotopy will immediately extend to a global homotopy of $\tilde{\gamma}_{p-1} \circ \theta_p$ and $l_p \circ \tilde{\gamma}_p$ on all of $\Omega^p \times \partial\Delta_{p-1} \cup_{\sigma_p} F_p \times \Delta_{p-1}$, thanks to this last property and to (6.90).

But, given (6.100), i.e. given the fact that g and g' differ at most by the values of the concentrations $(\lambda_1, \ldots, \lambda_{p-1})$ in $(\Lambda_0, +\infty)^{p-1}$ and given the convexity of $(\Delta_0, +\infty)$ we may define a barycentric homotopy from g' to g, i.e.

$$H: [0,1] \times \Omega^p \times \partial\Delta_{p-1} \cup_{\sigma_p} F_p \times \Delta_{p-1} \to B_{p-1} \tag{6.101}$$

$$(s, x, t) \to (x, ((1-s)\lambda_i' + s\lambda_1, \ldots, (1-s)\lambda_p' + s\lambda_p), t)$$

where $(x, (\lambda_1', \ldots, \lambda_p'), t) = g'(x,t)$ and $(x, (\lambda_1, \ldots, \lambda_p), t) = g(x,t)$.

Hence the proof of Lemma 6.4 reduces the existence of g and g' satisfying (6.98), (6.99) and (6.100).

Observe now that by the way γ_p (hence also γ_{p-1}) has been defined in (5.117)-(5.130), there are natural maps:

$$d_p: \Omega^p \to (\lambda_0, +\infty)^p, \tag{6.102}$$

$$d_{p-1}: \Omega^p \to (\lambda_0, +\infty)^{p-1}, \tag{6.103}$$

associated with these maps. d_p and d_{p-1} define the concentrations $(\lambda_1, \ldots, \lambda_p)$ and $(\lambda_1', \ldots, \lambda_p')$ which are used in defining $\gamma_p((x, \alpha))$ as $\sum_{i=1}^{p} \alpha_i P\delta(x_i, \lambda_i)$ and $\gamma_{p-1}(x, \alpha)$ as $\sum_{i=1}^{p-1} \alpha_i P\delta(x_i, \lambda_i')$. d_p and d_{p-1} are unambiguously defined in (5.117)-(5.130) and are continuous; they are also, respectively, σ_p and σ_{p-1} invariant.

Consider now (x,t) in U. There are then three possible cases.

<u>1st case:</u> x has only distinct components; t has only one zero component $(t_i = 0)$.

<u>2nd case:</u> x has two and only two equal components $x_i = x_j$; t has no zero component.

3rd case: x has two and only two equal components $x_i = x_j$; t_i (or t_j) is zero and it is the only zero component at t.

Furthermore, θ_p is well defined on U with values in $\Omega^{p-1} \times_{\sigma_{p-1}} \Delta_{p-2}$.

$$\theta_p : U \to \Omega^{p-1} \times_{\sigma_{p-1}} \Delta_{p-2} \tag{6.104}$$

We then set:

$$g'((x,t)) = \text{class in } B_{p-1} \text{ of } (X, T, d_{p-1}(X)) \text{ where } (X,T) \tag{6.105}$$

is a representative of $\theta_p(x,t)$ in $\Omega^{p-1} \times_{\sigma_{p-1}} \Delta_{p-2}$.

We clearly have by (6.103) and the definition $\tilde{\gamma}_{p-1}$:

$$\tilde{\gamma}_{p-1} \circ \theta_p |U = \phi_{p-1} \circ g'. \tag{6.106}$$

Observe that:

$$n \circ g' = \theta_p. \tag{6.107}$$

By the invariance of d_{p-1} under σ_{p-1}, g' is well defined and continuous. To define g, observe that, by (5.128):

$$d_p(x,t) = \underbrace{(\lambda, \dots, \lambda)}_{\text{p-uple}} \text{ as soon as } x \text{ has two neighbouring} \tag{6.108}$$

components.

For (x,t) in U being in the second or third case, it is then easy to define $g(x,t)$ to be, after a choice of a representative $(x_1, \dots, x_p, t_1, \dots, t_p)$ of (x,t):

$$g(x,t) = \tag{6.109}$$
$$((x_1, \dots, x_{i-1}, x_{i+1}, \dots, x_p), \underbrace{(\lambda, \dots, \lambda)}_{\text{(p-1)-uple}}, (t_1, \dots, t_{i-1}, t_i, \dots, t_{j-1}, t_i + t_j, t_{j+1}, \dots, t_p)).$$

Observe, here again, that for such (x, t), we have by (6.81):

$$n \circ g = \theta_p . \tag{6.110}$$

For (x, t) in U being in the first case, t_i is the only zero component of t. We then set, after choosing a representative $(x_1, \ldots, x_p, t_1, \ldots, t_p)$ in $\Omega^p \times \Delta_{p-1}$:

$$g(x, t) = \tag{6.110'}$$
$$((x_1, \ldots, x_{i-1}, x_{i+1}, \ldots, x_p), (\lambda_1, \ldots, \lambda_{i-1}, \lambda_{i+1}, \ldots, \lambda_p), (t_1, \ldots, t_{i-1}, t_{i+1}, \ldots, t_p))$$

where

$$(\lambda_1, \ldots, \lambda_p) = d_p(x_1, \ldots, x_p) . \tag{6.111}$$

(6.109) and (6.110) define g unambiguously, due to the invariance of d_p under σ_p. Observe that, by (6.80), we have:

$$n \circ g = \theta_p \text{ on such } (x, t) \text{ also.} \tag{6.112}$$

Thus (6.100) is established through (6.107), (6.110) and (6.112). The first identity of (6.99) has been seen to hold. For the second identity, we have, if (x, t) is in the second or third case:

$$\tilde{\gamma}_p(x, t) = \sum_{k=1}^{p} t_k P(x_k, \lambda) = \sum_{k=1}^{i=1} t_k P(x_k, \lambda) + \sum_{i+1}^{j-1} t_k P(x_k, \lambda) + \sum_{j+1}^{p} t_k P(x_k, \lambda) \tag{6.113}$$

$$+ (t_i + t_j) P(x_j, \lambda) = \phi_{p-1} \circ g(x, t)$$

as $x_i = x_j$.

In the first case, we have, with $(\lambda_1, \ldots, \lambda_p) = d_p(x_1, \ldots, x_p)$,

$$\tilde{\gamma}_p(x, t) = \sum_{k=1}^{p} t_k P(x_k, \lambda_k) = \sum_{k=1}^{i-1} t_k P(x_k, \lambda_k) + \sum_{i+1}^{p} t_k P(x_k, \lambda_k) \tag{6.114}$$

$$= \phi_{p-1} \circ g(x, t)$$

following (6.111). Hence the proof of (6.99).

It remains to show that g is continuous. When two components of x become neighbouring, then by (5.128), $d_p(x_1,\dots,x_p)$ is constant equal to $(\lambda,\dots,\lambda)$. If furthermore t_i (or t_j) is zero, then the two determinations of g through (6.109) and (6.110) coincide. Hence (6.109) and (6.110) define a continuous map. The proof of Lemma 6.4 is thereby complete.

Lemma 6.5: $H_\star(W_p, W_{p-1})$, if J has no critical point under the energy level b_p, and $H_\star(\Omega^p \times_{\sigma_p} \Delta_{p-1}, \Omega^p \times \partial\Delta_{p-1} \cup_{\sigma_p} F_p \times \Delta_{p-1})$, are modules over $H^\star(\Omega^p/\sigma_p)$ and the map $\tilde{\gamma}_{p\star}$ is $H^\star(\Omega^p/\sigma_p)$-linear from $H_\star(\Omega^p \times_{\sigma_p} \Delta_{p-1}, \Omega^p \times \partial\Delta_{p-1} \cup_{\sigma_p} F_p \times \Delta_{p-1})$ into $H_\star(W_p, W_{p-1})$.

Proof of Lemma 6.5: $H_\star(\Omega^p \times_{\sigma_p} \Delta_{p-1}, \Omega^p \times \partial\Delta_{p-1} \cup_{\sigma_p} F_p \times \Delta_{p-1})$ is a module over $H^\star(\Omega^p \times_{\sigma_p} \Delta_{p-1})$ hence over $H^\star(\Omega^p/\sigma_p)$.

The pair (W_p, W_{p-1}) injects into $(\overline{W}_p, \overline{W}_{p-1})$ and this injection is a homotopy equivalence if J has no critical point at the energy level b_{p-1}. Indeed, in such a case, both W_{p-1} and $\overline{W}_{p-1}$ retract by deformation onto a set $W'_{p-1} \subset W_{p-1}$. This retraction by deformation is induced by the flow of $-\partial J$. Hence $H_\star(W_p, W_{p-1}) = H_\star(\overline{W}_p, \overline{W}_{p-1})$. By Proposition 5.11, $(\overline{W}_p, \overline{W}_{p-1})$ retracts on $(R_p^{b_{p-1}}, \overline{W}_{p-1})$. Hence, $H_\star(\overline{W}_p, \overline{W}_{p-1}) = H_\star(R_p^{b_{p-1}}, \overline{W}_{p-1})$.

By the same proposition, $R_p^{b_{p-1}} - \overline{W}_{p-1}$ is contained in $V(p, \varepsilon_1)$. Hence:

$$H_\star(W_p, W_{p-1}) = H_\star(R_p^{b_{p-1}} \cap V(p, \varepsilon_1), \overline{W}_{p-1} \cap V(p, \varepsilon_1)).$$

Now, by Proposition 5.2, there is a continuous σ_p-invariant map from $V(p, \varepsilon_1)$ into Ω^p/σ_p; hence from $R_p^{b_{p-1}} \cap V(p, \varepsilon_1)$ into Ω^p/σ_p. This map provides $H_\star(W_p, W_{p-1})$ with a structure of module over $H^\star(\Omega^p/\sigma_p)$. Finally, consider for $\rho \in (0,1]$, the pair:

$$(A_\rho, B_\rho) = (\Omega^{p-1} \times_{\sigma_p} \rho\Delta_{p-1},\ \Omega^p \times \rho\,\partial\Delta_{p-1} \cup_{\sigma_p} F_p \times \rho\Delta_{p-1}) \tag{6.115}$$

where

$$\rho\,\Delta_{p-1} = \{(t_1', \ldots, t_p');\ \text{s.t.}\ (\frac{1}{p}+\frac{1}{\rho}(t_1'-\frac{1}{p}), \ldots, \frac{1}{p}+\frac{1}{\rho}(t_p'-\frac{1}{p}) \in \Delta_{p-1}\} \tag{6.116}$$

$\rho\ \partial\Delta_{p-1}$ is the boundary of $\rho\,\Delta_{p-1}$. $\rho\,\Delta_{p-1}$ is the p-simplex obtained by contracting Δ_{p-1} around the point $(\frac{1}{p}, \ldots, \frac{1}{p})$ at order ρ. Observe that

$$A_\rho \subset \Omega^{p-1} \times_{\sigma_p} \Delta_{p-1}\,. \tag{6.117}$$

When ρ runs in $(\rho_0, 1]$, this is an isotopy of the pair $(\Omega^{p-1} \times_{\sigma_p} \Delta_{p-1},\ \Omega^p \times \Delta_{p-1} \cup_{\sigma_p} F_p \times \Delta_{p-1})$ to (A_{ρ_0}, B_{ρ_0}) in $\Omega^{p-1} \times_{\sigma_p} \Delta_{p-1}$. Let

$$\rho\colon (\Omega^{p-1} \times_{\sigma_p} \Delta_{p-1},\ \Omega^p \times \partial\Delta_{p-1} \cup_{\sigma_p} F_p \times \Delta_{p-1}) \to (A_\rho, B_\rho) \tag{6.118}$$

be the associated map. As A_ρ is contained in $\Omega^p \times_{\sigma_p} \Delta_{p-1}$, the map $\tilde{\gamma}_p$ is defined on A_ρ, with values in W_p. On the other hand, consider the expansion of J provided by Proposition 5.10. This expansion, which holds under the sole hypothesis that the ε_{ij} are small yields:

$$J(\sum_{i=1}^{p} t_i P(x_i, \lambda_i)) = \tag{6.119}$$

$$\frac{(\sum_{i=1}^{p} t_i^2)^{n/(n-2)} (\int \delta^{2n/(n-2)})^{2n/(n-2)}}{\sum_{i=1}^{p} t_i^{2n/(n-2)}} \left(1 + 0\left(\sum_{i=1}^{p} \frac{1}{\lambda_i^{n-2}}\right)\right).$$

We also know, by Proposition 5.5, that $J(\sum_{i=1}^{p} t_i P(x_i, \lambda_i))$ belongs to W_{p-1} if one of the ε_{ij} is not small. These two facts that for any given $\rho_0 \in (0,1]$, one can choose λ large enough in (5.117), i.e. imply one can construct γ_p (hence $\tilde{\gamma}_p$

through a filtration of Ω by compact sets) so that:

$$\tilde{\gamma}_p(B_\rho) \subset W_{p-1} \quad \forall \rho \in (\rho_0, 1]. \tag{6.120}$$

Indeed, given (x,t') in B_ρ and considering $\gamma_p(x,t') = \sum_{i=1}^{p} t_i' P\delta(x_i, \mu_i)$ we use, assuming the ε_{ij} are small, the expansion provided in (6.119). This yields:

$$J\left(\sum_{i=1}^{p} t_i' P\delta(x_i,\mu_i)\right) = \frac{\left(\sum_{i=1}^{p} t_i'^2\right)^{n/(n-2)}}{\sum_{i=1}^{p} t_i'^{2n/(n-2)}} \left(\int \delta^{\frac{2n}{n-2}}\right)^{\frac{2}{n-2}} \left(1 + 0\left(\frac{1}{\lambda^{n-2}}\right)\right) \tag{6.121}$$

where λ has been introduced in (5.117) and $(\mu_1, \ldots, \mu_n)$ are the concentrations associated with γ_p and defined throughout (5.118)-(5.130). (In particular $\sum_{i=1}^{p} \frac{1}{\mu_i^{n-2}} = 0\left(\frac{1}{\lambda^{n-2}}\right)$).

Clearly, on $\rho\, \partial\Delta_{p-1}$, we have, assuming the ε_{ij} to be small,

$$J\left(\sum_{i=1}^{p} t_i' P\delta(x_i,\mu_i)\right) \leq \left[p^{\frac{2}{n-2}}\left(\int \delta^{\frac{2n}{n-2}}\right)^{\frac{2}{n-2}} - \phi(\rho_0)\right]\left(1+0\left(\frac{1}{\lambda^{n-2}}\right)\right); \tag{6.122}$$

for $\rho_0 \leq \rho \leq 1$,

where $\phi : \mathbb{R}^{+\star} \to \mathbb{R}^{+\star}$ is a strictly positive continuous function. Hence, for those points, $\gamma_p(x,t')$ belongs to W_{p-1}, provided λ is chosen large enough depending on ρ_0. The other points $\gamma_p(x,t')$ belong then to W_{p-1} by Proposition 5.5 and our claim (6.120) is thereby proved.

Thus the map $\tilde{\gamma}_p$:

$$\tilde{\gamma}_p : (\Omega^p \times_{\sigma_p} \Delta_{p-1},\ \Omega^p \times \partial\Delta_{p-1} \cup_{\sigma_p} F_p \times \Delta_{p-1}) \to (W_p, W_{p-1}) \tag{6.123}$$

is for any given $\rho_0 \in (0,1]$ homotopic to $\tilde{\gamma}_p \circ \rho_0$.

Now for any given compact set A in Ω, the map $\gamma_p \circ \rho_0$:

$$\gamma_p \circ \rho_0 : (A^p \times_{\sigma_p} \Delta_{p-1}, A^p \times \partial\Delta_{p-1} \cup_{\sigma_p} (A^p \cap F_p) \times \Delta_{p-1}) \to (W_p, W_{p-1}) \tag{6.124}$$

s defined as a map of pairs from $(A^p \times_{\sigma_p} \Delta_{p-1}, A^p \times \partial\Delta_{p-1} \cup_{\sigma_p} (A^p \cap V_p) \times \Delta_{p-1})$ nto (W_p, W_{p-1}), where V_p is a σ_p-invariant neighbourhood of F_p.

Using a filtration of Ω by compact sets A, as for passing from γ_p to $\tilde{\gamma}_p$, ve obtain a map:

$$\tilde{\gamma}_p \circ \rho_0 : (\Omega^p \times_{\sigma_p} \Delta_{p-1}, \Omega^p \times \partial\Delta_{p-1} \cup_{\sigma_p} V_p \times \Delta_{p-1}) \to (W_p, W_{p-1}) \tag{6.125}$$

greeing with $\tilde{\gamma}_p \circ \rho_0$ in $(\Omega^p \times_{\sigma_p} \Delta_{p-1}, \Omega^p \times \partial\Delta_{p-1} \cup_{\sigma_p} F_p \times \Delta_{p-1})$, for a suitable σ_p-invariant neighbourhood V_p of F_p. Thus, we have:

$$(\Omega^p \times_{\sigma_p} \Delta_{p-1}, \Omega^p \times \partial\Delta_{p-1} \cup_{\sigma_p} F_p \times \Delta_{p-1}) \xrightarrow{\text{inclusion}} \tag{6.126}$$

$$(\Omega^p \times_{\sigma_p} \Delta_{p-1}, \Omega^p \times \partial\Delta_{p-1} \cup_{\sigma_p} V_p \times \Delta_{p-1}) \xrightarrow[\tilde{\gamma}_p \circ \rho_0]{} (W_p, W_{p-1})$$

By (6.121), the inclusion in (6.126) is a homotopy equivalence.

We are then left with proving that the map $(\tilde{\gamma}_p \circ \rho_0)_\star$ from $H_\star(\Omega^p \times_{\sigma_1} \Delta_{p-1}, \Omega^p \times \partial\Delta_{p-1} \cup_{\sigma_p} V_p \times \Delta_{p-1})$ into $H_\star(W_p, W_{p-1})$ is $H^\star(\Omega^p / \sigma_p)$-linear. For this, we observe that the statement is equivalent, by excision, to the $H^\star(\Omega^p/\sigma_p)$-linearity of the map $(\tilde{\gamma} \circ \rho_0)_\star$ from $H_\star((\Omega^p)\star \times_{\sigma_p} \Delta_{p-1}, (\Omega^p)\star \times \partial\Delta_{p-1} \cup_{\sigma_p} \partial V_p \times \Delta_{p-1})$ into $H_\star(W_p, W_{p-1})$, or $H_\star(\overline{W}_p, \overline{W}_{p-1})$.

Consider now the map $\tilde{\gamma}_p \circ \rho_0$ from $((\Omega^p)\star \times_{\sigma_p} \Delta_{p-1}, (\Omega^p)\star \times \partial\Delta_{p-1} \cup_{\sigma_p} \partial V_p \times \Delta_{p-1})$ into $(\overline{W}_p, \overline{W}_{p-1})$. We claim that this map, with a proper choice of ρ_0 and $\tilde{\gamma}_p$, takes its value in $(R_p^{b\,p-1} \cap V(p, \varepsilon_1), \overline{W}_{p-1} \cap V(p, \varepsilon_1))$, where R_p and ε_1 are defined in Proposition 5.11. For this, consider a compact set A in Ω. Consider $(A^p \cap (\Omega^p)\star \times_{\sigma_p} \Delta_{p-1}, A^p \cap (\Omega^p)\star \times \partial\Delta_{p-1} \cup_{\sigma_p} (\partial V_p \cap$

$A^p) \times \Delta_{p-1})$ and the map $\gamma_p \circ \rho_0$ on it; for any x in $A^p \times (\Omega^p)^\star$, we have:

$$\underset{i \neq j}{\text{Min}} |x_i - x_j| \geq \tilde{\eta} > 0; \text{ with a uniform } \tilde{\eta}. \qquad (6.127)$$

We may assume η_p to be chosen in (5.118) so that:

$$\eta_p < \tilde{\eta}. \qquad (6.128)$$

We then have by (5.120) :

$$\lambda^2 \tilde{\eta}^2 \gg 1, \qquad (6.129)$$

and by (5.121) :

$$A^p \cap (\Omega^p)^\star \subset A_1^p. \qquad (6.130)$$

Hence the map γ_p is equal on $A_1^p \times \rho_0 \Delta_{p-1}$ to:

$$\forall (x, t) \in A_1^p \times \rho_0 \Delta_{p-1},\ \gamma_p(x, t) = \sum_{i=1}^{p} t_i P\delta(x_i, \bar{\lambda}_i), \qquad (6.131)$$

where $(\bar{\lambda}_1, \ldots, \bar{\lambda}_p)$ is defined in (5.124). Observe now that, by (5.136) :

there exists γ independent of λ, $\gamma > 0$ such that: (6.132)

$$\bar{\lambda}_i / \bar{\lambda}_j \leq \gamma \quad \forall\, i \neq j, \quad \forall\, x \in A_1^p.$$

As $\sum_{i=1}^{p} 1/\bar{\lambda}^{n-2} = p/\lambda^{n-2}$ by (5.126), we derive from (6.128) and (6.129) :

$$\bar{\lambda}_i \bar{\lambda}_j \tilde{\eta}^2 \gg 1 \qquad (6.133)$$

Thus, by (6.127)

$$\bar{\lambda}_i \bar{\lambda}_j |x_i - x_j|^2 \gg 1 \quad \forall\, i \neq j; \qquad (6.134)$$

hence the ε_{ij} are small for $i \neq j$, provided λ is large enough. Furthermore, on $\rho_0 \Delta_{p-1}$, we have:

$$\forall\, t' \in \rho_0 \Delta_{p-1},\ \left|t'_i - \frac{1}{p}\right| \leq 2\rho_0 . \tag{6.135}$$

Hence, with ρ_0 small enough, we have:

$$\left|\frac{t'_i}{t'_j} - 1\right| = o(1). \tag{6.136}$$

(6.136) and the fact that ε_{ij} are small imply then that

$$\gamma_p(x, t') \in V(p, \varepsilon) \quad \forall (x, t') \in A^p \cap (\Omega^p)^\star \times \rho_0 \Delta_{p-1}, \tag{6.137}$$

with $\varepsilon > 0$ as small as we want depending on the choice of λ in (5.117) large enough and ρ_0 small enough. Thus, the map $\gamma_p \circ \rho_0$ maps:

$$\begin{aligned} \gamma_0 \circ \rho_0 : (A^p \cap (\Omega^p)^\star \times_{\sigma_p} \Delta_{p-1},\ A^p \cap (\Omega^p)^\star \times \partial \Delta_{p-1} \cup_{\sigma_p} \partial V_p \times \Delta_{p-1}) \\ \to (\overline{W}_p \cap V(p, \varepsilon),\ \overline{W}_{p-1} \cap V(p, \varepsilon)) . \end{aligned} \tag{6.138}$$

Observe now that, by the definition of R_p (see (5.178)-(5.181)).

$$R_p(u) = J(u) - \mu(|\partial J(u)|^2) \tag{6.139}$$

hence

$$R_p(\gamma_p \circ \rho_0(x, t)) = J(\gamma_p \circ \rho_0(x, t)) - \mu(|\partial J(\gamma_p \circ \rho_0(x, t))|^2), \tag{6.140}$$

with $\mu(0) = \overline{\varepsilon}_p$.

Now, when ε goes to zero, we have:

$$\begin{cases} \lim\limits_{\varepsilon \to 0} \ \sup\limits_{u \in V(p, \varepsilon)} J(u) = b_{p-1} \\ \lim\limits_{\varepsilon \to 0} \ \sup\limits_{u \in V(p, \varepsilon)} |\partial J(u)|^2 = b_{p-1} \end{cases} \tag{6.141}$$

Thus, by (6.140) :

$$\lim_{\substack{\lambda\to+\infty\\ \rho_0\to 0}} \sup_{(x,t)\in A^p\cap(\Omega^p)^\star\times\Delta_{p-1}} J(\gamma_p\circ\rho_0(x,t)) = b_{p-1} \tag{6.142}$$

$$\lim_{\substack{\lambda\to+\infty\\ \rho_0\to 0}} \sup_{(x,t)\in A^p\cap(\Omega^p)^\star\times\Delta_{p-1}} \mu(|\partial J(\gamma_p\circ\rho_0(x,t))|^2) = \bar{\varepsilon}_p \tag{6.143}$$

Hence

$$\lim_{\substack{\lambda\to+\infty\\ \rho_0\to 0}} \sup_{(x,t)\in A^p\cap(\Omega^p)^\star\times\Delta_{p-1}} R_p(\gamma_p\circ\rho_0(x,t)) = b_{p-1}-\bar{\varepsilon}_p < b_{p-1}, \tag{6.144}$$

and our claim is proved; i.e. with a proper choice of ρ_0 and γ_p we have on any compact set A:

$$\gamma_p\circ\rho_0 : (A^p\cap(\Omega^p)^\star\times_{\sigma_p}\Delta_{p-1},\ A^p\cap(\Omega^p)^\star\times\partial\Delta_{p-1}\cup_{\sigma_p}\partial V_p \tag{6.145}$$
$$\cap A^p\times\Delta_{p-1}) \to (R_p^{b_{p-1}}\cap V(p,\varepsilon_1/2),\quad \overline{W}_{p-1}\cap V(p,\varepsilon_1/2)).$$

Consequently, using a filtration, we may assume that $\tilde{\gamma}_p\circ\rho_0$ is:

$$\tilde{\gamma}_p\circ\rho_0 : ((\Omega^p)^\star\times_{\sigma_p}\Delta_{p-1},\ (\Omega^p)^\star\times\partial\Delta_{p-1}\cup_{\sigma_p}\partial V_p\times\Delta_{p-1}) \tag{6.146}$$
$$\to (R_p^{b_{p-1}}\cap V(p,\varepsilon_1),\quad \overline{W}_{p-1}\cap V(p,\varepsilon_1)) \to (\overline{W}_p, \overline{W}_{p-1}).$$

By Proposition 5.11, the inclusion of the last arrow is a homological isomorphism which in fact defines the structure of $H^\star(\Omega^p/\sigma_p)$ module of $H_\star(\overline{W}_p, \overline{W}_{p-1})$ via the map from $V(p,\varepsilon_1)$ into Ω^p/σ_p provided by Proposition 5.2.

Now the diagram:

$$\begin{array}{ccc} ((\Omega^p)^\star\times_{\sigma_p}\Delta_{p-1},(\Omega^p)^\star\times\partial\Delta_{p-1}\cup_{\sigma_p}\partial V_p\times\Delta_{p-1}) & \to & (R_p^{b_{p-1}}\cap V(p,\varepsilon_1),\ \overline{W}_{p-1}\cap V(p,\varepsilon_1)) \\ \downarrow & & \downarrow \\ \Omega^p/\sigma_p & \to & \Omega^p/\sigma_p \end{array} \tag{6.147}$$

commutes.

Indeed, by the uniqueness of the solution to (5.12), the right vertical arrow maps $\sum_{i=1}^{p} t_i P\delta_i(x_i, \lambda_i)$ on $(x_1, \ldots, x_p)$, provided $\sum_{i=1}^{p} t_i P\delta_i(x_i, \lambda_i)$ is in $V(p, \varepsilon_1)$ with ε_1 small enough.

Hence this arrow maps $\gamma_p(x, t)$ onto x and the diagram commutes. The $H^*(\Omega^p/\sigma_p)$-linearity of γ_{p*} follows; and the proof of Lemma 6.5 is thereby complete.

We consider in the sequel a connected manifold V, compact and without boundary, of dimension $m \geq 1$. Consider the symmetric product:

$$V^p/\sigma_p \tag{6.148}$$

and the projection maps

$$\bar{q}_V : V^p \to V^p/\sigma_p; \quad q_{1,V} : V^p \to V \times V^{p-1}/\sigma_{p-1}. \tag{6.149}$$

Let

$\underline{D}_p$ be the subset of V^p where the action of σ_p degenerates. (6.150)

$\underline{V}_p$ be a σ_p-invariant tubular neighbourhood of $\underline{D}_p$ (6.151)

$$(V^p)^*/\sigma_p = \overline{V^p - \underline{V}_p/\sigma_p}. \tag{6.152}$$

Let

x_0 be a point of V (V is assumed to be connected). (6.153)

Let

$$I_{x_0}^V : V^{p-1}/\sigma_{p-1} \to V^p/\sigma_p \tag{6.154}$$

$$[(x_1, \ldots, x_{p-1})] \to [(x_1, \ldots, x_{p-1}, x_0)]$$

Let

$a_1, \ldots, a_{p-1}$ be p-1 distinct points in V, all of them distinct from x_0. (6.155)

Let

$$J^V : V \to V^p/\sigma_p \qquad (6.156)$$
$$x \to [(a_1, \ldots, a_{p-1}, x)]$$

Let

$$r^V : V \times V^{p-1}/\sigma_{p-1} \to V \qquad (6.157)$$

be the projection.

On V^p, σ_p acts; and so does σ_{p-1} which we consider as a subgroup of σ_p. The action of σ_{p-1} fixes the first factor of V^p and permutes the others. In this situation, we have (see Bredon [4]), well defined transfer maps $\mu^V_\star$ and $\mu^\star_V$:

$$\mu^V_\star : H_\star(V^p/\sigma_p) \to H_\star(V \times V^{p-1}/\sigma_{p-1}) \qquad (6.158)$$

$$\mu^\star_V : H^\star(V \times V^{p-1}/\sigma_{p-1}) \to H^\star(V^p/\sigma_p). \qquad (6.159)$$

The transfer maps are defined currently in the literature from a G-space X into X/G, where G is a finite group and X is paracompact finistic (see Bredon [4]). Here, we are considering a σ_p-space (manifold) V^p, together with a subgroup $H = \sigma_{p-1}$ of G and there is a transfer also in this situation between X/G and X/H (see Bredon [4]). Let

o_V be the generator of $H_m(V)$, (6.160)

$o^\star_V$ be the generator of $H^m(V)$. (6.161)

We then have

<u>Lemma 6.6</u>:

1) $H_i(V^p/\sigma_p) = 0$ for $i > pm$,

2) $H_{pm}(V^p/\sigma_p) = H_{pm}((V^p)\star/\sigma_p, \partial \underline{V}_p/\sigma_p) = \mathbb{Z}_2$

3) $H_l(\underline{D}_p/\sigma_p) = 0$ for $k > (p-1)m$.

Let then:

$$\omega_p^V \text{ be the generator of } H_{pm}(V^p/\sigma_p). \tag{6.162}$$

We have:

<u>Lemma 6.7</u>: $\mu_V^\star \circ r_V^\star(o_V^\star) \cap \omega_p^V = I_{x_0\star}^V(\omega_{p-1}^V)$

where $\cap$ is the cap-product.

(We start with proving Lemma 6.7.)

<u>Proof of Lemma 6.7</u>: First observe that $I_{x_0}^V(V^{p-1}/\sigma_{p-1})$ and $J^V(V)$ intersect transversely in the nonsingular part of $V^p/\sigma_p = (V^p - \underline{D}_p)\backslash\sigma_p$, at precisely one point. Namely, they intersect at:

$$[(a_1, \ldots, a_{p-1}, x_0)] \tag{6.163}$$

and the intersection is clearly transversal. Denoting then

$$\cdot \text{ the intersection pairing in } H_\star(V^p/\sigma_p), \tag{6.164}$$

we have:

$$I_{x_0\star}^V(\omega_{p-1}^V) \cdot J_\star^V(o_V) = 1 \in H_0(V^p/\sigma_p). \tag{6.165}$$

Thus,

$$I_{x_0\star}^V(\omega_{p-1}^V) \text{ is nonzero; as well as } J_\star^V(\omega_1). \tag{6.166}$$

Next, consider:

$$\phi : H_m(V) \to H_m(V)$$
$$\phi = r^V_\star \circ \mu^V_\star \circ J^V_\star \,. \tag{6.167}$$

We claim that

$$\phi = \mathrm{Id}\,. \tag{6.168}$$

Indeed, consider the generator o_V of $H_m(V)$. This generator, we may view as a chain map c in $S_m(V)$ (chain group of order m of V; with $\mathbb{Z}$-coefficients), which is decomposed on simplices $s_1, \ldots, s_n$:

$$o_V = [c] \;; \quad c = s_1 + \ldots + s_n \,. \tag{6.169}$$

Consider the simplices:

$$\hat{s}_i = J^V \circ s_i \in S_m(V^p/\sigma_p) \tag{6.170}$$

The transfer map μ^V works then as follows:

One considers a simplex $\tilde{s}_i$ in $S_m(V^p)$ over $\hat{s}_i$; that is a simplex in V^p whose projection on V^p/σ_p yields $\hat{s}_i$. Here, introducing:

$$\begin{aligned} h_V &: V \to V^p \\ &x \to (x, a_1, \ldots, a_{p-1}) \end{aligned} \tag{6.171}$$

we can use:

$$\tilde{s}_i = h_V \circ s_i \,. \tag{6.172}$$

Then, one considers representatives in σ_p of σ_p/σ_{p-1}. There are p of them and we may see them, acting on V^p, as:

$\tau_1, \ldots, \tau_p$; where τ_1 is the identity element and τ_k, $k \geq 2$, is the transposition between the first and the k-th factor. (6.173)

We then sum:

$$\sum_{j=1}^{p} \tau_j(\tilde{s}_i), \tag{6.174}$$

and then, project onto $V \times V^{p-1}/\sigma_{p-1}$, using $q_{1,V}$. Thus:

$$\mu_\star^V(J^V(c)) = \sum_{i=1}^{n}\left(\sum_{j=1}^{p} q_{1,V} \circ \tau_j \circ h_V(s_i)\right), \tag{6.175}$$

and consequently:

$$r_V \circ \mu^V \circ T^V(c) = \sum_{i=1}^{n}\left(\sum_{j=1}^{p} r_V \circ q_{1,V} \circ \tau_j \circ h_V(s_i)\right) \tag{6.176}$$

$$= \sum_{j=2}^{p} r_V \circ q_{1,V} \circ \tau_j \circ h_V(c)$$

Now, for $j \geq 2$, we have:

$$r_V \circ q_{1,V} \circ \tau_j \circ h_V : V \to V, \quad x \to a_j, \qquad j \geq 2 \tag{6.177}$$

Thus, passing to $\mathbb{Z}_2$-homology:

$$(r_V \circ q_{1,V} \circ \tau_j \circ h_V)_\star : H_m(V) \to H_m(V) \tag{6.178}$$

is zero. Therefore:

$$(r_V \circ \mu^V \circ J^V)_\star([c]) = [c], \tag{6.179}$$

or else

$$(r_V \circ \mu^V \circ J^V)_\star(o_V) = o_V, \tag{6.180}$$

and our claim (6.168) is proved.

This implies at once that the dual cohomological map:

$$\psi : H^m(V) \to H^m(V) \tag{6.181}$$

is also the identity. We then have a nonzero cohomological class in $H^m(V^p/\sigma_p)$:

$$\mu_V^\star \circ r_V^\star(o_V^\star) . \tag{6.182}$$

We now have the following formula, where q_V is defined as the projection map $V \times V^{p-1}/\sigma_{p-1} \to V^p/\sigma_p$:

$$\mu_V^\star \circ r_V^\star(o_V^\star) \cap \omega_p^V = q_{\star V}(r^\star(o_V^\star) \cap \mu_\star^V(\omega_p^V)) . \tag{6.183}$$

This formula is justified later on. Now, in (6.183), $\mu_\star^V(\omega_p^V)$ is the generator of $H_{mp}(V \times V^{p-1}/\sigma_{p-1})$. This is indeed a simple consequence of the definition of the transfer map $\mu_\star^V$: Considering a regular point z for the projection map:

$$q_V : V \times V^{p-1}/\sigma_{p-1} \to V^p/\sigma_p , \tag{6.184}$$

we have

$$q_V^{-1}(z) = \{y_1, \ldots, y_p\}. \tag{6.184}$$

Consider then U a disc, neighbourhood of z, where q_V is regular. We then have:

$$q_V^{-1}(U) = U_1 \cup \ldots \cup U_p, \tag{6.185}$$

where $U_1, \ldots, U_p$ are neighbourhoods of $y_1, \ldots, y_p$, which we may assume to be disjoint if U is small enough. We may also assume, choosing U to be a closed disc, the U_i to be closed discs

$$U_i \cap U_j = \phi, \quad i \neq j, \quad U_i = \overline{U}_i; \quad U_i \text{ a disc.} \tag{6.186}$$

We then have the following commutative diagram:

$$\begin{array}{ccc} H_\star(V \times V^{p-1}/\sigma_{p-1}) & \xrightarrow{\ \iota_\star\ } & H_\star(V \times V^{p-1}/\sigma_{p-1} - \bigcup_{i=1}^{p} U_i) \\ \mu_\star^V \uparrow & & \uparrow \mu_\star^V \\ H_\star(V^p/\sigma_p) & \xrightarrow{\ s_\star\ } & H_\star(V^p/\sigma_p, V^p/\sigma_p - U) \end{array} \tag{6.187}$$

On the other hand, the U_i being disjoint, we have:

$$H_{mp}(V \times V^{p-1}/\sigma_{p-1}, V \times V^{p-1}/\sigma_{p-1} - \bigcup_{i=1}^{p} U_i) \tag{6.188}$$

$$= \bigoplus_{i=1}^{p} H_{mp}(V \times V^{p-1}/\sigma_{p-1}, V \times V^{p-1}/\sigma_{p-1} - U_i).$$

Each of these $H_{mp}(V \times V^{p-1}/\sigma_{p-1}, V \times V^{p-1}/\sigma_{p-1} - U_i)$ is uniquely generated. We denote

$$o_i \text{ the generator of } H_{mp}(V \times V^{p-1}/\sigma_{p-1}, V \times V^{p-1}/\sigma_{p-1} - U_i) \tag{6.189}$$
$$= H_{mp}(U_i, \partial U_i)$$

The top arrow $l_\star$ in (6.187) is just localization. If we denote:

$$0 \text{ the generator of } H_{mp}(V \times V^{p-1}/\sigma_{p-1}), \tag{6.190}$$

we have:

$$l_\star(0) = \sum_{i=1}^{p} e_i . \tag{6.191}$$

The argument is similar for the bottom arrow. Thus, denoting:

$$\hat{e} \text{ the generator of } H_{mp}(V^p/\sigma_p, V^p/\sigma_p - U) = H_{mp}(U, \partial U), \tag{6.192}$$

we have:

$$s_\star(\omega_p^V) = \hat{e} . \tag{6.193}$$

Now, following the definition of $\mu_\star^V$, we have, through a localization argument on $(\bigcup_{i=1}^{p} U_i, \bigcup_{i=1}^{p} \partial U_i) \to (U, \partial U)$.

$$\mu_\star^V(\hat{e}) = \sum_{i=1}^{p} e_i . \tag{6.194}$$

Thus

$$l_\star \circ \mu^V_\star(\omega_p) = \mu^V_\star \circ s_\star(\omega_p) = \sum_{i=1}^{p} e_i . \tag{6.195}$$

Thus

$$\mu^V_\star(\omega^V_p) \text{ is nonzero and generates } H_{mp}(V \times V^{p-1}/\sigma_{p-1}). \tag{6.196}$$

This generator, we may view as:

$$J^V_{1\star}(o_V) \otimes I^V_{1x_0\star}(\omega_{p-1}), \tag{6.197}$$

where:

$$J^V_1 : V \to V \times V^{p-1}/\sigma_{p-1} \tag{6.198}$$

$$x \to (x, [(a_1, \ldots, a_{p-1})]),$$

$$J^V_{1x_0} : V^{p-1}/\sigma_{p-1} \to V \times V^{p-1}/\sigma_{p-1} \tag{6.199}$$

$$x \to (x_0, x)$$

Observe that:

$$q_V \circ J^V_1 = J^V ; \quad q_V \circ I^V_{1x_0} = I^V_{x_0} ; \quad r_V \circ J^V_1 = Id_V . \tag{6.200}$$

Thus, by (6.196) :

$$\mu^V_\star(\omega^V_p) = J^V_{1\star}(o_V) \otimes I^V_{1x_0\star}(\omega^V_{p-1}). \tag{6.201}$$

Hence, using (6.183), we have:

$$(\mu^\star_V \circ r^\star_V)(o^\star_V) \cap \omega^V_p = q_{\star V}(r^\star_V(o^\star_V) \cap \mu^V_\star(\omega^V_p)) \tag{6.202}$$

$$= q_{\star V}(r^\star_V(o^\star_V) \cap (J^V_{1\star}(o_V) \otimes I^V_{1x_0\star}(\omega_{p-1}))$$

$$= q_{\star V}(\langle r^\star_V(o^\star_V), J^V_{1\star}(o_V)\rangle I^V_{1x_0\star}(\omega_{p-1})).$$

Now:

$$\langle r^{\star}_V(o^{\star}_V), \overset{V}{J}_{1\star}(o_V)\rangle = \langle o^{\star}_V, r_{\star V} \circ \overset{V}{J}_{1\star}(o_V)\rangle = 1 \tag{6.203}$$

$$r_V \circ \overset{V}{J}_1 = Id_V; \quad \text{hence} \quad (r_V \circ \overset{V}{J}_1)_\star = Id_\star$$

(see (6.200)). Hence:

$$\mu^{\star}_V \circ r^{\star}_V(o^{\star}_V) \cap \overset{V}{\omega}_p = q_{\star V} \circ \overset{V}{I}_{1x_0 \star}(\overset{V}{\omega}_{p-1}) = \overset{V}{I}_{x_0 \star}(\overset{V}{\omega}_{p-1}), \tag{6.204}$$

and Lemma 6.7 is thereby proven up to (6.183) which we justify now. (6.183) comes from a more general formula which we prove now; namely:

$$q_{\star V}(\alpha \cap \overset{V}{\mu}_\star(y)) = \mu^{\star}_V(\alpha) \cap y. \tag{6.205}$$

To prove (6.205), we consider, via a natural diagonal map (see Dold [13]) a decomposition of y in $S(V^p/\sigma_p) \otimes S(V^p/\sigma_p)$:

$$y = \sum_{u,v} u \otimes v \tag{6.206}$$

u and v being simplices in $S(V^p/\sigma_p)$. By the definition of the transfer map $\mu^{\star}_V$, which is dual to $\overset{V}{\mu}_\star$ we have:

$$\langle \mu^{\star}_V(\alpha), u\rangle = \langle \alpha, \overset{V}{\mu}_\star(u)\rangle \tag{6.207}$$

Now, as explained in (6.168)-(6.175), if

$$\hat{u} \text{ is a chain in } S(V^p) \text{ such that } \bar{q}_V(\hat{u}) = u, \tag{6.208}$$

(i.e. $\hat{u}$ projects on u); we have:

$$\overset{V}{\mu}_\star(u) = \sum_{g \in \sigma_p/\sigma_{p-1}} q_{1V} \circ g(\hat{u}). \tag{6.209}$$

Thus:

$$\mu^{\star}_V(\alpha) \cap y = \sum_{u,v} \Big(\sum_{g \in \sigma_p/\sigma_{p-1}} \alpha(q_{1V} \circ g(\hat{u}))\, v. \tag{6.210}$$

On the other hand, we compute $\mu^V_\star(y)$ on the decomposition (6.206). We have, as can be checked through the definition of the transfer and that of the natural diagonals:

$$\mu^V_\star(y) = \sum_{u,v} \sum_{g\in\sigma_p/\sigma_{p-1}} [q_{1V} \circ g(\hat{u}) \otimes q_{1V} \circ g(\hat{v})], \tag{6.211}$$

where $\hat{u}$ is defined in (6.208) and $\hat{v}$ is the corresponding element for v. Thus:

$$\alpha \cap \mu^V_\star(y) = \sum_{u,v} \left(\sum_{g\in\sigma_p/\sigma_{p-1}} \alpha(q_{1V} \circ g(\hat{u}))\, q_{1V} \circ g(\hat{v}) \right) \tag{6.212}$$

Hence, observing that:

$$q_V \circ q_{1V} \circ g(\hat{v}) = \bar{q}_V(g(\hat{v})) = \bar{q}_V(\hat{v}) = v \tag{6.213}$$

($\bar{q}_V$ is the projection map $V^p \to V^p/\sigma_p$; hence $\bar{q}_V = q_V \circ q_{1V}$ and $\bar{q}_V(gx) = \bar{q}_V(x)$). We have:

$$q_{\star V}(\alpha \cap \mu^V_\star(y)) = \sum_{u,v} \left(\sum_{g\in\sigma_p/\sigma_{p-1}} \alpha(q_{1V} \circ g(\hat{u}))\right) v \tag{6.214}$$

Comparing with (6.210), we have (6.205). Lemma 6.7 is thereby proved.

Proof of Lemma 6.6: We give the proof for $m \geq 2$. The case $m = 1$ needs only minor modifications. First observe that $(V^p)\star/\sigma_p$ is a manifold of dimension pm, with boundary $\partial\underline{V}_p/\sigma_p$. As $\underline{D}_p$ does not disconnect V^p for $m \geq 2$, $(V^p)\star/\sigma_p$ is connected in this case. Thus:

$$H_{mp}(V^p/\sigma_p, \underline{D}_p/\sigma_p) = H_{pm}((V^p)\star/\sigma_p, \partial\underline{V}_p/\sigma_p) = \mathbb{Z}_2 \tag{6.215}$$

for $m \geq 2$.

Next, consider the filtration:

$$\underline{D}^1_p = V^p/\sigma_p \supset \underline{D}^2_p/\sigma_p = \underline{D}_p/\sigma_p \supset \underline{D}^3_p/\sigma_p \ldots \supset \underline{D}^p_p/\sigma_p \simeq V, \tag{6.216}$$

where $\underline{D}_p^i/\sigma_p$ is the subset of V^p/σ_p of p-uples where i points are identical. Let

$$\underline{V}_p^i/\sigma_p \text{ be a neighbourhood of } \underline{D}_p^i \text{ in } \underline{D}_p^{i+1}/\sigma_p . \tag{6.217}$$

We then have, as in (6.215),

$$H_{(p-i+1)m}(\underline{D}_p^i/\sigma_p, \underline{D}_p^{i-1}/\sigma_p) = \mathbb{Z}_2 \quad \text{for } m \geq 2. \tag{6.218}$$

Indeed, $\underline{D}_p^i/\sigma_p - \underline{D}_p^{i-1}/\sigma_p$ is a (p-i+1) m-dimensional manifold, $\left(\underline{D}_p^i/\underline{V}_p^{i-1}\right)/\sigma_p$ is a (p-i+1) m-dimensional manifold, with boundary $\partial\underline{V}_p^{i-1}/\sigma_p$; and the conclusion follows from the argument used in (6.215). On the other hand:

$$H_m(V) = \mathbb{Z}_2 = H_m(\underline{D}_p^p/\sigma_p); \quad H_i(V) = 0 \text{ for } i > m. \tag{6.219}$$

If $m \geq 2$, Lemma 6.6 follows by induction from the following arguments: We have:

$$H_r(\underline{D}_p^{i-1}/\sigma_p) \to H_r(\underline{D}_p^i/\sigma_p) \to H_r(\underline{D}_p^i/\underline{D}_p^{i-1}/\sigma_p) \to H_{r-1}(\underline{D}_p^{i-1}/\sigma_p). \tag{6.220}$$

Now, assume we have:

$$H_r(\underline{D}_p^{i-1}/\sigma_p) = \begin{cases} 0, & \text{if } r > (p-i)m \\ \mathbb{Z}_2 & \text{if } r = (p-i)m \end{cases} . \tag{6.221}$$

Then, (6.220) and (6.221) yield:

$$0 \to H_{(p-i+1)m}(\underline{D}_p^i/\sigma_p) \to \mathbb{Z}_2 = H_{(p-i+1)m}(\underline{D}_p^i/\sigma_p/\underline{D}_p^{i-1}/\sigma_p) \to 0, \tag{6.222}$$

under the assumption $m \geq 2$.

On the other hand, $\underline{D}_p^i/\sigma_p - \underline{D}_p^{i-1}/\sigma_p$ being a manifold of dimension (p-i+1) m, we also have:

$$H_r(\underline{D}_p^i/\sigma_p, \underline{D}_p^{i-1}/\sigma_p) = 0 \quad \text{for } r > (p-i+1)m . \tag{6.223}$$

this, with (6.222) yields:

$$0 = H_r(\underline{D}_p^i/\sigma_p, \underline{D}_p^{i-1}/\sigma_p) \to H_{r-1}(\underline{D}_p^{i-1}/\sigma_p) \to H_{r-1}(\underline{D}_p^i/\sigma_p) \to 0$$

$$= H_{r-1}(\underline{D}_p^i/\sigma_p, \underline{D}_p^{i-1}/\sigma_p) \tag{6.224}$$

as soon as $r-1 > (p-i+1)m$. Thus, from (6.224) and (6.272),

$$H_{(p-i+1)m}(\underline{D}_p^i/\sigma_p) = \mathbb{Z}_2; \; H_k(\underline{D}_p^i/\sigma_p) = 0 \tag{6.225}$$

for $k > (p-i+1)m$,

and the induction is established. This proves Lemma 6.6 in the case $m \geq 2$.

Consider now the isomorphism ${}^V u_2^p \cap .$ of Proposition 6.3, applied to $\Omega = V$,

$$H_{\star-p+1}(V^p/\sigma_p, \underline{D}_p/\sigma_p) \xleftarrow{{}^V u_2^p \cap .} H_\star((V^p)^\star \times_{\sigma_p} \Delta_{p-1},$$

$$\partial \underline{V}_p \times \Delta_{p-1} \cup_{\sigma_p} (V^p)^\star \times \partial \Delta_{p-1})$$

$$\simeq H_\star(V^p \times_{\sigma_p} \Delta_{p-1}, \underline{V}_p \times \Delta_{p-1} \cup_{\sigma_p} V^p \times \partial \Delta_{p-1}).$$

Consider also:

$$\theta_p^V, \tag{6.226}$$

the map defined in what follows Lemma 6.4, in particular (6.80) and (6.81), here again with $\Omega = V$. We thus have:

$$(\theta_p^V \circ \delta)_\star : H_\star(V^p \times_{\sigma_p} \Delta_{p-1}, V^p \times \partial\Delta_{p-1} \cup_{\sigma_p} \underline{D}_p \times \Delta_{p-1}) \tag{6.227}$$

$$\to H_{\star-1}(V^{p-1} \times_{\sigma_{p-1}} \Delta_{p-2}, V^{p-1} \times \partial\Delta_{p-2} \cup_{\sigma_{p-1}} \underline{D}_{p-1} \times \Delta_{p-2}.$$

Consider also the quotient homomorphism:

$$e_p^V : V^p/\sigma_p \to (V^p/\sigma_p)/(\underline{D}_p/\sigma_p) \tag{6.228}$$

We then have:

Lemma 6. 8: The diagram shown on the next page is commutative.

Proof of Lemma 6. 8: Let

$a_1, \dots, a_{p-1}$ be p-1 distinct points in V^{p-1}, (6. 229)

all of them distinct from x_0.

Let

U be a disc neighbourhood around $[(a_1, \dots, a_{p-1})]$ in V^{p-1}/σ_{p-1}, (6.230)

U is assumed to be small so that $U \cap \underline{D}_{p-1}/\sigma_{p-1} = \phi$

and $U \cap C^p_{x_0}/\sigma_p = \phi$,

where

$$C^p_{x_0} = \{(x_1, \dots, x_p) \text{ is } V^p \text{ such that } x_i = x_0 \text{ for a certain } i\}. \quad (6. 231)$$

Let

$$\underline{D}_p^{x_0} = \underline{D}_p \cap C^p_{x_1} . \quad (6. 232)$$

$0(a, x_0)$ = orbit of $(a_1, \dots, a_{p-1}, x_0)$ (6. 233)

under the action of σ_p in V^p,

$0(a)$ = orbit of $(a_1, \dots, a_{p-1})$ (6. 234)

under the action of σ_{p-1} in V^{p-1}.

$$(V^p)\bigstar_{x_0} = (V^p)\bigstar \cap C^p_{x_0} , \quad (6. 235)$$

$$(\partial \underline{V}_p)_{x_0} = \partial \underline{V}_p \cap C^p_{x_0} , \quad (6. 236)$$

$U_i = \{(x_1, \dots, x_p)$ in V^p such that $x_i = x_0$ and (6. 237)

$[(x_1, \dots, x_{i-1}, x_{i+1}, \dots, x_p)] \in U\}$ for $i = 1, \dots, p$.

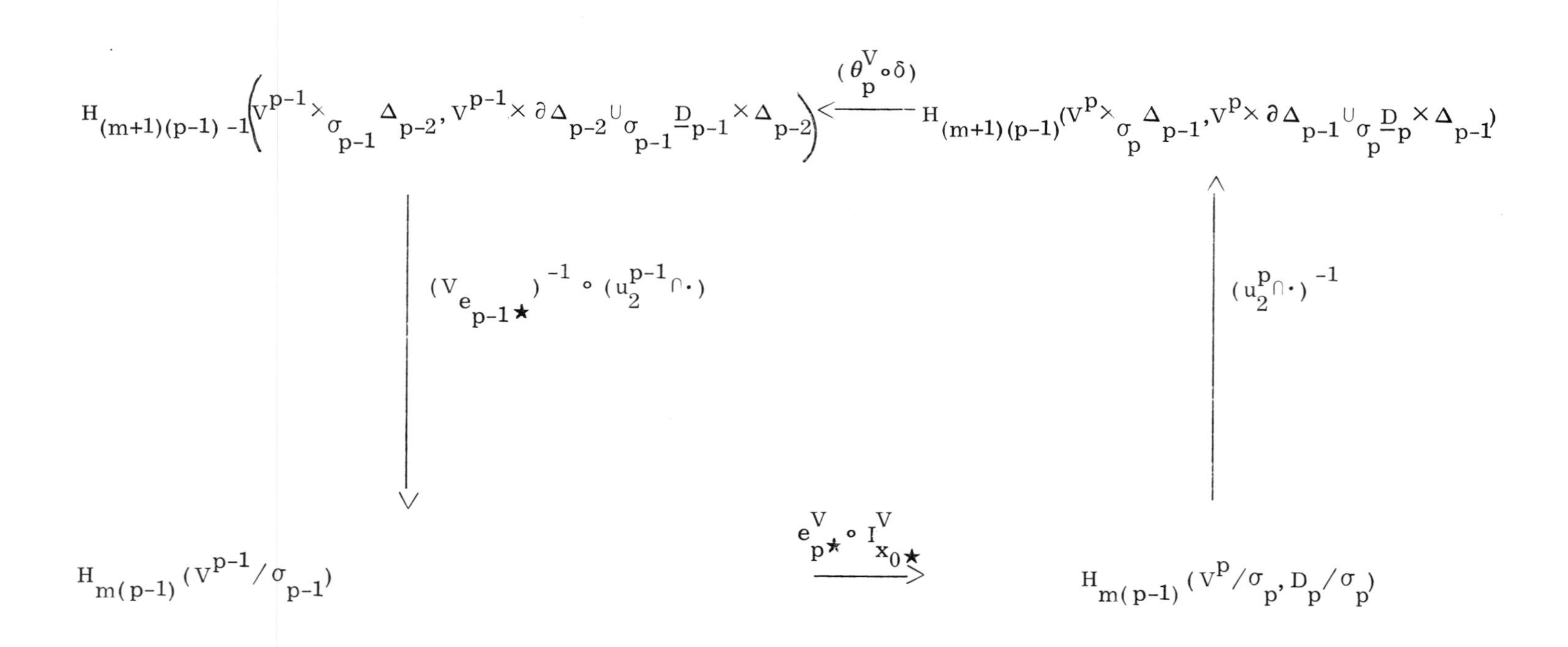
$H_{(m+1)(p-1)-1}\left(V^{p-1}\times_{\sigma_{p-1}}\Delta_{p-2}, V^{p-1}\times\partial\Delta_{p-2}\cup_{\sigma_{p-1}}\underline{D}_{p-1}\times\Delta_{p-2}\right)$
$(\theta^V_p\circ\delta)$
$H_{(m+1)(p-1)}(V^p\times_{\sigma_p}\Delta_{p-1}, V^p\times\partial\Delta_{p-1}\cup_{\sigma_p}\underline{D}_p\times\Delta_{p-1})$
$(V_{e_{p-1\star}})^{-1}\circ(u_2^{p-1}\cap\cdot)$
$(u_2^p\cap\cdot)^{-1}$
$H_{m(p-1)}(V^{p-1}/\sigma_{p-1})$
$e^V_{p\star}\circ I^V_{x_0\star}$
$H_{m(p-1)}(V^p/\sigma_p, D_p/\sigma_p)$

$(V^p)^{\star}_{x_0}$ is a manifold with boundary $(\partial \underline{V}_p)_{x_0}$. If U is small enough, the U_i are pairwise disjoint as $a_1, \ldots, a_{p-1}, x_0$ are pairwise distinct. We have well defined isomorphisms:

$$H_{\star}(V^{p-1}/\sigma_{p-1}, V^{p-1}/\sigma_{p-1} - \mathring{U}) \tag{6.238}$$

$$\leftarrow H_{\star+p-2}(V^{p-1} \times_{\sigma_{p-1}} \Delta_{p-2}, (V^{p-1} - \mathring{\hat{U}}) \times \Delta_{p-2} \cup_{\sigma_{p-1}} V^{p-1} \times \partial \Delta_{p-2}),$$

$$H_{\star}(I^V_{x_0}(V^{p-1}/\sigma_{p-1}), I^V_{x_0}(V^{p-1}/\sigma_{p-1} - \mathring{U})) \tag{6.239}$$

$$\leftarrow H_{\star+p-1}(C^p_{x_0} \times_{\sigma_p} \Delta_{p-1}) C^p_{x_0} \times \partial \Delta_{p-1} \cup_{\sigma_p} (C^p_{x_0} / \bigcup_{i=1}^{p} \mathring{U}_i) \times \Delta_{p-1})$$

$$H_{\star}(I^V_{x_0}(V^{p-1}/\sigma_{p-1}), \underline{D}_p^{x_0}/\sigma_p) \tag{6.240}$$

$$\leftarrow H_{\star+p-1}(C^p_{x_0} \times_{\sigma_p} \Delta_{p-1}, C^p_{x_0} \times \partial \Delta_{p-1} \cup_{\sigma_p} \underline{D}_p^{x_0} \times \Delta_{p-1})$$

$$H_{\star}(I^V_{x_0}(U), I^V_{x_0}(\partial U)) \tag{6.241}$$

$$\leftarrow H_{\star+p-1}(\bigcup_{i=1}^{p} U_i \times_{\sigma_p} \Delta_{p-1}, \bigcup_{i=1}^{p} U_i \times \partial \Delta_{p-1} \cup_{\sigma_p} \bigcup_{i=1}^{p} \partial U_i \times \Delta_{p-1}),$$

$$H_{\star}(U, \partial U) \leftarrow H_{\star+p-2}(\hat{U} \times_{\sigma_{p-1}} \Delta_{p-2}, \hat{U} \times \partial \Delta_{p-2} \cup_{\sigma_{p-1}} \partial \hat{U} \times \Delta_{p-2}), \tag{6.242}$$

where

$$\hat{U} \text{ is the converse image of } U \text{ under the projection map } V^{p-1} \to V^{p-1}/\sigma_{p-1}. \tag{6.243}$$

All the isomorphisms (6.238) - (6.242) are obtained through cap-product, following the same arguments used in the proof of Proposition 6.3. For the sake of completeness, we sketch here the proof of the existence of these isomorphisms.

Basically, we are facing in each of these situations, up to retraction by deformation, a pair (A, B), with B closed in A, the triad $(A; B; \overline{A-B})$ being excisive; and a disc-bundle over $\overline{A-B}$. The disc-bundle over $\overline{A-B}$ is given through a map:

$$q: X \to A; \quad q^{-1}(B) = A_2 \xrightarrow{q} B, \tag{6.244}$$

which, when restricted to $q^{-1}(\overline{A-B})$ defines a disc-fibering over $\overline{A-B}$:

$$X' \xrightarrow{q} \overline{A-B} \ ; \ X'' = q^{-1}(B \cap \overline{A-B}) \xrightarrow{q} B \cap \overline{A-B} \, . \tag{6.245}$$

Let

$$\mathring{X}', \mathring{X}'' \text{ be the associated sphere bundles,} \tag{6.246}$$

and let

$$u', \ u'' \text{ be their Thom class.} \tag{6.247}$$

We then have isomorphisms through cap-product:

$$H_{\star+l}(X', \mathring{X}') \xrightarrow{u' \cap \cdot} H_{\star}(\overline{A-B}), \tag{6.248}$$

$$H_{\star+l}(X'', \mathring{X}'') \xrightarrow{\ddot{u} \cap \cdot} H_{\star}(B, \ \overline{A-B}) \tag{6.249}$$

l is the dimension of the disc. Denoting:

$$k'': (X'', \mathring{X}'') \to (X', \mathring{X}') \text{ the injection,} \tag{6.250}$$

we have,

$$k''(u') = u'' , \tag{6.251}$$

by the uniqueness of this Thom class (see D. Husemoller [15]). Therefore the diagram:

$$\begin{array}{ccc} H_{\star+l}(X'',\overset{\circ}{X}') & \rightarrow & H_{\star+l}(X',\overset{\circ}{X}') \\ u''\cap. \downarrow & & \downarrow u'\cap. \\ H_\star((B,\overline{A-B}) & \rightarrow & H_\star(\overline{A-B}) \end{array} \qquad (6.251')$$

commutes.

Now, considering the pair $(X',\overset{\circ}{X}', X'')$, we have an action of u' on the homology of this pair by cap-product:

$$u'\cap. : H_{\star+l}(X',\overset{\circ}{X}' \cup X'') \rightarrow H_\star(X',X'') = H_\star(\overline{A-B}, \overline{A-B} \cap B). \qquad (6.252)$$

By naturality of the cap-product, we then have the commutative diagram:

$$\begin{array}{ccc} H_{\star+l}(X',\overset{\circ}{X}') & \rightarrow & H_{\star+l}(X',\overset{\circ}{X}' \cup X'') \\ u'\cap. \downarrow & & \downarrow u'\cap. \\ H_\star(\overline{A-B}) & \rightarrow & H_\star(\overline{A-B}, \overline{A-B} \cap B) \end{array} \qquad (6.253)$$

Finally, by stability of the cap-product, (see Dold [13]), we have the following commutative diagram:

$$\begin{array}{ccc} H_{\star+l}(X',\overset{\circ}{X}' \cup X'') & \overset{\delta}{\rightarrow} & H_{\star+l-1}(\overset{\circ}{X}' \cup X'', \overset{\circ}{X}') \simeq H_{\star+l-1}(X'',\overset{\circ}{X}') \\ \downarrow u'\cap. & & \downarrow u'\cap. \\ H_\star(\overline{A-B},\overline{A-B} \cap B) & \rightarrow & H_{\star-1}(\overline{A-B} \cap B) \end{array} \qquad (6.254)$$

gluing up (6.251), (6.253) and (6.254), we derive as in the proof of Proposition 6.3, through the five-lemma, an isomorphism:

$$H_{\star+l}(X, g^{-1}(B) \cup \overset{\circ}{X}') \qquad (6.255)$$

$$\simeq H_{\star+l}(X',\overset{\circ}{X}' \cup X'') \overset{u'\cap.}{\rightarrow} H_\star(\overline{A-B},\overline{A-B} \cap B) \simeq H_\star(A,B).$$

Hence (6.238)-(6.242). Notice also, that given two pairs:

$$(A_1,B_1), \ (A,B), \qquad (6.256)$$

with:

$$q_1 : X_1 \to A_1 \ ; \quad X_1' = q_1^{-1}(\overline{A_1 - B_1}) \ , \tag{6.257}$$

$$q: \ X \to A \ ; \quad X' = q^{-1}(\overline{A - B}) \ , \tag{6.258}$$

assuming that q_1 and q induced disc-bundles of the same dimension over $\overline{A_1 - B_1}$ and $\overline{A - B}$ respectively and given two maps $f, \hat{f}$, such that:

$$f: (A_1, B_1) \to (A, B) \ ; \quad \hat{f} : (X_1, q_1^{-1}(B_1)) \to (X, q^{-1}(B)) \ ; \tag{6.259}$$

$\hat{f}$ also maps $(X_1', \mathring{X}_1')$ into $(X', \mathring{X}')$ and $f \circ q_1 = q \circ \hat{f}$; we have, by naturality of the construction, the following commutative diagram:

$$\begin{array}{ccc} H_{\star+l}(X_1, q_1^{-1}(B_1) \cup \mathring{X}_1') & \xrightarrow{\hat{f}_\star} & H_{\star+l}(X, q^{-1}(B) \cup \mathring{X}') \\ \downarrow u_1' \cap \cdot & & \downarrow u' \cap \cdot \\ H_\star(A_1, B_1) & \xrightarrow{f_\star} & H_\star(A, B) \end{array} \tag{6.260}$$

Having justified (6.238)-(6.242) and pointed out (6.260), we turn now to the proof of Lemma 6.8.

First note that we have, by (6.260), the commutative diagram (6.261). The top and bottom arrows are induced by the natural injections. We denoted $u^{p-1} \cap$ the isomorphism of (6.238).

Next observe that we also have, through (6.260), the commutative diagram (6.262). The top and bottom arrows are induced again by natural inclusions. The vertical maps are the isomorphisms (6.239), (6.240). Likewise, we have the commutative diagram (6.263). In (6.263) again, the top and bottom arrows are induced by the natural inclusions. The left vertical arrow is (6.240). Finally, we define:

$$\Theta_p : C^p_{x_0} \times \partial\Delta_{p-1} \cup_{\sigma_p} (C^p_{x_0} / \bigcup_{i=1}^{p} \mathring{U}_i) \times \Delta_{p-1} \to \tag{6.264}$$

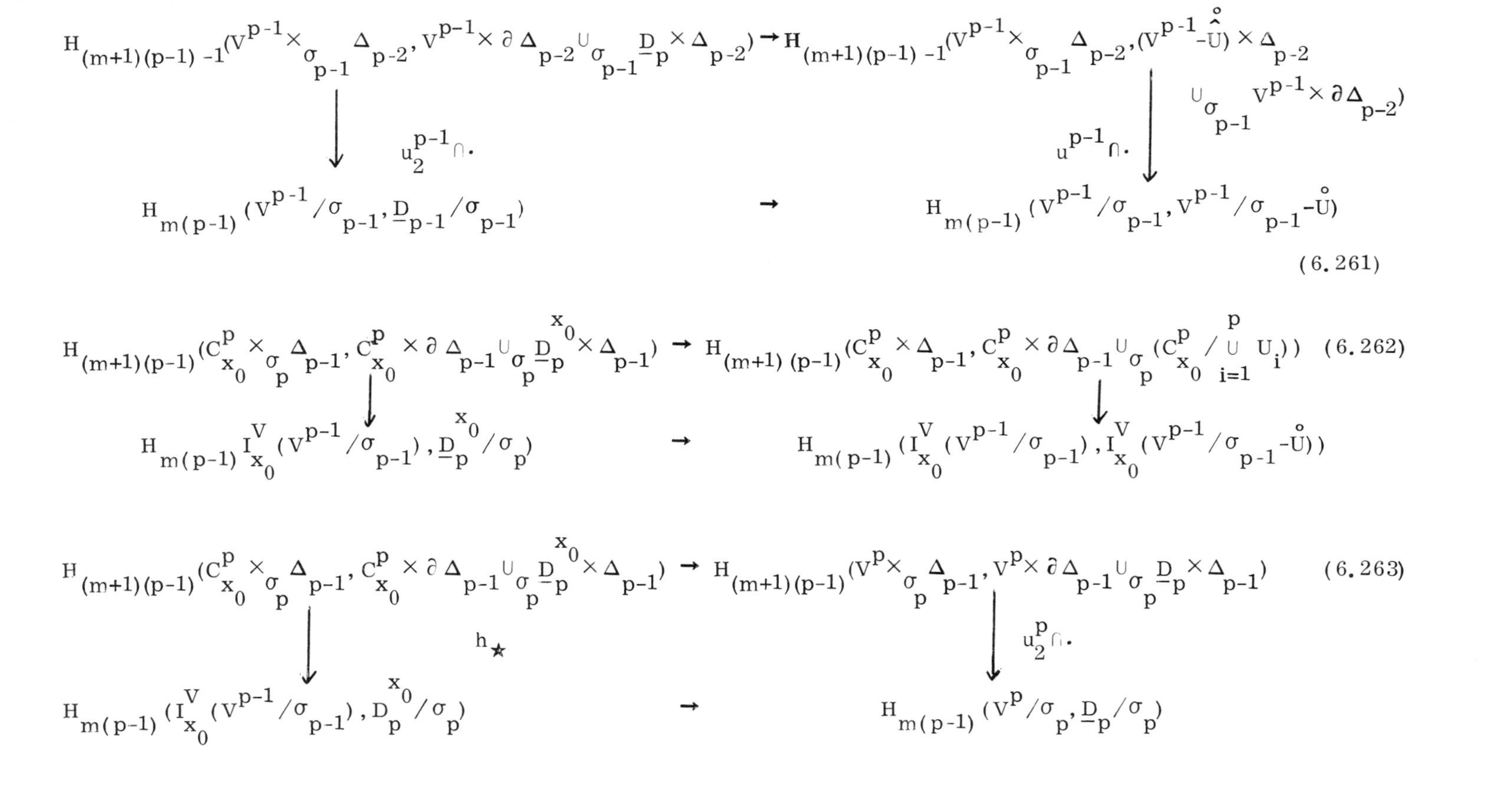

$$\begin{array}{ccc}
H_{(m+1)(p-1)-1}(V^{p-1}\times_{\sigma_{p-1}}\Delta_{p-2},\ V^{p-1}\times\partial\Delta_{p-2}\cup_{\sigma_{p-1}}\underline{D}_p\times\Delta_{p-2}) & \to & H_{(m+1)(p-1)-1}(V^{p-1}\times_{\sigma_{p-1}}\Delta_{p-2},\ (V^{p-1}-\hat{\mathring{U}})\times\Delta_{p-2}\cup_{\sigma_{p-1}}V^{p-1}\times\partial\Delta_{p-2}) \\
\downarrow u_2^{p-1}\cap\cdot & & \downarrow u^{p-1}\cap\cdot \\
H_{m(p-1)}(V^{p-1}/\sigma_{p-1},\ \underline{D}_{p-1}/\sigma_{p-1}) & \to & H_{m(p-1)}(V^{p-1}/\sigma_{p-1},\ V^{p-1}/\sigma_{p-1}-\mathring{U})
\end{array} \qquad (6.261)$$

$$\begin{array}{ccc}
H_{(m+1)(p-1)}(C^p_{x_0}\times_{\sigma_p}\Delta_{p-1},\ C^p_{x_0}\times\partial\Delta_{p-1}\cup_{\sigma_p}\underline{D}_p^{x_0}\times\Delta_{p-1}) & \to & H_{(m+1)(p-1)}(C^p_{x_0}\times\Delta_{p-1},\ C^p_{x_0}\times\partial\Delta_{p-1}\cup_{\sigma_p}(C^p_{x_0}/\cup_{i=1}^{p}U_i)) \\
\downarrow & & \downarrow \\
H_{m(p-1)}(I^V_{x_0}(V^{p-1}/\sigma_{p-1}),\ \underline{D}_p^{x_0}/\sigma_p) & \to & H_{m(p-1)}(I^V_{x_0}(V^{p-1}/\sigma_{p-1}),\ I^V_{x_0}(V^{p-1}/\sigma_{p-1}-\mathring{U}))
\end{array} \qquad (6.262)$$

$$\begin{array}{ccc}
H_{(m+1)(p-1)}(C^p_{x_0}\times_{\sigma_p}\Delta_{p-1},\ C^p_{x_0}\times\partial\Delta_{p-1}\cup_{\sigma_p}\underline{D}_p^{x_0}\times\Delta_{p-1}) & \to & H_{(m+1)(p-1)}(V^p\times_{\sigma_p}\Delta_{p-1},\ V^p\times\partial\Delta_{p-1}\cup_{\sigma_p}\underline{D}_p\times\Delta_{p-1}) \\
\downarrow & h_\star & \downarrow u_2^p\cap\cdot \\
H_{m(p-1)}(I^V_{x_0}(V^{p-1}/\sigma_{p-1}),\ D_p^{x_0}/\sigma_p) & \to & H_{m(p-1)}(V^p/\sigma_p,\ \underline{D}_p/\sigma_p)
\end{array} \qquad (6.263)$$

$$\to V^{p-1} \times_{\sigma_{p-1}} \Delta_{p-2} / (V^{p-1} \times \Delta_{p-2} - \hat{\mathring{U}}) \times \Delta_{p-2} \cup_{\sigma_{p-1}} V^{p-1} \times \partial\Delta_{p-2} \tag{6.264}$$

as follows:

$$\Theta_p([(x_0, x_1, \ldots, x_{p-1}, t_0, \ldots, t_{p-1})]) = \star \tag{6.265}$$

$$\text{if } [(x_0, x_1, \ldots, x_{p-1}, t_0, \ldots, t_{p-1})] \in (C^p_{x_0} / \bigcup_{i=1}^{p} \mathring{U}_i) \times_{\sigma_p} \Delta_{p-1}$$

where $\star$ is the image of $(V^{p-1}/\hat{\mathring{U}}) \times \Delta_{p-2} \cup_{\sigma_{p-1}} \times \partial\Delta_{p-2}$ under the quotient map:

$$V^{p-1} \times_{\sigma_{p-1}} \Delta_{p-2} \xrightarrow{r} V^{p-1} \times_{\sigma_{p-1}} \Delta_{p-2} / (V^{p-1} - \hat{\mathring{U}}) \times \Delta_{p-2} \cup_{\sigma_{p-1}} V^{p-1} \times \partial\Delta_{p-2}. \tag{6.266}$$

if one of the t_i is zero, we drop (x_i, t_i) from $(x_0, x_1, \ldots, x_{p-1},$ (6.267) $t_1, \ldots, t_{p-1})$, take the class of the element in $V^{p-1} \times_{\sigma_{p-1}} \Delta_{p-1}$ and the image of this class under r.

Θ_p is continuous and well defined: This is obvious on $(C^p_{x_0} / \bigcup_{i=1}^{p} \mathring{U}_i) \times_{\sigma_p} \Delta_{p-1}$. On $C^p_{x_0} \times_{\sigma_{p-1}} \partial\Delta_{p-1}$, Θ_p is equal (see (6.80)-(6.81)) to:

$$\Theta_p \Big|_{C^p_{x_0} \times_{\sigma_{p-1}} \partial\Delta_{p-1}} = r_1 \circ \theta_p \Big|_{C^p_{x_0} \times_{\sigma_{p-1}} \partial\Delta_{p-1}}, \tag{6.268}$$

where r_1 is the quotient map:

$$r_1 : V^{p-1} \times_{\sigma_{p-1}} \Delta_{p-2} / V^{p-1} \times \partial\Delta_{p-2} \cup_{\sigma_{p-1}} \underline{D}_{p-1} \times \Delta_{p-2} \tag{6.269}$$

$$\to V^{p-1} \times_{\sigma_{p-1}} \Delta_{p-2} / (V^{p-1} - \hat{\mathring{U}}) \times \Delta_{p-2} \cup_{\sigma_{p-1}} V^{p-1} \times \partial\Delta_{p-2}.$$

Observe that $\underline{D}_{p-1} \times \Delta_{p-2} \subset (V^{p-1}/\hat{\mathring{U}}) \times \Delta_{p-2}$, as $U \cap \underline{D}_{p-1}/\sigma_{p-1} = \phi$.) Therefore, Θ_p is well defined and continuous on $C^p_{x_0} \times_{\sigma_{p-1}} \partial\Delta_{p-1}$. Now, on

$(C^p_{x_0} / \overset{p}{\underset{j=1}{\cup}} \mathring{U}_j) \times_{\sigma_p} \partial\Delta_{p-1}$, according to (6.267), Θ_p acts by dropping a component (x_i, t_i) since that t_i is zero. Considering an element of $C^p_{x_0} / \overset{p}{\underset{j=1}{\cup}} \mathring{U}_j$, one of its components is x_0. The remaining (p-1)-components build up an element which does not belong to $\hat{\mathring{U}}$. Therefore, if Θ_p acts by dropping x_0 on this element, the image in $V^{p-1} \times_{\sigma_{p-1}} \Delta_{p-2}/(V^{p-1}/\hat{\mathring{U}}) \times \Delta_{p-2} \cup_{\sigma_{p-1}} V^{p-1} \times \partial\Delta_{p-1}$ is $\star$. Now, if Θ_p acts by dropping another component, the element left has x_0 as one of its components. But (see (6.230)), $U \cap (C^p_{x_0}|_{\sigma_p}) = \phi$. Thus, this element belongs to $(V^{p-1}/\hat{\mathring{U}}) \times \Delta_{p-2}$ and again the image is $\star$. Θ_p is therefore well defined globally and continuous as well.

Observe that we have the following commutative diagram:

$$\begin{array}{ccc} C^p_{x_0} \times \partial\Delta_{p-1} \cup_{\sigma_p} \underline{D}^{x_0}_p \times \Delta_{p-1} & \xrightarrow{i} & C^p_{x_0} \times \partial\Delta_{p-1} \cup_{\sigma_p} (C^p_{x_0} / \overset{p}{\underset{i=1}{\cup}} \mathring{U}_i) \times \Delta_{p-1} \\ \downarrow j & & \downarrow \Theta_p \\ V^p \times \partial\Delta_{p-1} \cup_{\sigma_p} \underline{D}_p \times \Delta_{p-1} & \xrightarrow{r_1 \circ \theta_p} & V^{p-1} \times_{\sigma_{p-1}} \Delta_{p-2}/(V^{p-1} - \hat{\mathring{U}}) \times \Delta_{p-2} \cup_{\sigma_{p-1}} V^{p-1} \\ & & \times \partial\Delta_{p-2} \end{array} \qquad (6.270)$$

i and j are the inclusion maps. (6.270) is commutative by (6.268) on $C^p_{x_0} \times_{\sigma_p} \partial\Delta_{p-1}$. On $\underline{D}^{x_0}_p \times \Delta_{p-1}$, $\Theta_p \circ i$ maps this set on $\star$, as $i(\underline{D}^{x_0}_p \times \Delta_{p-1})$ is contained in $(C^p_{x_0} / \overset{p}{\underset{i=1}{\cup}} \mathring{U}_i) \times_{\sigma_p} \Delta_{p-1} (U \cap \underline{D}_{p-1}/\sigma_{p-1} = \phi)$.

On the other hand, $j(\underline{D}^{x_0}_p \times_{\sigma_p} \Delta_{p-1})$ is contained in $\underline{D}_p \times_{\sigma_p} \Delta_{p-1}$. $\theta_j \circ j$, following (6.81), acts by dropping one of the x_i such that $x_i = x_j$ $(i \neq j)$. If x_0 is dropped, the resulting element still bears x_0 as one of its components (we are in $\underline{D}^{x_0}_p$ or in $\underline{D}_p$ more generally). Therefore, this element belongs to $(V^{p-1}/\hat{\mathring{U}}) \times_{\sigma_{p-1}} \Delta_{p-2}$ and is mapped on $\star$ by r_1. If, on the contrary, x_0 is not dropped, then x_0 remains a component of the resulting element and the same conclusion holds. Therefore

$$\epsilon_{p \circ i}\Big|_{\underline{D}_p^{x_0} \times_{\sigma_p} \Delta_{p-1}} = r_1 \circ \theta_p \circ j\Big|_{\underline{D}_p^{x_0} \times_{\sigma_p} \Delta_{p-1}} .$$

Consequently, (6.270) commutes and yields the commutative diagram (6.271). This diagram yields in turn the commutative diagram (6.272). j is the natural inclusion; δ' is the connecting homomorphism of the pair $(C^p_{x_0} \times_{\sigma_p} \Delta_{p-1}, C^p_{x_0} \times \partial \Delta_{p-1} \cup_{\sigma_p} \underline{D}_p^{x_0} \times \Delta_{p-1})$.

Now, as $H_{m(p-1)}(V^{p-1}/\sigma_{p-1})$ is by Lemma 6.6 uniquely generated and we have isomorphisms:

$$e^V_{p-1\star} : H_{m(p-1)}(V^{p-1}/\sigma_{p-1}) \to H_{m(p-1)}(V^{p-1}/\sigma_{p-1}, \underline{D}_{p-1}/\sigma_{p-1}) \quad (6.273)$$

$$l_\star : H_{m(p-1)}(V^{p-1}/\sigma_{p-1}, \underline{D}_{p-1}/\sigma_{p-1}) \to H_{m(p-1)}(V^{p-1}/\sigma_{p-1}, (V^{p-1}/\sigma_{p-1})/\mathring{U}), \quad (6.274)$$

the latter being induced by the inclusion map, the commutativity of the diagram of Lemma 6.8 reduces through (6.261) to the commutativity of (6.275). Indeed, by (6.261), we have:

$$l_\star \circ u_2^{p-1} \cap \cdot = (u^{p-1} \cap \cdot) \circ r_{1\star}. \quad (6.276)$$

$l_\star$ is an isomorphism on $H_{m(p-1)}(V^{p-1}/\sigma_{p-1}, \underline{D}_{p-1}/\sigma_{p-1})$. Therefore:

$$u_2^{p-1} \cap \cdot = l_\star^{-1} \circ (u^{p-1} \cap \cdot) \circ r_1^\star \quad (6.277)$$

on $H_{(m+1)(p-1)}(V^{p-1} \times_{\sigma_{p-1}} \Delta_{p-1}, V^{p-1} \times \partial\Delta_{p-2} \cup_{\sigma_{p-1}} \underline{D}_{p-1} \times \Delta_{p-2})$

and we are reduced to proving the commutativity of (6.275).

Next, observe that $I^V_{x_0 \star}$ factorizes through:

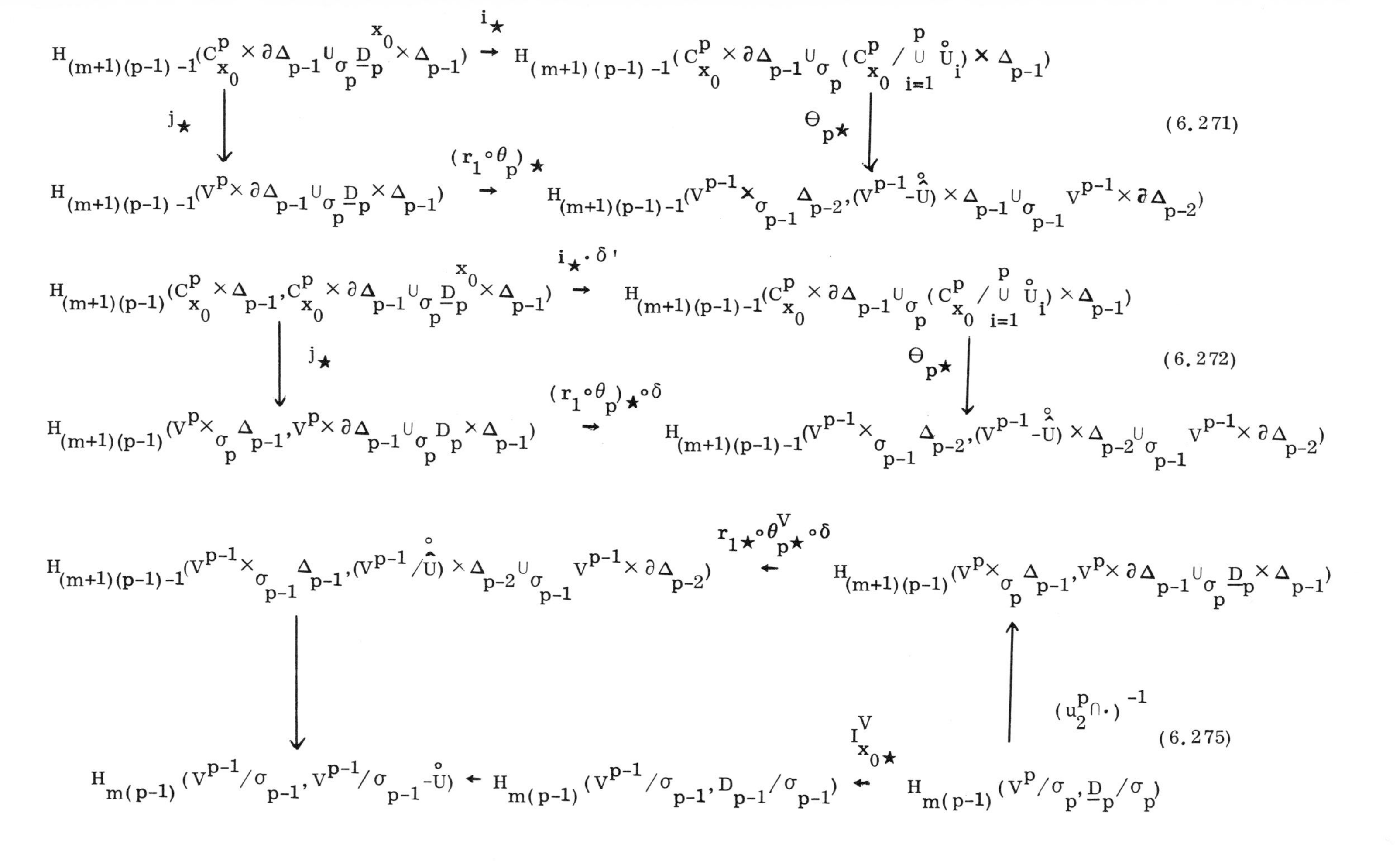

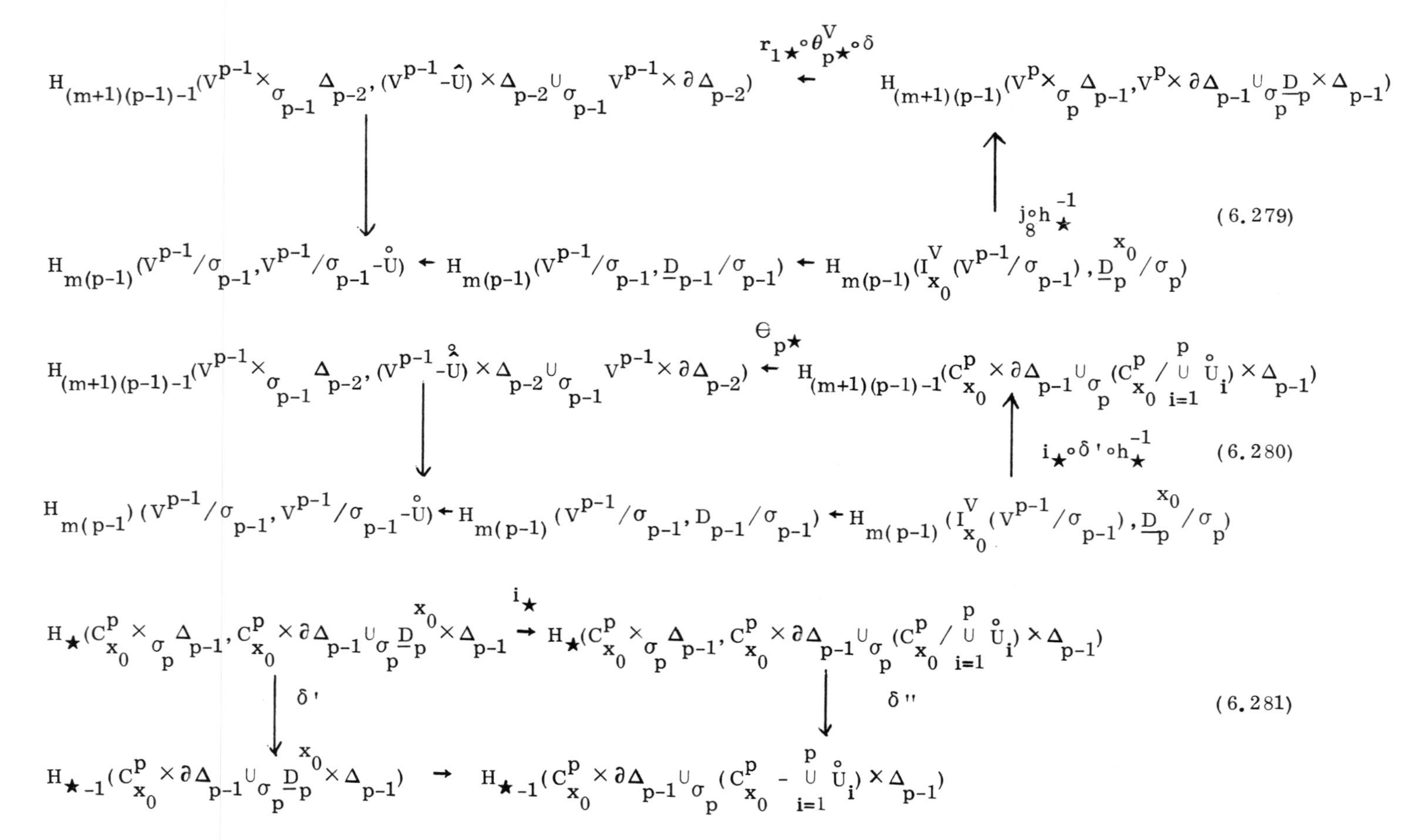

$$H_{m(p-1)}(V^{p-1}/\sigma_{p-1}, \underline{D}_{p-1}/\sigma_{p-1}) \to H_{m(p-1)}(I^V_{x_0}(V^{p-1}/\sigma_{p-1}), \underline{D}_p^{x_0}/\sigma_p), \tag{6.278}$$

and we have a commutative diagram associated to this factorization in (6.263). $h_\star$ is in (6.263) the isomorphism of (6.240). We may thus replace (6.275) by (6.279). Now, using (6.272), we may replace $r_{1\star} \circ \theta^V_{p\star} \circ \delta \circ j_\star$ by $\Theta_{p\star} \circ i_\star \circ \delta'$. Therefore, we derive (6.280). Finally, using (6.262) and the commutativity of the diagram (6.281), we are reduced to (6.281').

Observe now that:

$$H_{m(p-1)}(V^{p-1}/\sigma_{p-1}, V^{p-1}/\sigma_{p-1} - \overset{\circ}{U}) \tag{6.282}$$

$$\overset{t_\star \circ n_\star^{-1}}{\to} H_{m(p-1)}(I^V_{x_0}(V^{p-1}/\sigma_{p-1}), I^V_{x_0}(V^{p-1}/\sigma_{p-1}) - \overset{\circ}{U}))$$

is just $I^V_{x_0\star}$ as all the groups in the bottom arrow of (6.281) are uniquely generated and as $n_\star$ is an isomorphism. We are thus left with (6.283), which, by excision, yields (6.284). This excision is allowed by the fact that Θ_p is well defined on the excised set.

We are now left with trivial bundles and free actions. Introducing

$$\tilde{U} \text{ a disc neighbourhood of } (a_1, \ldots, a_{p-1}) \text{ in } V^{p-1} \text{ which projects homeomorphically on } U, \tag{6.285}$$

$$\tilde{I}^V_{x_0} : V^{p-1} \to V^p \tag{6.286}$$

$$(v_1, \ldots, v_{p-1}) \to (x_0, v_1, \ldots, v_{p-1})$$

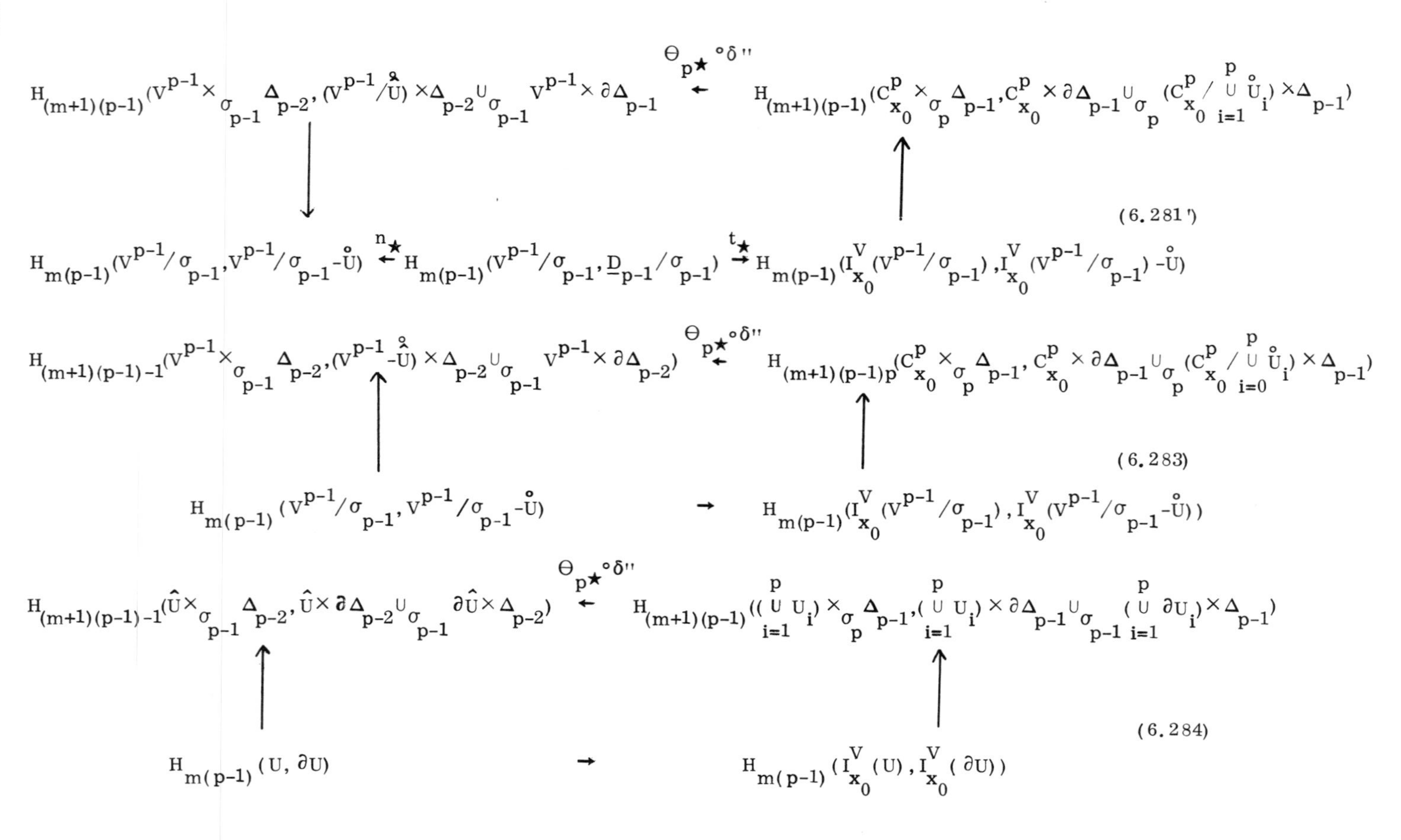

observing that:

$$U_1 = \{(y_1,\ldots,y_p);\, y_1 = x_0;\ (y_2,\ldots,y_p) \in \tilde{U}\} \tag{6.287}$$

$$\left(\bigcup_{i=1}^{p} U_i \times_{\sigma_p} \Delta_{p-1},\ \bigcup_{i=1}^{p} U_i \times \partial\Delta_{p-1} \cup_{\sigma_p} \left(\bigcup_{i=1}^{p} \partial U_i\right) \times \Delta_{p-1}\right) \tag{6.288}$$

$$= (U_1 \times \Delta_{p-1},\ U_1 \times \partial\Delta_{p-1} \cup \partial U_1 \times \Delta_p)$$

$$(\hat{U} \times_{\sigma_{p-1}} \Delta_{p-2},\ \hat{U} \times \partial\Delta_{p-2} \cup_{\sigma_{p-1}} \partial\hat{U} \times \Delta_{p-2}) \tag{6.289}$$

$$= (\tilde{U} \times \Delta_{p-2},\ \tilde{U} \times \partial\Delta_{p-2} \cup \partial\tilde{U} \times \Delta_{p-2});$$

identifying Θ_p:

$$\Theta_p : (U_1 \times \partial\Delta_{p-1} \cup \partial U_1 \times \Delta_{p-1}) \to \tilde{U} \times \Delta_{p-2}/\tilde{U} \times \partial\Delta_{p-2} \cup \partial\tilde{U} \times \Delta_{p-2} \tag{6.290}$$

being the map which acts by dropping x_0 and the corresponding coefficient t_0, we are left with the diagram (6.291). In (6.291) δ''' is the connecting homomorphism of $(U_1 \times \Delta_{p-1},\ U_1 \times \partial\Delta_{p-1} \cup \partial U_1 \times \Delta_{p-1})$. The vertical arrows are clearly tensor-products with the generators of $H_{p-2}(\Delta_{p-2}, \partial\Delta_{p-2})$ (for the left one) and $H_{p-1}(\Delta_{p-1}, \partial\Delta_{p-1})$ (for the right one), respectively. The commutativity of (6.291) reads off now easily and the proof of Lemma 6.8 is complete.

Now let:

$$f: V \to \Omega \tag{6.292}$$

be such that:

$$f^\star : H^m(\Omega) \to H^m(V) \tag{6.293}$$

is onto; V is an m-dimensional manifold as previously. Let

$$o^\star_\Omega \text{ in } H^m(\Omega) \text{ be such that } f^\star(o^\star_\Omega) = o^\star_V. \tag{6.294}$$

$$
\begin{array}{ccc}
H_{m(p-1)}(\tilde{U}, \partial\tilde{U}) & \rightarrow & H_{m(p-1)}(\tilde{I}^{V}_{x_0}(\tilde{U}), \tilde{I}^{V}_{x_0}(\partial\tilde{U})) = H_{m(p-1)}(U_1, \partial U_1) \\
\downarrow & & \downarrow \\
H_{(m+1)(p-1)-1}(\tilde{U}\times\Delta_{p-2}, \tilde{U}\times\partial\Delta_{p-2}\cup\partial\tilde{U}\times\Delta_{p-2}) & \underset{\Theta_{p\star}\circ\delta'''}{\leftarrow} & H_{(m+1)(p-1)}(U_1\times\Delta_{p-1}, U_1\times\partial\Delta_{p-1}\cup\partial U_1\times\Delta_{p-1})
\end{array}
\qquad (6.291)
$$

Diagram to Lemma 6.9:

$$
\begin{array}{ccc}
H_\star(V^p\times_{\sigma_p}\Delta_{p-1}, V^p\times\partial\Delta_{p-1}\cup_{\sigma_p} \underline{D}_p\times\Delta_{p-1}) & \overset{\tilde{f}^p_\star}{\rightarrow} & H_\star(\Omega^p\times_{\sigma_p}\Delta_{p-1}, \Omega^p\times\partial\Delta_{p-1}\cup_{\sigma_p} F_p\times\Delta_{p-1}) \\
\downarrow (\theta_p\circ\delta)_\star & & \downarrow (\theta_p\circ\delta)_\star \\
H_\star(V^{p-1}\times_{\sigma_{p-1}}\Delta_{p-2}, V^{p-1}\times\partial\Delta_{p-2}\cup_{\sigma_{p-1}} \underline{D}_{p-1}\times\Delta_{p-2}) & \overset{\tilde{f}^{p-1}_\star}{\rightarrow} & H_\star(\Omega^{p-1}\times_{\sigma_{p-1}}\Delta_{p-2}, \Omega^{p-1}\times\partial\Delta_{p-2}\cup_{\sigma_p} F_{p-1}\times\Delta_{p-2})
\end{array}
$$

$$
\begin{array}{ccc}
V^p\times\partial\Delta_{p-1}\cup_{\sigma_p} \underline{D}_p\times\Delta_{p-1} & \overset{\tilde{f}^p}{\rightarrow} & \Omega^p\times\partial\Delta_{p-1}\cup_{\sigma_p} F_p\times\Delta_{p-1} \\
\downarrow \theta^V_p & & \downarrow \theta_p \\
V^{p-1}\times_{\sigma_{p-1}}\Delta_{p-2}/V^{p-1}\times\partial\Delta_{p-2}\cup_{\sigma_{p-1}} \underline{D}_{p-1}\times\Delta_{p-2} & \overset{\tilde{f}^{p-1}}{\rightarrow} & \Omega^{p-1}\times_{\sigma_{p-1}}\Delta_{p-2}/\Omega^{p-1}\times\partial\Delta_{p-2}\cup_{\sigma_{p-1}} F_{p-1}\times\Delta_{p-2}
\end{array}
\qquad (6.303)
$$

Considering the point $f(x_0)$, we define (similarly to (6.154)):

$$I_{x_0}^{\Omega} : \Omega^{p-1}/\sigma_{p-1} \to \Omega^p/\sigma_p \tag{6.295}$$
$$[(x_1, \ldots, x_{p-1})] \to [(x_1, \ldots, x_{p-1}, f(x_0))]$$

Consider then the points

$$f(a_1), \ldots, f(a_{p-1}) \tag{6.296}$$

where $a_1, \ldots, a_{p-1}$ have been defined in (6.155). Let (compare (6.156)):

$$J^{\Omega} : \Omega \to \Omega^p/\sigma_p \tag{6.297}$$
$$x \to [(f(a_1), \ldots, f(a_{p-1}), x)]$$

Let (compare (6.157))

$$r^{\Omega} : \Omega \times \Omega^{p-1}/\sigma_{p-1} \to \Omega \text{ be the projection;} \tag{6.298}$$

and similarly to (6.158) and (6.259), let:

$$\mu_{\star}^{\Omega} : H_{\star}(\Omega^p/\sigma_p) \to H_{\star}(\Omega \times \Omega^{p-1}/\sigma_{p-1}) \tag{6.299}$$

$$\mu_{\Omega}^{\star} : H^{\star}(\Omega \times \Omega^{p-1}/\sigma_{p-1}) \to H^{\star}(\Omega^p/\sigma_p) \tag{6.300}$$

Let

$$f^p : V^p/\sigma_p \to \Omega^p/\sigma_p \tag{6.301}$$

be the map induced by f,

$$\tilde{f}^p : (V^p \times_{\sigma_p} \Delta_{p-1}, V^p \times \partial\Delta_{p-1} \cup_{\sigma_p} \underline{D}_p \times \Delta_{p-1}) \tag{6.302}$$
$$\to (\Omega^p \times_{\sigma_p} \Delta_{p-1}, \Omega^p \times \partial\Delta_{p-1} \cup_{\sigma_p} F_p \times \Delta_{p-1})$$

also be the map induced by f. We then have:

Lemma 6.9: (1) The diagram on page 280 is commutative.

(2) $H_*(V^p \times_{\sigma_p} \Delta_{p-1}, V^p \times \partial\Delta_{p-1} \cup_{\sigma_p} \underline{D}_p \times \Delta_{p-1})$ is an $H^*(V^p/\sigma_p)$ module.

Let $x \cap y$ denote the action of x in $H^*(V^p/\sigma_p)$ on y in $H_*(V^p \times_{\sigma_p} \Delta_{p-1}, V^p \times \partial\Delta_{p-1} \cup_{\sigma_p} \underline{D}_p \times \Delta_{p-1})$.

Let also $x' \cap y'$ denote the action of x' in $H^*(\Omega^p/\sigma_p)$ on y' in $H_*(\Omega^p \times_{\sigma_p} \Delta_{p-1}, \Omega^p \times \partial\Delta_{p-1} \cup_{\sigma_p} F_p \times \Delta_{p-1})$.

(3) We have: $\tilde{f}^p_*(\mu^*_V \circ r^*_V(o^*_V) \cap y) = \mu^*_\Omega \circ r^*_\Omega(o^*_\Omega) \cap \tilde{f}^p_*(y)$ for any y in $H_*(V^p \times_{\sigma_p} \Delta_{p-1}, V^p \times \partial\Delta_{p-1} \cup_{\sigma_p} \underline{D}_p \times \Delta_{p-1})$.

Corollary 6.10: The equation

$$\begin{cases} \Delta u + u^{\frac{n+2}{n-2}} = 0 \\ u = 0 \big|_{\partial\Omega} \end{cases}, \quad u > 0, \ \Omega \text{ connected, bounded, regular in } \mathbb{R}^n$$

has a solution as soon as the reduced $\mathbb{Z}_2$-homology of Ω is nonzero.

Proof of Lemma 6.9: Item (1) of Lemma 6.9 follows from the commutativity of diagram (6.303), which can be checked easily. Item (2) of Lemma 6.9 follows from the definition of the cap-product which provides $H_*(V^p \times_{\sigma_p} \Delta_{p-1}, V^p \times \partial\Delta_{p-1} \cup_{\sigma_p} \underline{D}_p \times \Delta_{p-1})$ with a structure of $H^*(V^p \times_{\sigma_p} \Delta_{p-1}) = H^*(V^p/\sigma_p)$ module. Observe that it is also the cap-product which provides $H_*(\Omega^p \times_{\sigma_p} \Delta_{p-1}, \Omega^p \times \partial\Delta_{p-1} \cup_{\sigma_p} F_p \times \Delta_{p-1})$ with a structure of the $H^*(\Omega^p/\sigma_p)$ module.

Now, combining $\tilde{f}^p$, we have, by naturality of the cap-product (see Dold [13]):

$$\tilde{f}^p_*(\tilde{f}^{p*}(x') \cap y) = x' \cap \tilde{f}^p_*(y) \tag{6.304}$$

for any y in $H_*(V^p \times_{\sigma_p} \Delta_{p-1}, V^p \times \partial\Delta_{p-1} \cup_{\sigma_p} F_p \times \Delta_{p-1})$ and any x' in $H^*(\Omega^p \times_{\sigma_p} \Delta_{p-1}) = H^*(\Omega^p/\sigma_p)$. Thus, in order to prove (3) of Lemma 6.9, we are left with proving:

$$\tilde{f}^{p\star}(\mu^\star_\Omega \circ r^\star_\Omega(o^\star_\Omega)) = \mu^\star_V \circ r^\star_V(o^\star_V) \; . \tag{6.305}$$

In (6.305), $\tilde{f}^{p\star}$ is the map induced by $\tilde{f}^p$ as acting from $V^p \times_{\sigma_p} \Delta_{p-1}$ into $\Omega^p \times_{\sigma_p} \Delta_{p-1}$. Thus:

$$\begin{array}{ccc} \tilde{f}^{p\star} : H^\star(\Omega^p \times_{\sigma_p} \Delta_{p-1}) & \to & H^\star(V^p \times_{\sigma_p} \Delta_{p-1}) \\ \| & & \| \\ H^\star(\Omega^p/\sigma_p) & & H^\star(V^p/\sigma_p) \end{array} \tag{6.306}$$

Clearly, $\tilde{f}^{p\star} = f^{p\star}$, where f^p is defined in (6.301), and (6.305) is then equivalent to:

$$f^{p\star}(\mu^\star_\Omega \circ r^\star_\Omega(o^\star_\Omega)) = \mu^\star_V \circ r^\star_V(o^\star_V) . \tag{6.307}$$

Now, by naturality of the transfer maps (see Bredon [4]) and by the commutativity of the following diagrams,

$$\begin{array}{ccc} V \times V^{p-1}/\sigma_{p-1} & \xrightarrow{(f, f^{p-1})} & \Omega \times \Omega^{p-1}/\sigma_{p-1} \\ r_V \downarrow & & \downarrow r_\Omega \\ V & \xrightarrow{f} & \Omega \end{array} \quad ; \tag{6.308}$$

$$\begin{array}{ccc} V^p/\sigma_p & \xrightarrow{f^p} & \Omega^p/\sigma_p \\ \uparrow & & \uparrow \\ V \times V^{p-1}/\sigma_{p-1} & \xrightarrow{(f, f^{p-1})} & \Omega \times \Omega^{p-1}/\sigma_{p-1} \end{array}$$

the following diagram commutes:

$$\begin{array}{ccc} H^\star(V) & \xleftarrow{f^\star} & H^\star(\Omega) \\ \downarrow r_V^\star & & \downarrow r_\Omega^\star \\ H^\star(V \times V^{p-1}/\sigma_{p-1}) & & H^\star(\Omega \times \Omega^{p-1}/\sigma_{p-1}) \\ \downarrow \mu_V^\star & & \downarrow \mu_\Omega^\star \\ H^\star(V^p/\sigma_p) & \xleftarrow{f^{p\star}} & H^\star(\Omega^p/\sigma_p) \end{array} \tag{6.309}$$

By (6.293), we have:

$$o_V^\star = f^\star(o_\Omega^\star). \tag{6.310}$$

Thus:

$$\mu_V^\star \circ r_V^\star(o_V^\star) = \mu_V^\star \circ r_V^\star \circ f^\star(o_\Omega^\star) = f^{p\star} \circ \mu_\Omega^\star \circ r_\Omega^\star(o_\Omega^\star), \tag{6.311}$$

and (6.307) is thereby proved. The proof of Lemma 6.9 is thereby complete.

Proof of Corollary 6.10: Let

$$w_p = \tilde{\gamma}_{p\star} \circ \tilde{f}^p_\star((u_2^p \cap \cdot)^{-1}(\omega_p^V)) \tag{6.312}$$

where

$$\omega_p^V \text{ generates } H_{mp}(V^p/\sigma_p) \tag{6.313}$$

and

$(u_2^p \cap \cdot)^{-1}$ is the inverse of the isomorphism defined in Proposition 6.3. (6.314)

We claim that ω_p is nonzero as long as J has no critical point under the energy level b_p; thus contradicting (5.132) of Proposition 5.10 with $A = f(V)$. In order to prove the claim, observe that, by Lemma 6.7, we have:

$$\mu_V^\star \circ r_V^\star(o_V^\star) \cap \omega_p^V = I^V_{x_0\star}(\omega^V_{p-1}). \tag{6.315}$$

Thus:

$$(u_2^p \cap \cdot)^{-1}(\mu_V^\star \circ r_V^\star(o_V^\star) \cap \omega_p^V) = (u_2^p \cap \cdot)^{-1}(I_{x_0 \star}^V(\omega_{p-1}^V)). \quad (6.316)$$

Using Lemma 6.8, we derive:

$$(u_2^{p-1} \cap \cdot)^{-1}(\omega_{p-1}^V) = \theta_{p\star}^V \circ \delta \circ (u_2^p \cap \cdot)^{-1}(\mu_V^\star \circ r_V^\star(o_V^\star) \cap \omega_p^V). \quad (6.317)$$

Using Lemma 6.9, we derive:

$$\tilde{f}_\star^{p-1} \circ (u_2^{p-1} \cap \cdot)^{-1}(\omega_{p-1}^V) = ((\theta_p)_\star \circ \delta) \circ \tilde{f}_\star^p \circ (u_2^p \cap \cdot)^{-1}(\mu_V^\star \circ r_V^\star(o_V^\star \cap \omega_p^V). \quad (6.318)$$

Thus, comparing with $\tilde{\gamma}_{p-1\star}$ and using Lemma 6.4, we derive:

$$w_{p-1} = \partial(\tilde{\gamma}_{p\star} \circ \tilde{f}_\star^p \circ (u_2^p \cap \cdot)^{-1}(\mu_V^\star \circ r_V^\star(o_V^\star) \cap \omega_p^V)). \quad (6.319)$$

Now, we have:

$$(u_2^p \cap \cdot)^{-1}(\mu_V^\star \circ r_V^\star(o_V^\star) \cap \omega_p^V) = \mu_V^\star \circ r_V^\star(o_V^\star) \cap [(u_2^p \cap \cdot)^{-1}(\omega_p^V)). \quad (6.320)$$

Indeed, with $\cup$ being the cup-product, we have:

$$\begin{aligned}
&(u_2^p \cap \cdot)(\mu_V^\star \circ r_V^\star(o_V^\star) \cap ((u_2^p \cap \cdot)^{-1}(\omega_p^V))) \quad (6.321)\\
&= (u_2^p \cup \mu_V^\star \circ r_V^\star(o_V^\star)) \cap (u_2^p \cap \cdot)^{-1}(\omega_p^V)\\
&= (\mu_V^\star \circ r_V^\star(o_V^\star) \cup u_2^p) \cap (u_2^p \cap \cdot)^{-1}(\omega_p^V)\\
&= (\mu_V^\star \circ r_V^\star(o_V^\star) \cap (u_2^p \cap (u_2^p \cap \cdot)^{-1}(\omega_p^V)\\
&= (\mu_V^\star \circ r_V^\star)(o_V^\star) \cap \omega_p^V,
\end{aligned}$$

which implies (6.320) by composing with $(u_2^p \cap \cdot)^{-1}$. Formulas (6.319) and (6.320) now yield:

$$w_{p-1} = \partial(\tilde{\gamma}_{p\star} \circ \tilde{f}_\star^p(\mu_V^\star \circ r_V^\star(o_V^\star) \cap (u_2^p \cap \cdot)^{-1}(\omega_p^V))). \quad (6.322)$$

Using (3) of Lemma 6.9, we derive:

$$w_{p-1} = \partial(\tilde{\gamma}_{p\star}(\mu^\star_\Omega \circ r^\star_\Omega(o^\star_\Omega) \cap \tilde{f}^p_\star((u_2^p \cap \cdot)^{-1}(\omega^V_p)))). \tag{6.323}$$

Using Lemma 6.5 and the $H^\star(\Omega^p/\sigma_p)$-linearity of $\tilde{\gamma}_{p\star}$, we then have:

$$w_{p-1} = \partial(\mu^\star_\Omega \circ r^\star_\Omega(o^\star_\Omega) \cap w_p). \tag{6.324}$$

Therefore, w_p is nonzero by induction, as long as J has no critical level under the energy level b_p. As previously said, there is then a contradiction with (5.132) of Proposition 5.10 with $A = f(V)$. The corollary follows.

7 The Palais-Smale condition on flow-lines

Definition 7.1: We will say that the Palais-Smale condition holds on flow-lines in the $V(p,\varepsilon)$ if, taking an initial data u_0 in $V(p,\varepsilon)$, with ε_0 small enough (but fixed), the solution $u(s,u_0)$ of the differential equation $\frac{\partial u}{\partial s} = -\partial J(u)$ with initial data u_0 remains outside a $V(p,\varepsilon_1)$, $\varepsilon_1 > 0$, which depends only on u_0.

We then have:

Proposition 7.2: (i) Assume $\Omega_- = S^n$ and $K = 1$. The Palais-Smale condition holds on flow-lines in $V(p,\varepsilon)$ for any p. In particular, every flow-line in $V(1,\varepsilon_0)$, $\varepsilon_0 > 0$ small enough, is compact.

(ii) Assume $\Omega = S^3$; K arbitrary with isolated critical points; the Palais-Smale condition holds along flow-lines in $V(p,\varepsilon)$; $p \geq 2$.

In $V(1,\varepsilon)$, let $x_1(s)$ be the solution of (5.11). Then, if a flow-line remains in $V(1,\varepsilon)$, ε going to zero with the time s on the line, then we have:

$$DK(x_1(s)) \to 0; \; x_1(s) \text{ converges to a critical point of } K. \tag{7.1}$$

$$\overline{\lim} \frac{-LK(x_1(s))}{K(x_1(s))} \geq 0; \quad \text{where } -L \text{ is the Yamabe operator on } S^3. \tag{7.2}$$

Proof of Proposition 7.2: We complete a stereographic projection and we therefore work on $\mathbb{R}^3$. In order to prove Proposition 7·2, we turn back to the form of the dynamical system near infinity given in Chapter 4.

Turning back to (4.53), the matrix A' may be simplified in:

$$A' = \left|\begin{array}{c|c|c} \begin{matrix}\ddots & 0(\varepsilon_{ij}) \\ \frac{1}{C}+o(1) & \ddots\end{matrix} & \begin{matrix}\ddots & 0(\frac{1}{\lambda_j}\varepsilon_{ij}) \\ o(\frac{1}{\lambda_j}) & \ddots\end{matrix} & \begin{matrix}\ddots & 0(\lambda_j\varepsilon_j) \\ o(\lambda_j) & \ddots\end{matrix} \\ \hline \begin{matrix}\ddots & 0(\varepsilon_{ij}) \\ 0(\Sigma\varepsilon_{jk}^2) & \ddots\end{matrix} & \begin{matrix}\ddots & \frac{1}{\lambda_j}0(\varepsilon_{ij}) \\ \frac{1}{\lambda_j}[C_{n/\alpha_j}+0(1)] & \ddots\end{matrix} & \begin{matrix}\ddots & \lambda_j 0(\varepsilon_{ij}) \\ \lambda_j o(1) & \ddots\end{matrix} \\ \hline \begin{matrix}\ddots & 0(\varepsilon_{ij}) \\ 0(\Sigma\varepsilon_{jk}^2) & \ddots\end{matrix} & \begin{matrix}\ddots & 0(\frac{\varepsilon_{ij}}{\lambda_j}) \\ \frac{1}{\lambda_j}0(|v|+\operatorname{Sup}\varepsilon_{jk}^2) & \ddots\end{matrix} & \begin{matrix}\ddots & 0(\lambda_j\varepsilon_{ij}) \\ \lambda_j(\frac{1}{\alpha_j}C'+0(1)) & \ddots\end{matrix} \end{array}\right| \tag{7.3}$$

Turning back to (4.7), the term S_i simplifies when $\Omega = S^n$, yielding:

$$S_i = o(\varepsilon_{ij}) + 0_\varepsilon\Big(\int_{|v|\geq\varepsilon\Sigma\alpha_j\delta_j} \delta_i |v|^{\frac{n+2}{n-2}}\Big) \tag{7.4}$$

$$+ 0_{\varepsilon'}\Big(\int_{|v|\leq\varepsilon\Sigma\alpha_j\delta_j,\ |v|\leq\varepsilon'\alpha_i\delta_i} \delta_i^{4/(n-2)} v^2\Big)$$

$$+ 0_K\Big(\Big(\int|\nabla v|^2\Big)^{1/2}\Big(\frac{|DK(x_i)|}{\lambda_i} + \frac{1}{\lambda_i^2}\Big)\Big).$$

Using then Lemma 4.2, we derive:

$$S_i = o(\varepsilon_{ij}) + 0_K\left(|\bar{w}|_H\Big(\frac{|DK(x_i)|}{\lambda_i} + \frac{1}{\lambda_i^2}\Big) + \frac{|DK(x_i)|^2}{\lambda_i^2} + \frac{1}{\lambda_i^4}\right) \tag{7.5}$$

$$+ 0\left(\int_{|\bar{w}|\leq 2\varepsilon\Sigma\alpha_j\delta_j,\ |\bar{w}|\leq 2\varepsilon'\alpha_i\delta_i} \delta_i^{\frac{4}{n-2}}\bar{w}^2\right.$$

$$\left. + \int_{|\bar{w}|\geq\frac{\varepsilon}{2}\alpha_i\delta_i} \delta_i|\bar{w}|^{\frac{n+2}{n-2}}\right) + 0(|\partial J(\Sigma\alpha_j\delta_j+v)|^2)$$

Observe now that:

$$0\left(\int_{|\bar{w}|\geq\frac{\varepsilon}{2}\alpha_i\delta_i}\delta_i|\bar{w}|^{\frac{n+2}{n-2}}\right)+0\left(\int_{|\bar{w}|\leq 2\varepsilon\Sigma\alpha_j\delta_j,\ |\bar{w}|\leq 2\varepsilon'\alpha_i\delta_i}\delta_i^{\frac{4}{n-2}}\bar{w}^2\right) \tag{7.6}$$

$$\leq C_\varepsilon|\bar{w}|_H^2$$

Thus, using (4.61):

$$S_i = o(\Sigma\,\varepsilon_{kr}) + 0_K\left(\frac{|DK(x_i)|^2}{\lambda_i^2}+\frac{1}{\lambda_i^4}\right)+0(|\partial J(\Sigma\,\alpha_j\delta_j+v)|^2) \tag{7.7}$$

Next, we turn to simplifying the second member terms in (4.5), (4.8) and (4.10). Due to the definition (5.10) of $V(p,\varepsilon)$, we have on these sets:

$$\alpha_i = \frac{1}{K(x_i)^{(n-2)/4}}\ \frac{(\Sigma 1/K(x_r)^{(n-2)/2})^{-1/2}}{(\int\delta^{2n/(n-2)})^{1/2}}, \tag{7.8}$$

$$\frac{(\Sigma\,\alpha_r^2)^{n/(n-2)}}{\Sigma\,\alpha_r^{2n/(n-2)}K(x_r)} = \left(\Sigma\,\frac{1}{K(x_r)^{(n-2)/2}}\right)^{\frac{2}{n-2}}+o(1). \tag{7.9}$$

Also: as $\Omega = S^n$, the remainder terms in (4.6), (4.9) and (4.11) simplify yielding, for (4.6),

$$S_i + 0_K\left(\Sigma\,\frac{1}{\lambda_k^{n-2}}+\sum_{k\neq j}\varepsilon_{kj}+\frac{1}{\lambda_k^2}\right) \tag{7.10}$$

for (4.9), taking also into account that $\frac{1}{\lambda_i}\left|\frac{\partial\varepsilon_{ij}}{\partial x_i}\right| = 0(\varepsilon_{ij})$,

$$\lambda_iS_i + 0_K\left(\Sigma\,\frac{1}{\lambda_k^{n-2}}+\sum_{k\neq j}\varepsilon_{kj}+\frac{1}{\lambda_k^2}\right) \tag{7.11}$$

for (4.11), taking into account that $\left|\frac{\partial}{\partial\lambda_i}\varepsilon_{ij}\right| = \frac{1}{\lambda_i}0(\varepsilon_{ij})$;

$$\frac{1}{\lambda_i}S_i + 0_K\left(\frac{1}{\lambda_i^3}\left(\Sigma\,\frac{1}{\lambda_k^{n-2}}+\Sigma\,\frac{1}{\lambda_k^2}\right)\right). \tag{7.12}$$

Observing now that:

$$\frac{(\Sigma \alpha_r^2)^{n/(n-2)}}{\Sigma \alpha_r^{2n/(n-2)} K(x_r)} \left(\int \delta^{\frac{2n}{n-2}}\right)^{\frac{2}{n-2}} \alpha^{\frac{4}{n-2}} K(x_i) = 1 + o(1) \text{ on } V(p, \varepsilon), \qquad (7.13)$$

and that, by (7. 8) and by the definition of the $V(p, \varepsilon)$

$$\lambda(u) K(x_i)^{\frac{n-2}{4}} \alpha_i = 1 + o(1) \quad \text{on} \quad V(p, \varepsilon), \qquad (7.14)$$

$$\varepsilon_{ij} = o(1). \qquad (7.15)$$

We derive that the second member of (4. 5) is (using also (7.10)) :

$$o(1). \qquad (7.15')$$

For the second member of (4. 8), we get, taking into account (7. 11) :

$$\lambda(u) \left[\frac{1}{K(x_i)^{(n+2)/4}} C_1 DK(x_i) + C_2 \frac{\Delta(DK)}{\lambda_i^2}(x_i) \right. \qquad (7.16)$$

$$+ C_3 \left(\sum_{j \neq i} \frac{1}{K(x_j)^{(n-2)/4}} \frac{\partial \varepsilon_{ij}}{\partial x_i} \right) + 0 \left(\sum_{j \neq i} \varepsilon_{ij} \right) + o \left(\sum_{j \neq i} \left| \frac{\partial \varepsilon_{ij}}{\partial x_i} \right| \right)$$

$$\left. + o(|DK(x_i)|) + \lambda_i S_i + 0_K \left(\Sigma \frac{1}{\lambda_k^{n-2}} + \sum_{k \neq j} \varepsilon_{kj} + \frac{1}{\lambda_k^2} \right) \right]$$

where

$$C_1 = c_0^{\frac{2n}{n-2}} \times \frac{n-2}{2n} \times \int \frac{r^{n-1} dr}{(1+r^2)^n} \times \frac{1}{\left(\int \delta^{2n/(n-2)} \times \Sigma \frac{1}{K(x_r)^{(n-2)/2}}\right)^{1/2}} \qquad (7.17)$$

$$C_2 = C_1 \left(\int \frac{|x|^2 dx}{(1+|x|^2)^n} \right) \times \left(\int \frac{r^{n-1} dr\, d\sigma}{(1+r^2)^n} \right)^{-1}. \qquad (7.18)$$

$$C_3 = \frac{2n}{n-2} C_1 \left(\int \frac{r^{n-1} dr}{(1+r^2)^n} \right)^{-1}. \qquad (7.19)$$

For the second member of (4. 10), we get, taking into account (7. 12) :

$$\lambda(u)\left[\frac{1}{K(x_i)^{(n+2)/4}}\left(-\frac{C_2}{2}\frac{\Delta K(x_i)}{\lambda_i^3}\right)+C_3\left(\sum_{j\neq i}\frac{1}{K(x_j)^{(n-2)/4}}\frac{\partial\varepsilon_{ij}}{\partial\lambda_i}\right)\right.$$

$$\left.+o\left(\sum_{j\neq i}\left|\frac{\partial\varepsilon_{ij}}{\partial\lambda_i}\right|\right)+o_K\left(\frac{1}{\lambda_i^3}\right)+\frac{1}{\lambda_i}S_i\right] \qquad (7.20)$$

Observe also that, using Lemma 6.2, in particular (4.61), we have:

$$|v|_H\leq|\overline{w}|_H+0(|\partial J(\Sigma\alpha_k\delta_k+v)|)=o(\sum_{k\neq r}\varepsilon_{kr}) \qquad (7.21)$$

$$+0(|\partial J(\Sigma\alpha_k\delta_k+v)|)$$

which allows us to replace the matrix

$$\begin{bmatrix} 0(\frac{\varepsilon_{ij}}{\lambda_j}) \\ \frac{1}{\lambda_j}0(|v|+\operatorname{Sup}\varepsilon_{jk}^2) \end{bmatrix} \text{ by } \begin{bmatrix} 0(\frac{\varepsilon_{ij}}{\lambda_j}) \\ \frac{1}{\lambda_j}[o(\sum_{k\neq r}\varepsilon_{kr})+0(|\partial J(\Sigma\alpha_k\delta_k+v)|) \end{bmatrix}$$

in A' (see (7.3)).

We then have, using (7.3) with the improvement of (7.21), (7.16), (7.20) and (7.15), as well as (7.7), the following simplified normal form of the dynamical system near infinity under the hypothesis of Proposition 7.2 (in order to derive (7.12), we also use that $\frac{1}{\lambda_i}\frac{\partial\varepsilon_{ij}}{\partial x_j}=0(\varepsilon_{ij})$; $\lambda_i\frac{\partial\varepsilon_{ij}}{\partial\lambda_i}=0(\varepsilon_{ij})$).

$$\frac{\dot\lambda_i}{\lambda_i}=\lambda(u)C''_n\left(-\frac{1}{2}\frac{\Delta K(x_i)}{K(x_i)\lambda_i^2}+a_n\left(\sum_{j\neq i}\alpha_j\lambda_i\frac{\partial\varepsilon_{ij}}{\partial\lambda_i}\right)\right)+o(\sum_{k\neq r}\varepsilon_{kr}) \qquad (7.22)$$

$$+0_K\left(\frac{|DK(x_i)|^2}{\lambda_i^2}+\frac{1}{\lambda_i^4}\right)+0(|\partial J(\Sigma\alpha_j\delta_j+v)|^2),$$

with C''_n and a_n suitable positive constants depending only on n.

$$\dot{x}_i = \lambda(u)\frac{C'''_n}{\lambda_i^2}\frac{DK(x_i)}{K(x_i)} + b_n \frac{\Delta(DK)}{K(x_i)\lambda_i^2}(x_i) \tag{7.23}$$

$$+ o\left(\frac{|DK(x_i)|}{\lambda_i^2}\right) + \frac{1}{\lambda_i} o(\Sigma\, \varepsilon_{kr}) + o_K\left(\frac{1}{\lambda_i^3}\right) + \frac{1}{\lambda_i} 0(|\partial J(\Sigma\, \alpha_j \delta_j + v)|^2)$$

C''_n and a_n are again suitable positive constants depending only on n. We now need the following Proposition, whose proof is provided in the Appendix:

Proposition 7.3: Let $u = \sum_{i=1}^{P} \alpha_i \delta_i + v$, where v satisfies (G0). Then

$$\frac{|DK(x_i)|}{\lambda_i} + \frac{1}{\lambda_i^2} \leq C(|\partial J(u)| + \sum_{j\neq i} \varepsilon_{ij}).$$

We are now ready to prove Proposition 7.2. Indeed, assume first that $K = 1$. Then, (7.22) and Proposition 7.3 yields:

$$\frac{\dot{\lambda}_i}{\lambda_i} = \lambda(u)\, C''_n a_n \sum_{j\neq i} \lambda_i \frac{\partial \varepsilon_{ij}}{\partial \lambda_i} + o(\sum_{k\neq r} \varepsilon_{kr}) \tag{7.24}$$

$$+ 0(|\partial J(\Sigma\, \alpha_j \delta_j + v)|^2)$$

We have also

$$\lambda_i \frac{\partial \varepsilon_{ij}}{\partial \lambda_i} = -\frac{n-2}{2} \varepsilon_{ij}^{\frac{n}{n-2}} \left(\frac{\lambda_i}{\lambda_j} - \frac{\lambda_j}{\lambda_i} + \lambda_i \lambda_j |x_i - x_j|^2\right). \tag{7.25}$$

Observe that, if $\lambda_i \geq \lambda_j$ and ε_{ij} is small, then $\lambda_i/\lambda_j + \lambda_j/\lambda_i + \lambda_i\lambda_j|x_i - x_j|^2$ is large, we have:

$$-\lambda_i \frac{\partial \varepsilon_{ij}}{\partial \lambda_i} \geq \frac{n-2}{4} \varepsilon_{ij}, \quad \lambda_i \geq \lambda_j; \quad \varepsilon_{ij} \text{ small.} \tag{7.26}$$

Observe also that:

$$\lambda_i \frac{\partial \varepsilon_{ij}}{\partial \lambda_i} + \lambda_j \frac{\partial \varepsilon_{ij}}{\partial \lambda_j} = -\frac{2}{n-2} \varepsilon_{ij}^{\frac{n}{n-2}} \lambda_i \lambda_j |x_i - x_j|^2 < 0. \tag{7.27}$$

These inequalities will also be used later on, for the proof of (ii).

We now classify our concentrations λ_i by increasing order, which we assume for sake of simplicity to be:

$$\lambda_1 \leq \lambda_2 \leq \ldots \leq \lambda_p \tag{7.28}$$

This order may change when s increases, so that none of these functions, as defined, is differentiable. They are however continuous; and their derivatives take at most p-values. Consider then the function:

$$\text{Log}\,\lambda_1 + 2\,\text{Log}\,\lambda_2 + 4\,\text{Log}\,\lambda_3 + \ldots + 2^{p-1}\text{Log}\,\lambda_p = \psi(s). \tag{7.29}$$

This is again a continuous function, which might not be differentiable if there is an interchange in the order of the λ_i; but has in any case a finite number of derivatives at such an interchange.

We compute one of them. Using formula (7.24), we have:

$$\dot{\psi}(s) = \lambda(u)\,C''_n a_n \left(\sum_{j\neq 1} \lambda_1 \frac{\partial \varepsilon_{1j}}{\partial \lambda_1} + 2 \sum_{j\neq 2} \lambda_2 \frac{\partial \varepsilon_{2j}}{\partial \lambda_2} + \ldots + 2^p \sum_{j\neq p} \lambda_p \frac{\partial \varepsilon_{pj}}{\partial \lambda_p} \right. \tag{7.30}$$

$$+ o\Big(\sum_{k\neq r} \varepsilon_{kr}\Big) + 0\big(\,|\,\partial J(\Sigma\, \alpha_j \delta_j + v)\,|^2\big)$$

$$= \lambda(u)\,C''_n a_n \quad \sum_{j\neq 1} \left(\lambda_1 \frac{\partial \varepsilon_{1j}}{\partial \lambda_1} + \lambda_j \frac{\partial \varepsilon_{1j}}{\partial \lambda_j}\right) + \left[2 \sum_{\substack{j\neq 2\\ j\neq 1}} \left(\lambda_2 \frac{\partial \varepsilon_{2j}}{\partial \lambda_2} + \lambda_j \frac{\partial \varepsilon_{2j}}{\partial \lambda_j}\right)\right.$$

$$\left. + \lambda_2 \frac{\partial \varepsilon_{12}}{\partial \lambda_2}\right] + \ldots + \left[2^{k-1} \sum_{j>k} \left(\lambda_k \frac{\partial \varepsilon_{kj}}{\partial \lambda_k} + \lambda_j \frac{\partial \varepsilon_{kj}}{\partial \lambda_j}\right)\right.$$

$$\left. + 2^{k-2}\lambda_k \frac{\partial \varepsilon_{k,k-1}}{\partial \lambda_k} + 2^{k-1}\left(1 - \frac{1}{2^2}\right) \lambda_k \frac{\partial \varepsilon_{k,k-2}}{\partial \lambda_k} + \ldots + (2^{k-1}-1)\lambda_k \frac{\partial \varepsilon_{1k}}{\partial \lambda_k}\right]$$

$$+ \ldots + \left[2^{p-2}\lambda_p \frac{\partial \varepsilon_{p,p-1}}{\partial \lambda_p} + 2^{p-1}\left(1 - \frac{1}{2^2}\right) \lambda_p \frac{\partial \varepsilon_{p,p-2}}{\partial \lambda_p}\right.$$

$$\left. + \ldots + (2^{p-1}-1)\,\lambda_p \frac{\partial \varepsilon_{1p}}{\partial \lambda_p}\right] + o\Big(\sum_{k\neq r} \varepsilon_{kr}\Big) + 0\big(\,|\,\partial J(\Sigma\, \alpha_j \delta_j + v)\,|^2\big).$$

Using now (7.26) and (7.27), we derive:

$$\dot{\psi}(s) \leq C\left(-\frac{n-2}{4}\sum_{\substack{i,j\\ i\neq j}} \varepsilon_{ij} + o\Big(\sum_{i\neq j} \varepsilon_{ij}\Big) + 0\Big(\big|\partial J(\Sigma\, \alpha_j \delta_j + v)\big|^2\Big)\right) \tag{7.31}$$

(7.31) holds with any determination we could take for the ordering in (7.28) in case of an interchange. (7.31) now implies:

$$\dot{\psi}(s) \leq 0\big(\,|\partial J(\Sigma\, \alpha_j \delta_j + v)|^2\big) \tag{7.32}$$

Thus:

$$\lambda_1(s)\lambda_2^2(s)\ \ldots\ \lambda_p^{2p}(s) \leq e^{c\int_{s_0}^{s} |\partial J(\Sigma\, \alpha_j \delta_j + v)|^2 + \psi(s_0)}\ . \tag{7.33}$$

Now

$$\int_{s_0}^{s} |\partial J(\Sigma\, \alpha_j \delta_j + v)|^2 = J(u(s_0)) - J(u(s)) \leq J(u(s_0)), \tag{7.34}$$

as

$$u = \Sigma\, \alpha_j \delta_j + v \quad\text{and}\quad \frac{\partial u}{\partial s} = -\partial J(u).$$

Thus,

$$e^{\psi(s)} \text{ is bounded.} \tag{7.35}$$

Therefore, the concentrations cannot blow up, all of them to $+\infty$; and the Palais-Smale condition holds on the flow-lines for $p \geq 2$. For $p = 1$, we are left with a single differential equation:

$$\frac{\dot{\lambda}_i}{\lambda_i} = 0\big(\,|\partial J(\alpha_i \delta_i + v)|^2\big). \tag{7.36}$$

Hence, through the same argument, using (7.34), the condition (i) is then proved.

We turn now to the proof of (ii). Observe that, for $n = 3$, we have:

$$\varepsilon_{ij} = \frac{1}{\left(\frac{\lambda_i}{\lambda_j}+\frac{\lambda_j}{\lambda_i}+\lambda_i\lambda_j|x_i-x_j|^2\right)^{1/2}} \tag{7.37}$$

(7.14) implies the existence of two constants m and M, depending only on p s.t

$$0 < m \leq \alpha_j \leq M \tag{7.37'}$$

Considering now the new function:

$$\phi(s) = \text{Log}\,\lambda_2 + \frac{2M}{m}\,\text{Log}\,\lambda_3 + \ldots + \left(\frac{2M}{m}\right)^{p-2}\,\text{Log}\,\lambda_p, \tag{7.38}$$

we can do the same computations as in (7.30) and (7.31). Due to the presence of the K-term, we end up with:

$$\dot{\phi}(s) \leq C\sum_{i\neq 1}\frac{1}{\lambda_i^2} - \frac{c}{4}\sum_{\substack{i,j;\, i\neq j\\ i,j\neq 1}}\varepsilon_{ij} + o(\sum_{i\neq j}\varepsilon_{ij}) + 0(|\partial J(\Sigma\,\alpha_j\delta_j+v)|^2) \tag{7.39}$$

$$\leq C\sum_{i\neq 1}\frac{1}{\lambda_i^2} - \frac{c}{2}\sum_{\substack{i,j\\ i\neq j}}\varepsilon_{ij} + o(\sum_{i\neq j}\varepsilon_{ij}) + 0(|\partial J(\Sigma\,\alpha_j\delta_j+v)|^2)$$

$$\leq C\sum_{i\neq 1}\frac{|\Delta K(x_i)|}{\lambda_i^2} - \frac{c}{4}\sum_{i\neq j}\varepsilon_{ij} + 0(|\partial J(\Sigma\,\alpha_j\delta_j+v)|^2).$$

Now for any i different from 1, we have:

$$\frac{1}{\lambda_i^2} = o(\varepsilon_{i1}), \quad i \neq 1. \tag{7.40}$$

Indeed, as $\lambda_1 \leq \lambda_i$, we have:

$$\varepsilon_{i1} \sim C\,\text{Min}\left(\sqrt{\frac{\lambda_1}{\lambda_i}},\ \frac{1}{\sqrt{\lambda_i\lambda_1}\,|x_i-x_1|}\right). \tag{7.41}$$

We may always assume, due to the fact that the problem is on S^3 and that we transformed it into a problem on $\mathbb{R}^3$, that we choose the stereographic projection, at the time when we are computing the derivative so that $|x_i-x_1|$ is bounded by a uniform constant C. Then (7.40) follows readily and yields:

$$\dot{\phi}(s) \leq -\frac{c}{8}\sum_{i\neq j}\varepsilon_{ij} + 0(|\partial J(\Sigma\,\alpha_j\delta_j+v)|^2), \tag{7.42}$$

and the conclusion follows that the Palais-Smale condition holds on flow-lines for $p \geq 2$.

We are thus left with a single max δ_1. We then have:

$$\frac{\dot{\lambda}_1}{\lambda_1} = \lambda(u)\, C''_n \left(-\frac{1}{2}\frac{\Delta K(x_1)}{K(x_1)\lambda_1^2}\right) + 0\left(\frac{|DK(x_1)|^2}{\lambda_1^2} + \frac{1}{\lambda_1^4}\right) \tag{7.43}$$

$$+ 0(|\partial J(\alpha_1\delta_1+v)|^2)$$

$$\dot{x}_1 = \lambda(u)\, C'''_3 \frac{DK(x_1)}{\lambda_1^2 K(x_1)} + o\left(\frac{1}{\lambda_1^3}\right) + \frac{1}{\lambda_1} 0(|\partial J(\alpha_1\delta_1+v)|^2). \tag{7.44}$$

Let

$$x'_1(s) = x_1(s) - \int_{s_0}^{s} \frac{0(|\partial J(\alpha_1\delta_1+v)|^2)}{\lambda_1} dt, \tag{7.45}$$

where the function we are integrating is the $0\left(\frac{|\partial J(\alpha_1\delta_1+v)|^2}{\lambda_1}\right)$ of (7.44).

We have, using the mean value theorem:

$$\dot{x}'_1(s) = \lambda(u)\, C''_3 \frac{1}{\lambda_1^2} \frac{DK(x_1)}{K(x_1)^{5/4}} + o\left(\frac{1}{\lambda_1^3}\right) \tag{7.46}$$

$$= \lambda(u)\, C''_3 \frac{1}{\lambda_1^2} \frac{DK(x'_1)}{K(x'_1)^{5/4}} + \frac{1}{\lambda_1^2} 0\left(\int_{s_0}^{s} \frac{1}{\lambda_1} |\partial J(\alpha_1\delta_1+v)|^2 ds\right) + o\left(\frac{1}{\lambda_1^3}\right)$$

Consider now $y_1, \dots, y_m$ the critical points of the functions K. Assume they are ordered as noted:

$$K(y_1) \geq \dots \geq K(y_m). \tag{7.47}$$

If $K(y_i) = K(y_j)$, there is no trajectory of the flow of K leading from y_i to y_j or conversely from y_j to y_i. Hence, the same feature holds on neighbourhoods of the y_i's.

Consider then $U_1, \dots, U_m$, neighbourhoods of $y_1, \dots, y_m$; which are arbitrary, up to the fact that they are small enough so that there is no flow trajectory from U_i to U_j if $K(y_i) = K(y_j)$. Consider $U'_1 \subset \overline{U}'_1 \subset U_1, \ldots,$ $U'_m \subset \overline{U}'_m \subset U_m$, U'_i neighbourhood of y_i. x_1 must enter one of the U_i's and remain there. For this we consider x'_1, with a choice of s_0 we will prescribe later on. Clearly, as $\int_0^{+\infty} |\partial J(\alpha_1 \delta_1 + v)|^2 dx < +\infty$, we have, for s_0 large enough:

$$\frac{1}{2}\frac{d}{ds} K(x'_1)^2 = \lambda(u) C''_n \frac{1}{\lambda_1^2} |DK(x'_1)|^2 + o(\frac{1}{\lambda_1^2}). \tag{7.48}$$

Thus, as $\lambda(u)$ is bounded below by the universal strictly positive constant $\operatorname{Inf}_{u\in\Sigma^+} J(u)^{\frac{n-2}{4}} = \operatorname{Inf}_{u\in\Sigma^+} J(u)^{1/4} (\lambda(u) = J(u)^{\frac{n-2}{4}})$, we have:

$$\frac{d}{ds} K(x'_1)^2 > 0 \text{ as soon as } x'_1 \text{ is not in the } U'_j\text{'s}. \tag{7.49}$$

Hence, when x'_1 is in one of the U'_j's or $K(x'_1)^2$ increases. Now, from (7.43), we derive:

$$|2\lambda_1 \dot{\lambda}_1 + 0(|\partial J(\alpha_1 \delta_1 + v)|^2) \lambda_1^2| \le C, \tag{7.50}$$

hence

$$\left|\frac{d}{ds}(\lambda_1^2 e^{\int_0^s 0(|\partial J(\alpha_1 \delta_1 + v)|^2)}\right| \le C e^{\int_0^s 0(|\partial J(\alpha_1 \delta_1 + v)|^2)}. \tag{7.51}$$

Hence, using the boundedness of $\int_0^{+\infty} |\partial J(\alpha_1 \delta_1 + v)|^2)$, we derive:

$$\lambda_1^2 \le C's + C''. \tag{7.52}$$

Thus

$$\int_0^s \frac{1}{\lambda_1^2} dt \text{ diverges}. \tag{7.53}$$

Setting

$$\frac{d}{dz} = \lambda_1^2(s) \frac{d}{ds}; \text{ or } z = \int_0^s \frac{1}{\lambda_1^2(\tau)} d\tau, \tag{7.54}$$

we then derive from (7.46) and (7.47):

$$\frac{d}{dz} x_1'(z) = \lambda(u)\, C_n'' \frac{DK(x_1')}{K(x_1')} + o(1), \tag{7.55}$$

$$\frac{1}{2}\frac{d}{dz} K(x_1')^2 = \lambda(u)\, C_n'' |DK(x_1')|^2 + o(1). \tag{7.56}$$

Notice that, because of (7.53), z runs from 0 to $+\infty$ when s runs from 0 to $+\infty$. If x_1' would remain outside of the U_j''s after a certain time z_0, (7.56) would imply:

$$\lim_{z \to +\infty} K(x_1')^2 = +\infty \tag{7.57}$$

which is impossible. Thus x_1' after having possibly left a U_i' must re-enter a U_j'. We claim that for z large enough, we must then have for such a process to take place:

$$K(y_j) > K(y_i), \tag{7.58}$$

with a strict inequality. Indeed with such consecutive U_i' and U_j', if such a process would take place, i.e. x_1' leaves U_i' and re-enters U_j', by (7.55), we would have an orbit of $\lambda(u)$ $C_3'' \frac{DK(x_1')}{K(x_1')^{5/4}} + o(1)$ leading from U_1' to U_j'. As there is no orbit of $DK(x)$ from U_i to U_j for $K(y_j) \le K(y_i)$ by the choice of the U_k, there is no orbit of $DK(x)$ from $\overline{U}_i'$ to $\overline{U}_j'$ for $K(y_j) \le K(y_i)$ and, by continuity, there will be no orbit of $\lambda(u)\, C_3'' \frac{DK(x)}{K(x)^{5/4}} + o(1)$ (taking z large enough) for $\overline{U}_i'$ to $\overline{U}_j'$ for $K(y_j) \le K(y_i)$. Hence (7.58) holds. Thus x_1' must enter one of the U_k''s after any given time and the index k must increase strictly. Therefore, x_1' has to rest after a certain time in one of the U_k', say U_{k_0}'.

Now, (7.45) implies, with s_0 large enough:

$$|x_1' - x_1| = o(1). \tag{7.59}$$

Thus, x_1 must then be in U_{k_0} for k large enough. As a corollary x_1 converges to y_{k_0} and $DK(x_1)$ goes to zero. Turning back then to (7.43), we have:

$$2\lambda_1\dot{\lambda}_1 + \lambda_1^2 0(|\partial J(\alpha_1\delta_1+v)|^2) = -\lambda(u)\, C_3'' \frac{\Delta K(y_{k_0})}{K(y_{k_0})} + o(1), \tag{7.60}$$

or

$$\lambda_1^2(s) = \left[\lambda_1^2(0) - \int_0^s e^{\int_0^\tau 0(|\partial J(\alpha_1\delta_1+v)|^2)} \lambda(u)\, C_3'' \frac{(\Delta K(y_{k_0})}{K(y_{k_0})} + o(1)\, d\tau\right] \times$$

$$\times\, e^{\int_0^s 0(|\partial J(\alpha_1\delta_1+v)|^2)} \tag{7.61}$$

The boundedness of $\int_0^{+\infty} 0(|\partial J(\alpha_1\delta_1+v)|^2)$ and the fact that $\lambda(u) \geq c > 0$, then implies that $\Delta K(y_{k_0})$ must be negative. Hence item (ii) of Proposition 7.2. Observe also that if $\Delta K(y_{k_0})$ is strictly negative, we have:

$$\lambda_1(s) \underset{s\to+\infty}{\sim} C(y_{k_0})\sqrt{s}. \tag{7.62}$$

Corollary 7.3: Assume K is a C^2 strictly positive function on S^3 with nondegenerate critical points $y_1, \ldots, y_m$. Let $k_1, \ldots, k_m$ be the Morse index of K at $y_1, \ldots, y_m$ respectively. Assume $\Delta K(y_i)$ is nonzero for all i, where Δ is the Laplacian operator on S^3. If $\sum_{i \mid \Delta K(y_i) < 0;} (-1)^{k_i}$ is different from -1, the equation

$$\begin{cases} Lu - K(x)\, u^{\frac{n+2}{n-2}} = 0 \\ u > 0 \ \text{ on } S^3 \end{cases}$$

has a solution.

Proof: Arguing by contradiction, we may assume there is no solution to the equation. The gradient vector field ∂J on Σ^+ has then only singularities at infinity, at those single points y_i such that $-\Delta K(y_i)/K(y_i)$ is positive; this, by Proposition 7.2.

Next, using the expansion provided by Proposition 5.3, we have:

$$J(\alpha_1 \delta_1(x_1, \lambda_1) + v) = \frac{1}{\alpha_1^6 K(x_1)} (\int \delta^6)^2 \times \tag{7.63}$$

$$\times \left(\frac{1}{K(x_1)} \frac{\Delta K(x_1)}{\lambda_1^2} + o\left(\frac{1}{\lambda_1^2}\right)\right)$$

$$+ (f, v) + Q(v) + o(|v|^2),$$

where Q is a positive definite quadratic form. With x_1 being close to y_i, such that $\Delta K(y_i)$ is strictly negative and α_1 being equal to:

$$\alpha_1 = \frac{1}{(\int |\nabla \delta_1)^2)^{1/2}} + o(|v|^2) = \frac{1}{(\int \delta^{2n/(n-2)})^{(n-2)/2n}} \tag{7.64}$$

$$+ o(|v|^2)$$

we have:

$$J(\alpha_1 \delta_1(x_1, \lambda_1) + v) = \left((\int \delta^6 \right)^{-\frac{1}{2}} \left(\frac{1}{K(x_1)} + \frac{\Delta K(y_i)}{K(y_i)^2} \cdot \frac{1}{\lambda_1^2}\right) + o\left(\frac{1}{\lambda_1^2}\right) \tag{7.65}$$

$$+ Q(v) + (f, v) + o(|v|^2).$$

Here, following Proposition 5.3, we have:

$$(f, v) = \int_{S^3} K(x) \delta_1^5 v = \int_{S^3} (K(x) - K(x_1)) \delta_1^5 v = o\left(\frac{1}{\lambda_1} |x_1 - y_i|\right) |v|. \tag{7.66}$$

Therefore, as $|x_1 - y_i| = o(1)$,

$$(f, v) = o\left(|v|^2 + \frac{1}{\lambda_1^2}\right). \tag{7.67}$$

We thus have:

$$J(\alpha_1 \delta_1(x_1, \lambda_1) + v) = (\int \delta^6)^{-\frac{1}{2}} \left(\frac{1}{K(x_1)}\right) + -(\int \delta^6)^{-\frac{1}{2}} \frac{\Delta K(y_i)}{\lambda_1^2 K(y_i)^2} \cdot$$

$$+ Q(v) + o\left(\frac{1}{\lambda_1^2}\right) + o(|v|^2). \tag{7.68}$$

As $Q(v)$ is definite positive, the study of the difference of topology induced by such a singularity is easily brought back to the difference of topology in the variation space $U(y_i) \times (A, +\infty)$, A a large number, $U(y_i)$ a small neighbourhood of y_i, by the functional

$$\frac{1}{K(x_1)} \quad \frac{\Delta K(y_i)}{K(y_i)^2} \cdot \frac{1}{\lambda_1^2}$$

between the values

$$\frac{1}{K(y_i)} + \varepsilon \text{ and } \frac{1}{K(y_i)} - \varepsilon \; ; \quad \varepsilon > 0 \text{ small.}$$

This difference of topology, using a Morse reduction of K near y_i, is the same as the one induced on $U(0) \times (A, +\infty)$ by the functional

$$\frac{1}{K(y_i)} + \sum_{i=1}^{k_i} x_r^2 - \sum_{k_i+1}^{3} x_r^2 + \frac{a_i}{\lambda_1^2} \; ; \quad a_i = \frac{-\Delta K(y_i)}{K(y_i)^2} \; ;$$

k_i, Morse index of K at y_i.

Considering this difference of topology between $c+\varepsilon$ and $c-\varepsilon$, i.e. considering the pair:

$$\left\{ (x_1, x_2, x_3, \lambda_1) \text{ s.t. } \sum_{i=1}^{k_i} x_r^2 - \sum_{k_i+1}^{3} x_r^2 + \frac{a_i}{\lambda_i^2} \leq \varepsilon \right\} ,$$

$$\left\{ (x_1, x_2, x_3, \lambda_1) \text{ s.t. } \sum_{i=1}^{k_i} x_r^2 - \sum_{k_i+1}^{3} x_r^2 + \frac{a_i}{\lambda_i^2} \leq -\varepsilon \right\} ,$$

this pair is nontrivial if and only if a_i is strictly positive. Therefore we must have:

$$-\Delta K(y_i) > 0 \tag{7.69}$$

This difference of topology is then given by the co-index of K at y_i, i.e. the pair is homotopically equivalent to (D^{3-k_i}, S^{2-k_i}). Therefore, $-\partial J$ has essential singularities at those y_i such that $-\Delta K(y_i) > 0$. The local index at these singularities of $-\partial J$ is $3-k_i$ and Corollary 7.3 follows.

Appendix

Proof of Proposition 7.3: Let

$$\psi = \frac{1}{\lambda_i}\frac{\partial \delta_i}{\partial x_i} \quad \text{or} \quad \psi = \lambda_i \frac{\partial \delta_i}{\partial \lambda_i} \tag{A1}$$

In both cases, we have:

$$|\psi| \leq C\delta_i \ ; \qquad |\psi|_{-L} \leq C \tag{A2}$$

with a suitable constant C.
Observe also that:

$$\int \nabla\delta_i \nabla\psi = 0 \tag{A3}$$

since $\int \nabla\delta_i \nabla\delta_i$ is independent of λ_i and x_i.
We compute

$$\partial J(u)\cdot\psi \tag{A4}$$

We have by (A3) and (G0):

$$\partial J(u)\cdot\psi = -\lambda(u)^5 \int_{S^3} K(x)\Big(\sum_{i=1}^{P} \alpha_j\delta_j + v\Big)^5\psi - \lambda(u)\sum_{j\neq i}\alpha_j\int_{S^3}\delta_j^5\psi$$

$$= -\lambda(u)^5\alpha_i^4\int_{S^3} K(x)\delta_i^5\psi \ + R$$

Using (A2) and (E3), we derive:

$$|R| \leq C\Big(\sum_{j\neq i}\int\delta_j^5\delta_i + \sum_{j\neq i}\int\delta_j^4\delta_i|v| + \Big|\int K(x)\delta_i^4 v\psi\Big| + |v|_{-L}^2\Big) \leq \tag{A6}$$

$$\leq C\Big(\sum_{j\neq i}\varepsilon_{ij} + |v|_{-L}\ \Sigma\varepsilon_{ij}(\log\varepsilon_{ij}^{-1})^{1/3} + |v|_{-L}^2 + \Big|\int K(x)\delta_i^4 v\psi\Big|\Big)$$

By Lemma 4.1, we know that:

$$|v|_{-L} \leq |v-\bar{w}|_{-L} + |\bar{w}|_{-L} \leq C(\varepsilon_{ij}(\log\varepsilon_{ij}^{-1})^{1/3} + \tag{A7}$$

$$+ \left| \partial J(\Sigma \alpha_i \delta_i + v) \right| + \sum_{j=1}^{P} \left(\frac{|DK(x_j)|}{\lambda_j} + \frac{1}{\lambda_j^2} \right)) .$$

Since v satisfies (G0), $\int \nabla\psi\nabla v = 0$ and therefore

$$\int K(x)\delta_i^4 v\,\psi = \int (K(x) - K(x_i))\delta_i^4 v\psi \tag{A8}$$

Using (A1), we thus derive:

$$\left|\int K(x)\delta_i^4 v\,\psi\right| \leq C|v|_{-L}(\int |K(x)-K(x_i)|^{6/5}\delta_i^6)^{5/6} \leq \tag{A9}$$

$$\leq C|v|_{-L}\left(\frac{|DK(x_i)|}{\lambda_i} + \frac{1}{\lambda_i^2}\right)$$

Combining A6-A9, we derive:

$$\left|\int K(x)\delta_i^5\psi\right| \leq C\left(|\partial J(u)|_{-L} + \sum_{j\neq i} \varepsilon_{ij} + \frac{|DK(x_i)|^2}{\lambda_i^2} + \frac{1}{\lambda_i^4}\right). \tag{A10}$$

On another hand, using estimate (F14) and (F18), we derive:

$$\left|\int K(x)\delta_i^5 \frac{1}{\lambda_i}\frac{\partial\delta_i}{\partial x_i}\right| + \left|\int K(x)\delta_i^5\lambda_i\frac{\partial\delta_i}{\partial\lambda_i}\right| = \tag{A11}$$

$$= c\left|\frac{DK(x_i)}{\lambda_i} + O\left(\frac{1}{\lambda_i^2}\right)\right| + c'\left|\frac{\Delta K(x_i)}{\lambda_i^2} + o\left(\frac{1}{\lambda_i^2}\right)\right|$$

Since $K(x_i)$ does not vanish when $DK(x_i)$ vanishes, we have:

$$c\left|\frac{|DK(x_i)|}{\lambda_i} + O\left(\frac{1}{\lambda_i^2}\right)\right| + c'\left|\frac{\Delta K(x_i)}{\lambda_i^2} + o\left(\frac{1}{\lambda_i^2}\right)\right| \geq \tag{A12}$$

$$\geq c_1\left(\frac{|DK(x_i)|}{\lambda_i} + \frac{1}{\lambda_i^2}\right).$$

(A10), (A11) and (A12) imply:

$$\frac{|DK(x_i)|}{\lambda_i} + \frac{1}{\lambda_i^2} \leq C'\left(\left|\partial J\left(\sum_{j=i}^{P} \alpha_j\delta_j + v\right)\right| + \right. \tag{A13}$$

$$+ \sum_{j \neq i} \varepsilon_{ij} + \frac{|DK(x_i)|^2}{\lambda_i^2} + \frac{1}{\lambda_i^4}),$$

hence, with λ_i large enough:

$$\frac{|DK(x_i)|}{\lambda_i} + \frac{1}{\lambda_i^2} \leq C''(|J(\sum_{j=1}^{P} \alpha_j \delta_j + v)| + \sum_{j \neq i} \varepsilon_{ij}) \tag{A14}$$

Proposition 7.3 follows.

References

[1] A. BAHRI. Pseudo-orbits of contact forms. Proceedings of the NATO conference on Dynamical Systems, Il Ciocco, October 1986.

[2] A. BAHRI-P. L. LIONS. Elliptic differential equations in unbounded domains, to appear.

[3] A. BAHRI-J. M. CORON. On a non-linear elliptic equation involving the critical Sobolev exponent: the effect of the topology of the domain, to appear in Communications in Pure and Applied Mathematics.

[4] G. E. BREDON. Compact transformation groups. Pure and Applied Mathematics, Academic Press, 1972.

[5] H. BREZIS. Elliptic equations with limiting Sobolev exponents - the impact of topology. Proc. Symp. 50th Anniversary of the Courant Institute. Comm. Pure Appl. Math. 39 (1986), p. 517-539.

[6] H. BREZIS-J. M. CORON. Convergence of solutions of H-systems or how to blow bubbles. Arch. Rat. Mech. An., 89, 1 (1985), p. 21-56.

[7] P. L. LIONS. The concentration compactness principle in the calculus of variations, the limit case. Rev. Mat. Iberoamericana 1, 2 (1985), p. 145-201.

[8] J. SACKS-K. UHLENBECK. The existence of minimal immersions of 2-spheres. Ann. of Math. 113 (1981), p. 1-24.

[9] S. SEDLACEK. A direct method for minimizing the Yang-Mills functional over 4-manifolds. Comm. Math. Phys. 86 (1982), p. 515-527.

[10] Y. T. SIU-S. T. YAU. Compact Kähler manifolds of positive bisectional curvature. Inv. Mat. 59 (1980), p. 189-204.

[11] M. STRUWE. A global compactness result for elliptic boundary values problems involving nonlinearities. Math. Z. 187 (1984), p. 511-517.

[12] H. WENTE. Large solutions of the volume constrained Plateau problem, Arch. Rational Mech. Anal. 75 (1980), p. 59-77.

[13] A. DOLD. Lectures on Algebraic Topology. Springer Verlag, 1972.

[14] R. M. SWITZER. Algebraic topology. Homotopy and homology. Springer Verlag, 1975.

[15] D. HUSEMOLLER. Fibre bundles. New York, McGraw-Hill, 1966.